MATHEMATICAL MODELLING WITH CASE STUDIES

Using Maple™ and MATLAB®

Third Edition

TEXTBOOKS in MATHEMATICS

Series Editors: Al Boggess and Ken Rosen

TEXTBOOKS in MATHEMATICS

MATHEMATICAL MODELLING WITH CASE STUDIES

Using Maple™ and MATLAB®

Third Edition

B. Barnes and G. R. Fulford

CRC Press
Taylor & Francis Group
Boca Raton   London   New York

CRC Press is an imprint of the
Taylor & Francis Group, an informa business

A CHAPMAN & HALL BOOK

CRC Press
Taylor & Francis Group
6000 Broken Sound Parkway NW, Suite 300
Boca Raton, FL 33487-2742

© 2015 by Taylor & Francis Group, LLC
CRC Press is an imprint of Taylor & Francis Group, an Informa business

No claim to original U.S. Government works

Printed on acid-free paper
Version Date: 20150127

International Standard Book Number-13: 978-1-4822-4772-5 (Hardback)

Library of Congress Cataloging-in-Publication Data

Barnes, Belinda, 1959-
    Mathematical modelling with case studies : using Maple and Matlab / B. Barnes, G.R. Fulford. -- Third edition.
        pages cm
    "A CRC title."
    Includes bibliographical references and index.
    ISBN 978-1-4822-4772-5 (hardcover : alk. paper) 1. Differential equations--Data processing. 2. Mathematical models. 3. Maple (Computer file) I. Fulford, Glenn. II. Title.

QA371.3.B37 2015
515'.35028553--dc23   2015001599

Visit the Taylor & Francis Web site at
http://www.taylorandfrancis.com

and the CRC Press Web site at
http://www.crcpress.com

Contents

Preface

Aims and objectives

This book aims to provide the student with some basic modelling skills that will have application to a wide variety of problems.

The focus is on those mathematical techniques that are applicable to models involving differential equations, and which describe rates of change. We consider models in three different areas: growth and decay processes, interacting populations and heating/cooling problems. The main mathematical technique used is that of solving differential equations. However, the range of applications and mathematical techniques presented should provide a broad appreciation of the scope of this type of modelling, while the skills developed are transferable to many other areas.

Layout

The book is divided into three general sections: compartmental models, population models and heat transfer problems.

Within each of the modelling sections the process of constructing a model is carefully considered and presented in full detail. Because this is one of the most difficult and crucial aspects of modelling, we emphasise that the process followed is: start simple, evaluate and extend.

One very important skill is that of converting a seemingly complex problem into a much simpler one. To illustrate this, there is an emphasis on the development of simple models, where some reality may be sacrificed in order to gain a basic understanding of the problem being modelled. Flow diagrams and word equations are constructed, both to aid in the model building process and to develop the mathematical equations. The assumptions made and consequent limitations are discussed together with their ramifications in the interpretation of results. This structured and detailed approach has proved to be a successful learning strategy.

Analysis using a variety of techniques with theoretical, graphical and computational tools follows. Understanding the behaviour of the model under changing conditions (such as changes in the parameters and initial conditions) is part of this process. Graphical methods are encouraged, and it is the intention of this course to provide an interactive use of software. To this end, $\texttt{Maple}^{\text{TM}}$ and $\texttt{MATLAB}^{\circledR}$ code and ensuing results are included within the text.

Validation of the model is discussed, using observations or prior knowledge, and extensions to the models are suggested and developed or incorporated into the exercises at the end of each chapter. There is an emphasis on identifying and recognising the strengths and limitations of a model, as well as its relevance in context. A more detailed overview of the layout is presented in Section 1.2.

Features

Applications and case studies are integral to this text. Case studies have been included throughout to provide the reader with relevant applications of the skills learned and an understanding of how such skills are used and interpreted in current research. We have aimed to include models from a diverse range of disciplines, such as, environmental, biological, archaeological, etc. The authors strongly encourage students to read some of the research papers for which references are provided.

An important feature of this book is the use of compartmental diagrams and work equations as a way for students to help conceptualise the formulation of differential equations. In this book we require students to formulate variants of the various models as exercises. We believe it is important that a student attempts to formulate differential equations where they do not see the answer; this will be closer to real life. To this end, we have sometimes organised material into a formulation chapter and an analysis chapter, particularly for the later chapters. Also, we will usually explore a model numerically before attempting to solve symbolically or to use analytic techniques to analyze a model. This is sometimes the opposite way that models have been presented in traditional mathematics texts, but we think this is more beneficial to students studying mathematical modelling for the first time as it allows them to explore the models without getting too bogged down with algebra. Some of the simpler models introduced early in the text are further developed and extended and analysed in subsequent chapters of the book.

There are problems at the end of each chapter to practice the skills developed in formulating differential equation models as well as analysing and interpreting models. A solutions manual is available from the publisher for instructors.

Prerequisites

Some basic calculus, such as a first-year modelling/calculus course or advanced school mathematics, as well as fundamental computing skills, are required to make optimal use of this text. We do expect that students will have some basic familiarity with differential equations.

This book is not intended to teach techniques for solving differential equations from the start; there are plenty of other books that do a good job of this. However, many details of the mathematics required for differential equations and their analysis are covered in the appendices, as a reminder for those who wish some review of the basic techniques. Since much `Maple` and `MATLAB` code is included, prior detailed knowledge of any software package is not a prerequisite and basic computing skills should suffice.

Course suggestions

This book contains far more than was given by the authors in any one course. Chapters 2 and 3 provide a good introduction for students with little previous background in modelling. This could be followed by a choice in emphasis on either population models (Chapters 5, 6 and perhaps 8) or heat transport models (Chapters 9, 10 and perhaps 11).

The authors also suggest the inclusion of a project to complement the course and contextualise the modelling process. This may involve an in-depth examination of one of the case studies or other relevant research papers that incorporate the approaches and techniques of mathematical modelling learned throughout the various chapters of this book.

Software integration

The analysis of these models is carried out using an analytical approach as well as the integrated use of a software package, which provides both numerical solutions and an excellent means by which graphics are displayed. It allows us to obtain solutions to differential equations that are not always possible analytically, and provides a valuable tool in the analysis process. `Maple` and `MATLAB` code are provided in the text so that students can reproduce the results exactly. Other software packages could replace `Maple` (such as *Mathematica* or *Sage*) where students are more familiar with these, as the principles are the same and their use would in no way detract from the thrust of the book. Many universities have a site licence for `Maple` or `MATLAB` or one of the other packages, with cheap student versions available. Electronic copies of the code used in this book are available from the publisher's web-site.

For **MATLAB** product information, please contact:

The MathWorks, Inc.
3 Apple Hill Drive
Natick, MA 01760-2098 USA
Tel: 508-647-7000
Fax: 508-647-7001
E-mail: info@mathworks.com
Web: www.mathworks.com

For **Maple** product information, please contact:

Maplesoft,
615 Kumpf Drive
Waterloo, Ontario
Canada, N2V 1K8
Tel: 519-747-2373
Fax: 519-747-5284
E-mail: info@maplesoft.com
Web: http://www.maplesoft.com/

Authors

Belinda Barnes is currently a director in the Australian Government Research Bureau and a visiting fellow at the National Centre for Epidemiology and Population Health at the Australian National University (ANU), Canberra. She has worked at a number of research schools at the ANU and as a lecturer in the Department of Mathematics, establishing courses in mathematical modelling and bifurcation theory. Her work has focussed on the formulation of models that capture biological, ecological and epidemiological processes, while remaining practical to use and with the facility for theoretical analysis. She has published work in a number of applied areas such as bifurcation theory, population dynamics, carbon sequestration, biological processes and disease transmission.

Glenn Fulford was recently a research associate and senior lecturer in applicable mathematics at the Queensland University of Technology. He has taught at a number of universities in Australia, including the Queensland University of Technology, the University of NSW (ADFA) and La Trobe University and has developed a number of courses on mathematical modelling. He has published several other textbooks on mathematical modelling and industrial mathematics. His PhD was in the area of physiological fluid mechanics at the University of Wollongong and he has published in a variety of areas, including mucus transport, spermatozoa propulsion, infectious disease modelling, tuberculosis in possums, tear-flow dynamics in the eye, population genetics and industrial mathematics.

Acknowledgements

There are many friends and colleagues who have had input into this book, with typing, technical support, suggestions and encouragement, and we would like to offer them all our warmest thanks.

We are extremely grateful to Jill Smith, Sharon Eldridge and Annette Hughes at the Australian National University, and to Annabelle Lippiatt from University College, ADFA, who contributed significantly to the typing of this document. And to Tracey Hansen who undertook the copy editing. We would also like to thank James Gifford and Nick Guoth for the valuable technical advice they provided, as well as Harvinder Sidhu for his `Maple` code to generate the bifurcation diagram in Section 3.7. Furthermore, we thank Scott Rabuka and Waterloo `Maple` for their support.

One of the students taking the course, Cezary Niewiadomski, constructed some of the diagrams incorporated into the book, and we thank him warmly for this generous contribution. There are many other students who pointed out errors and provided suggestions, to all of whom we are most grateful. Willow Hart spent many tedious hours compiling the index and updating files, with our great appreciation.

We are very grateful for information on alcohol and drug absorption supplied to us by Kelly FitzGerald of Ainslie Pharmacy (Canberra, Australia) as well as the Alcohol and Other Drugs Council of Australia. Thanks are also due to Duck Australia, who provided details of Australian standards for hot water heaters. Many thanks also to Tony Richings for his background information on land mines — we greatly admire your heroic task of clearing Cambodia of mines.

We extend our thanks to those colleagues who read and commented on sections of the book, particularly Harvinder Sidhu, David Rowland, Daryl Daley and Joe Gani. Your comments were a highly valued contribution in the construction of the text.

For the second edition of this book we are very grateful to Jen Pestana for her careful and enthusiastic work on producing a solutions manual. We would also like to thank Mark Nelson for pointing out some typographical errors and making some excellent suggestions, and Charisse Farr for commenting on the new case study on economic modelling.

For the third edition of this book we would first like to again thank Mark Edwards who has continued to provide excellent feedback as he used this book for his students. Also, we would like to thank Sucheta Nayak for her useful comments on the 2nd edition. We are extremely grateful to Xing Lee who has read the book from cover to cover and kindly provided us with the majority of improvements for this edition.

Finally, we would like to thank our families for their support and encouragement, as such an enterprise always has an impact on family life. Our warmest thanks for this — it has made the experience all the more enjoyable for us.

Belinda Barnes and Glenn Fulford

mailto:belindagb2@gmail.com
glenn.fulford@gmail.com

Chapter 1

Introduction to mathematical modelling

Mathematical modelling has a diversity of applications and thus a range of possible approaches. Together with an outline of the contents, aims and scope of the book, this chapter provides an introduction to the cyclic nature of the modelling process, a brief description of some modelling approaches and a discussion of the wider ramifications of model application.

1.1 Mathematical models

Models of systems have become part of our everyday lives: they range from global decisions having a profound impact on our future, to local decisions about whether to cycle to university based on weather predictions. Together with their provision of a deeper understanding of the processes involved, this predictive nature of models, which aids in decision-making, is one of their key strengths.

In particular, many processes can be described with mathematical equations, that is, by mathematical models. Such models have use in a diverse range of disciplines.

There is an aesthetic use, for example, in constructing perspective in paintings or etchings such as is seen in the paradoxical work of Escher. The proportions of the golden mean and the Fibonacci series of numbers, occurring in many natural phenomena such as the arrangement of seed spirals in sunflowers, have been applied to methods of information storage in computers. This well-known mathematical series is also applied in models describing the growth nodes on the stems of plants, as well as in aesthetically pleasing proportions in painting and sculpture and the design of musical instruments. From a philosophical perspective, mathematical logic and rigour provide a model for the construction of argument.

In a more practical and analytical mode there is a plethora of applications. Mathematical optimisation theory has been applied in the clothing industry to minimise the required cloth for the maximum number of garments, and to the arrangement of odd-shaped chocolates in a box to minimise the number required to give the impression that the box is full! The mathematics of fractals has allowed the successful development of fractal image compression techniques, requiring little storage for extremely precise images.

Some other areas of application include the physical sciences (such as astronomy), medicine (such as the absorption of medication), and the social sciences (such as patterns in election voting). Mathematical models are used extensively in biology and ecology to examine population fluctuations, water catchments, erosion and the spread of pollutants, to name just a few. Fluid mechanics is another extensive area of research, with applications ranging from the modelling of evolving tsunamis across the ocean, to the flow of lolly mixture into moulds. (Mathematicians were consulted to establish the best entry points for the mixture to the mould in order to ensure a filled nose for a Mickey Mouse lollypop!)

The use and versatility of such mathematical models has been heightened by the power of computers. This trend is likely to continue, as such modelling results in an efficient and economical way of understanding, analysing and designing processes. Furthermore, the diverse range of applications, where mathematical models are used, implicates an appeal to many areas of mathematics as well as to many types of models. Some models can be extremely accurate in their predictions, whereas others may be more susceptible to a range of interpretations, particularly in the case of large systems with many interacting mechanisms.

The use of information from this modelling process to reach decisions is now very much in the public view, particularly in the case of environmental issues exemplified in the 1998 international meeting on Global Warming in Kyoto. We would hope that by encouraging the responsible use of models, we may improve public confidence in the role of science, which has been declining in recent years (Beck et al., 1993). It is well to remember that models are not reality and should not be promoted as such. They are considered by the authors to be a valuable and integral part of a multi-disciplinary approach towards decision-making on issues we face today, and we encourage future modellers to adopt and promote this perspective.

1.2   An overview of the book

This course aims to provide the student with some basic modelling skills that have application to a wide variety of problems. There is an emphasis on the initial development of simple models, which are then extended and improved by incorporating further features. As well, we have focused attention on the importance of declaring any assumptions made in the modelling process, and bearing these in mind when interpreting the predictions. We discuss various theoretical approaches as they arise, but the emphasis is on the modelling itself and the interpretation of the results.

Layout of the book

The analysis of these models is carried out using a theoretical approach as well as the integrated use of `Maple` or `MATLAB` (two mathematical software packages), both of which provide numerical solutions and an excellent means by which graphics are displayed. `Maple` and `MATLAB` code and output are presented throughout the text.

Furthermore, case studies have been included throughout to provide the reader with relevant applications of the skills learned and an understanding of how such skills are used and interpreted in current research. We have aimed to include models from a diverse range of subjects — environmental, biological, archaeological, etc. The authors strongly encourage students to read some of the research papers, for which references are provided.

In general, with such a diversity of modelling applications it is not surprising that many different areas of mathematics are used. These include differential and integral calculus, differential equations, matrices, statistics, probability, to name just a few.

This course focuses on those mathematical techniques that are applicable to differential equation models, in particular ordinary differential equations (ODEs), although partial differential equations (PDEs) are also briefly discussed. We consider models in three different areas: decay processes, interacting populations and heating/cooling problems. The main mathematical technique used is that of solving differential equations. However, this range of applications and the mathematical techniques presented should still provide an appreciation of the scope of this type of mathematical modelling because the skills developed are transferable to many other areas.

The book is divided into three general parts: compartmental models, population models and heat transfer problems.

Within each of the modelling sections, the process of constructing a model is carefully considered because this aspect is one of the most difficult and crucial parts of modelling. One very important skill is that of converting a seemingly complex problem into a much simpler one. There is an emphasis in this book on the development of simple models where some features of reality are sacrificed in order to gain a basic understanding of the underlying problem. With each model type, the text covers this building process in detail, with a critical evaluation of the assumptions made and their subsequent limitations on interpretation. Flow diagrams aid the understanding of this process and clarify any division of sub-sections within a model. Then 'word equations' are developed describing the process under consideration, and from these the differential equations follow. This structured approach has proved a successful learning strategy.

Once a model has been established, it is analysed using a variety of techniques with theoretical, graphical and computational tools. Understanding the behaviour of the model under a variety of conditions (changes in the parameters and different initial conditions) is discussed, with an awareness of possible limitations on the interpretations (or misinterpretations) of the results. Graphical methods are encouraged, and it is the intention of this course to provide an interactive use of `MATLAB` and `Maple` software. To this end, `MATLAB` and `Maple` code, and ensuing results are included within the text; they are easy to use and applicable to a variety of problems. Simple analytical techniques are developed and applied to solve the differential equations; typically this is confirmed with `Maple`. Again, all relevant code is included within the text, and a list of commonly used procedures can be found under the `Maple` and `MATLAB` entries in the index.

Validation of the model is discussed, using observations or prior knowledge. Extensions are suggested and developed or incorporated into the exercises at the end of each chapter. Thus the process followed is to start simple, evaluate and extend. Subsequent interpretations are critically evaluated in terms of the assumptions made, with an emphasis on the strengths, limitations and relevance of the models constructed.

We now outline two typical problems discussed in this book to provide an indication of how this process of simplifying a complex problem into its basic parts is approached in the text.

A typical problem in population growth

A major issue in the farming industry is dealing with insect pests that infest and destroy crops. To this end, we examine an example from the Citrus Industry in America that was threatened by an insect pest (introduced from Australia). One obvious approach is to spray the crops with a pesticide, such as DDT, to eradicate the insects. However, this is not necessarily the most efficient approach in that complicated ecological interactions between the pest and other animals or plants may be affected in unforeseen ways.

To simplify the problem, we consider this population of aphid-like insects (the pest) that interact with a type of ladybird beetle. The interaction is a predator-prey relationship since the aphids are eaten by the beetles. Thus the aphid population and the beetle population interact, in that the aphid numbers are reduced by the presence of the beetles and the beetle population is enhanced by the presence of the aphids. Furthermore, our model needs to incorporate the effect of the pesticide on both the aphid and beetle populations.

We formulate two differential equations: one describing the aphid population as a function of time and the other describing the beetle population as a function of time. Solving the differential equations gives us an unexpected result: the effect of the pesticide actually increases the aphid population, which is not what was expected or desired. This problem illustrates how simple models can provide improved insight into the way populations interact.

Other interacting populations that we examine in detail include competing species models, as well as models of populations in combat and the spread of epidemics.

A typical problem in cooling

Mathematical models can be useful in domestic situations. To illustrate this, we consider how to determine the optimal amount of insulation to wrap around a water pipe to prevent its freezing on a cold morning. As heat is lost from the water in the pipe to the outside, the temperature of the pipe will drop, and if the water freezes it may burst the pipe. The placement of a layer of insulating material around the pipe is designed to reduce the rate of heat loss and thus prevent the water from freezing as quickly. The aim is to determine the rate of heat loss, as a function of the thickness of the insulating material, in order to determine the specifications for sufficient and effective coverage of the pipe.

To produce a mathematical model, we envisage first a simplified situation. We consider an infinite pipe where the heat loss is in the radial direction from the centre to the outside of the pipe. We can also simplify the problem by ignoring the metal pipe and just considering the heat transfer through the insulation itself.

We aim to develop a differential equation that predicts the temperature at any point in the insulation, and then use this to find the rate of heat loss. Again a surprising result emerges from the mathematics. It turns out that for some (not all) insulating materials, there can be more heat loss than if there were no insulation at all.

Similar models examined are those of insulating the wall of a house and the effect of double glazing in windows. Further problems include determining the optimal length of heat fins that are used to reduce the temperature of computer chips, and establishing criteria to avoid spontaneous combustion in storage bins.

1.3   Some modelling approaches

There are many modelling approaches, and while this book is mainly concerned with mathematical modelling with differential equations it is essential to recognise that there may be other approaches with alternative conceptual frameworks that are more appropriate for different problems. A combination of approaches may provide the 'best' understanding, each approach providing insight into some different aspect.

Below we have briefly outlined some of the approaches that can be taken when formulating a mathematical model. In order to clarify each method, we apply it to the problem of determining a population size as a function of time.

Empirical models

The empirical approach is the most basic, but also the least useful. The idea is to fit a curve through a set of data and then to use this curve in order to predict outcomes where there are no data.

In terms of the population example, if t denotes time and $P(t)$ denotes the population at time t, then we might try curves of the form

$$P(t) = a + bt + ct^2 \qquad \text{and} \qquad P(t) = ae^{bt}$$

and use the method of least squares[1] to estimate the parameters a, b and c. The disadvantage with this approach is that we cannot be confident that the formula applies outside the range of data considered, and the parameters a, b and c have no specific meaning. Thus we cannot extend the model easily to a related problem. This method is often used to create a continuous data set for input to another model, such as completing a temperature profile for a water column to be used as data in an ecological model predicting water velocities.

Stochastic models

Another approach to modelling is the probabilistic or stochastic approach ('stochastic' comes from the Greek word to guess). Using this method we try to estimate the probability of certain outcomes based on the available data. In terms of the population example, we would aim to establish a formula for the probability of a population changing size on the basis of assumed formulae for the probability of population births or deaths. These assumed formulae may in turn be based on probability distributions. Quantities of interest to be calculated would be standard statistics such as the mean and the standard deviation of the population size.

These models can by extremely complicated, although this is not necessarily the case. They do have the advantage of incorporating a degree of uncertainty within them and ideally should be used when there is a high degree of variability in the problem. This method is typically used for models of small populations when reproduction rates need to be predicted over a time interval. They also have valuable application in many other areas such as economic fluctuations, insurance problems, telecommunications and traffic theory, and biological models.

Simulation models

In a simulation model one writes a computer program that applies a set of rules, or possibly even physically builds a scale model. It is intended to produce a set of data that mimic a real outcome including extreme events. In a population model, the rules would apply to each member of the population. Usually a simulation will involve some random components. The computer program can be run many times and statistical information gained in the process.

Typically such models are used in engineering applications as an aid to identifying problems that may arise during use or construction. They also provide a useful means by which data sets can be generated. One such example is rainfall data, which may be required as input to a model forecasting phosphorus concentrations in a stream network.

It might be argued that this simulation approach provides the most realistic models, but this does not mean it provides the best models. The best models are usually those which are simple and yet still provide results that are useful.

Deterministic models

Another approach to mathematical modelling is the deterministic approach. This is the approach we adopt in this book. Modelling in this manner, we ignore random variation and try to formulate mathematical equations describing the basic fundamental relationships between the variables of the problem.

For instance, in the population model we would aim to obtain an equation relating birth rates and death rates, which themselves are related through equations, to the population size at any given time. In a sense we are constructing a model made of several empirical

---

[1] In the method of least squares the total error is given as the sum of the squares of the difference between the curve and the data points, and the unknowns are estimated by minimising this error.

and inter-related sub-models, linking these sub-models together and then using the whole system to predict the outcome from a set of initial conditions.

This process is widely used and can be extremely accurate, such as in the case of predicting satellite orbits. It has the drawback that in other cases it is not possible to establish all the component mechanisms of a process or, even if this were possible, that including all known relationships renders the model unwieldy. However, as we shall see, even under these conditions this modelling technique can provide valuable insights into a process.

Statistical models

Statistical models concern the testing (referred to as hypothesis testing) of whether a set of empirical data is from one or another category. These categories are assumed to have particular distributions (with associated means and standard deviations), and the results suggest the data are drawn from one such category. This distribution can then be used to predict the outcome of further trials. In terms of our population, such a model may test whether a sample of heights was taken from a population with an average height of six feet (say category A), or one with an average height of five feet and ten inches (say category B). Our results would indicate a percentage error margin with which we have made this prediction.

Statistical testing is used widely in psychology, palaeontology and the biological sciences. One example application would be to establish spatial distributions of some ancient species from sets of fossilised teeth found in different locations.

1.4   Modelling for decision-making

Currently, modelling, and in particular mathematical modelling, provides a means by which many political and management planning decisions are made, both locally and globally. Such models have huge economic clout and wield social and environmental ramifications. There is a need to use these models wisely.

Modelling is an extremely powerful tool, a framework for research, debate and planning, which provides a valuable source of information for decision-making. However, while models aim to incorporate the main mechanisms of a process and have the capacity not only to describe and explain but to predict as well, they are not infallible and should not be considered in isolation, particularly in light of the above remarks. Decision makers need to be aware of the modelling process and its context-specific strengths and limitations in order to make best use of the interpretations.

The main objectives of any model are twofold: first, to provide a deeper understanding of a system or process, and second, to use this knowledge for predictions and decision-making. In the first instance, mathematical modelling can be extremely successful with, in many cases, a combination of models providing the best insight into a problem. Of course, any model is, at best, merely a means of mimicking reality and as such at each stage of the process, there is the possibility of error or introduced limitations. This does not render the process useless, but rather suggests a need for an understanding of the complexities involved.

No part of the process is necessarily simple. For example, in the decision-making process individuals invariably have different interpretations of the results, and differ in opinion as to what might most benefit a situation. They may disagree on a 'safe' level of pollution in a local river, or how 'green' the Sydney Green Olympics should be, or what constitutes an endangered species, or even whether a population facing extinction warrants concern at all.

Furthermore, the model needs to be sufficiently simple so that the available equipment is adequate for sensible and useful predictions. If it took until next week to predict tomorrow's weather we might as well not bother: clearly, accurate predictions are only useful before the event! Certain physical processes, such as the path of a satellite, can be accurately described, while others, such as ecological systems or the process of global warming, are far more complex and require modelling and interpretation of a very different sort. Thus the modelling process is not necessarily simple, and it needs to be well understood with all its strengths, simplifications, assumptions and limitations to utilise the results optimally.

When involved in decision-making processes, to take the results of any single model as absolute would be short-sighted. It constitutes one part of a multi-dimensional field, far more complex in most cases than our single mathematical model can hope to represent. But, for their part, decision makers need to be made aware of the limitations of the model predictions, as well as their strengths. They need to recognise the assumptions on which the model is based and the rationale behind them. They should consider a variety of interpretations. They should be as well informed as possible.

Furthermore, beyond the mathematics there may be, for example, historical, legal or cultural issues involved that are integral to the planning process and should be acknowledged. It is the modellers' responsibility to provide the best information possible, not to sell a 'perfect' model. It is well to remember that, ultimately, political and management decisions have an impact on us all.

In many instances, the reticence of scientists to work together with those in other 'softer' disciplines, and *vice versa*, has emphasised the distrust between them and the lack of respect for each others' research. There are many who ignore the imperfect nature of models, and emphasise only the accuracy of mathematics. There are those who deride quantitative research. The authors consider the contributions from each discipline to carry value and encourage their collaboration and cooperation in order to realise the full and powerful potential of modelling.

With a combination of the qualitative and the quantitative, a recognition of the complexity of systems and the true nature of models, we might aim to capture (to the best of our ability) more than just the physics of the system. If we are willing to recognise and acknowledge the complexities and limitations involved, cooperative and informed modelling has a truly enormous potential.

For further discussion on certain of the issues raised in this section, see Gani (1972), Gani (1980) and Beck et al. (1993).

Chapter 2

Compartmental models

The compartmental model framework is an extremely natural and valuable means with which to formulate models for processes that have inputs and/or outputs over time. In this chapter, we model radioactive decay processes, pollution levels in lake systems and the absorption of drugs into the bloodstream, using compartmental techniques.

2.1 Introduction

Many processes may be considered as compartmental models: that is, the process has inputs to and outputs from a 'compartment' over time.

Compartmental diagram

One example of this compartmental notion is the amount of carbon dioxide in the Earth's atmosphere. The compartment is the atmosphere. The input of CO_2 occurs through many processes, such as burning, and the output occurs through processes such as plant respiration. This is illustrated in Figure 2.1 in a diagram known as a *compartmental diagram*.

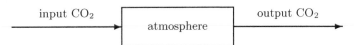

Figure 2.1: Input-output compartmental diagram for CO_2.

What we are interested in is the amount of CO_2 in the compartment at any given time, or whether over long periods the levels are increasing or decreasing. Currently, the input of CO_2 into the Earth's atmosphere appears to exceed the output, so that there is an overall increase in the level of CO_2 in the compartment.

Balance law

Suppose we are modelling the amount of CO_2 as above, or the size of a population, each of which is changing with time. We can think of the amount of a substance (CO_2 or the population) as occupying the compartment and the rate of change can be considered as the 'rate in' minus the 'rate out'. Thus we have what is called the *balance law*. In words, this is

$$\left\{ \begin{array}{c} \textit{net rate} \\ \textit{of change} \\ \textit{of a substance} \end{array} \right\} = \left\{ \begin{array}{c} \text{rate} \\ \text{in} \end{array} \right\} - \left\{ \begin{array}{c} \text{rate} \\ \text{out} \end{array} \right\}. \tag{2.1}$$

It could describe, for example:

- the decay process of radioactive elements;
- births and deaths in a population;
- pollution into and out of a lake or river, or the atmosphere; or
- drug assimilation into, and removal from, the bloodstream.

In the following sections, we use this balance law approach to formulate mathematical models of differential equations that describe such processes.

2.2   Exponential decay and radioactivity

The process of dating aspects of our environment is essential to the understanding of our history. From the formation of the Earth through the evolution of life and the development of mankind, historians, geologists, archaeologists, palaeontologists and many others use dating procedures to establish theories within their disciplines.

Compartmental diagram

While certain elements are stable, others (or their isotopes) are not, and emit α-particles, β-particles or photons while decaying into isotopes of other elements. Such elements are called *radioactive*.

The decay, or disintegration, of one nucleus is a random event and so for small numbers of nuclei one might apply probability functions. However, when dealing with large numbers of nuclei we can be reasonably certain that a proportion of the nuclei will decay in any time interval and thus we can model the process as continuous with some fixed rate of decay. We can consider the process in terms of a compartment without input but with output over time, as in Figure 2.2.

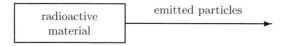

Figure 2.2: Input-output compartmental diagram for radioactive nuclei.

We make the following assumptions and then, based on these, develop a model to describe the process.

- We assume that the amount of an element present is large enough so that we are justified in ignoring random fluctuations.
- We assume the process is continuous in time.
- We assume a fixed rate of decay for an element.
- We assume there is no increase in mass of the body of material.

The first step is to determine an equation describing this disintegration process and, as with each subsequent model building in this book, we start with a *word equation* and then replace each 'word-term' with mathematical symbols. In a word equation, we can express the above decay process as

$$\left\{ \begin{array}{c} \textit{rate of change of} \\ \textit{radioactive material} \\ \textit{at time } t \end{array} \right\} = - \left\{ \begin{array}{c} \textit{rate amount of} \\ \textit{radioactive} \\ \textit{material} \\ \textit{decayed} \end{array} \right\}. \tag{2.2}$$

Formulating the differential equation

To convert the terms in the word equation into symbols let $N(t)$ be the number of radioactive nuclei at time t and let Δt be a small change in time. We know that the change in the number of nuclei is proportional to the number of nuclei at the start of the time period. Hence (2.2) translates to

$$\frac{dN}{dt} = -kN \tag{2.3}$$

where k is a positive constant of proportionality indicating the rate of decay per nucleus in unit time. We write N on the right-hand side for $N(t)$ as the dependence on t is implied by the derivative dN/dt. We assume k to be fixed although it will have a different value for different elements/isotopes.

Alternatively, we can obtain the differential equation from a limiting process of a consideration of the change in the number of particles in a small time interval Δt. Thus the number of particles at time $t + \Delta t$ is the number at time t minus the number lost,

$$N(t + \Delta t) = N(t) - kN(t)\Delta t.$$

Dividing through by Δt and then letting $\Delta t \to 0$ gives the instantaneous rate of change on the left-hand side, and so we obtain

$$\frac{dN}{dt} = -kN.$$

Note that $N(t)$ should be an integer (number of nuclei) while $-kN(t)\,\Delta t$ may not be. However, if instead we consider N as a mass in grams (from which we can find the number of nuclei), then we can overcome this problem. Note that we then have an equation for $N(t)$, which assumes that $N(t)$ is a continuous, differentiable function rather than an integer. This is an appropriate approximation, particularly when the number of nuclei is large.

Given a sample of a radioactive element at some initial time, say n_0 nuclei at t_0, we may want to predict the mass of nuclei at some later time t. We require the value of k for the calculations; it is usually found through experimentation. Then, with known k and an initial condition $N(0) = n_0$ we have an initial value problem (IVP)

$$\frac{dN}{dt} = -kN, \qquad N(0) = n_0. \tag{2.4}$$

Solving the differential equation

We can solve the differential equation numerically, using either MATLAB or Maple, and we can also solve this differential equation exactly by the technique of separation of variables. The techniques for numerical solutions are discussed in detail in Chapter 4. First, the numerical solution can be obtained using the MATLAB code in Listing 2.1. The graph of the solution is shown in Figure 2.3.

Listing 2.1: MATLAB code: c_cm_expdecay.m

```
function c_cp_expdecay
```

```
global k1; %make k1 available to function

k1=2.0; %set parameter value
tend=5; %endtime in hours
x0=10^5; %initial population
[tsol, xsol] = ode45(@rhs, [0, tend], x0); %solve DE
plot(tsol, xsol,'k');    %plot solution

function xdot = rhs(t, x) %RHS of DE
global k1;      %make k1 available from main
xdot = -k1*x;  %RHS of DE
```

A similar numerical solution can also be easily obtained using `Maple`, and in particular, the `detools` command. Some sample code is given in Listing 2.2.

Listing 2.2: Maple code: c_cm_expdecay.mpl

```
restart; with(plots): with(DEtools):
de := diff(N(t),t) = -k*N(t);
inits := N(0) = 10^5;
k := 2.0;
plot1 := DEplot(de, N, t = 0 .. 5, [inits], arrows = none, scene = [t, N]);
display(plot1);
```

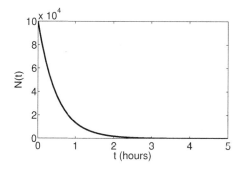

Figure 2.3: Numerical solution of the exponential decay differential equation with $n_0 = 10^5$ and $k = 2.0$.

Exact solution

We can also solve this equation theoretically using the separation of variables technique as in the following example. (See Appendix A.3 for a review of this technique for solving differential equations.)

Example 2.1: *Solve the initial value problem (IVP) in equation (2.4) with initial condition* $N(t_0) = n_0$.

Solution: *Since the differential equation is separable,*

$$\int \frac{1}{N} \frac{dN}{dt}\, dt = \int -k\, dt,$$

$$\int \frac{1}{N}\, dN = -k \int dt,$$

and hence, since N *is a positive quantity,*

$$\ell n\, N = -kt + C,$$

where C is an arbitrary constant. Taking exponentials of both sides we obtain

$$N(t) = Ae^{-kt}, \qquad \text{where} \quad A = e^C,$$

and note that $N \geq 0$.

Using the initial condition $N(t_0) = n_0$ we get

$$n_0 = Ae^{-kt_0} \qquad \text{and} \quad A = n_0 e^{kt_0}.$$

Thus

$$N(t) = n_0 e^{-k(t-t_0)}, \tag{2.5}$$

which is the solution of the IVP.

Note that above we considered the indefinite integral, when technically we should specify the integral over a particular interval $[0, t]$. In the next example we show that this has no impact on the final solution, and thus for the many integrations carried out in this book the interval is often not given explicitly, but understood from the context.

Example 2.2: *Solve the IVP in equation (2.4) on the interval $[0, t]$.*

Solution: *Since the differential equation is separable,*

$$\int_0^t \frac{1}{N} \frac{dN}{dt} \, dt = \int_0^t -k \, dt,$$

$$\int_{n_0}^N \frac{1}{N} \, dN = -k \int_0^t dt,$$

and hence, since N and n_0 are positive,

$$\ell n\, N - \ell n\, n_0 = -kt + 0,$$

$$\ell n \frac{N}{n_0} = -kt.$$

Taking exponentials of both sides, we obtain

$$N(t) = n_0 e^{-kt},$$

which is the solution of the initial value problem (IVP).

We can also obtain the analytic solution of the differential equation using `Maple`, or some other symbolic software package. We include many inserts of `Maple` code throughout the book, which are listed in the index and which allow the reader to reproduce the analytic results and some illustrated graphs. For a brief introduction to `Maple` see Appendix C.1. The code to establish the analytic solution is given in Listing 2.3

Listing 2.3: Maple code: c_cm_expdecay_sym.mpl

```
restart;
de := diff(N(t),t) = -k*N(t);
dsolve({de, N(0)=n0}, N(t));
```

`MATLAB`, with the symbolic toolbox, can also be used to obtain the exact symbolic solution. (The symbolic toolbox comes with the student version of `MATLAB`.) The code to obtain the exact solution in `MATLAB` is given in Listing 2.4. We also include many inserts of `MATLAB` code throughout this book. For a brief introduction to `MATLAB` see Appendix C.3.

Listing 2.4: MATLAB code: c_cm_expdecay_sym.m

```
dsolve('DN = -k*N', 'N(0)=N0', 't');
```

Consider a graph of the solution $N = n_0 e^{-k(t-t_0)}$, from equation (2.5). First, since $t - t_0 > 0$, we have that $e^{-k(t-t_0)}$ is a negative exponential. We also have that

$$\lim_{t \to \infty} n_0 e^{-k(t-t_0)} = 0$$

and $N' = -kN < 0$ (since N and k are positive). This implies that the function $N(t)$ is monotonically decreasing and has no turning points. This information allows us to construct a graph of the solution illustrated in Figure 2.4.

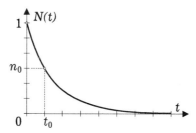

Figure 2.4: Graph of solution (2.5) showing exponential decay.

Equally, `Maple` could have been used to construct the graph with the following code in Listing 2.5.

Listing 2.5: Maple code: c_cm_expdecay_symplot.mpl

```
restart; with(plots);
de := diff(N(t), t) = k*N(t);
soln := dsolve({de, N(t0) = n0}, N(t));
k := -.5; n0 := 1.; t0 := 0;
plot1 := plot(rhs(soln), t = 0 .. 15);
display(plot1, view = [0 .. 15, 0 .. 1]);
```

Often experimental data provides information concerning $N(t)$ at a given time t_1 and what is required is to estimate the constant k and thus predict future or past values of N. This can be done algebraically, or using the substitution procedure of `Maple` (see Exercise 2.3) with code in Listing 2.6.

Listing 2.6: Maple code: c_cm_expdecay_findt.mpl

```
restart:
de:=diff(N(t),t)=k*N(t);
soln:=n0->dsolve({de,N(t0)=n0},N(t)):
t0:=0;n0:=1;
ans:=soln(n0);
get_k:=solve(subs(t=2,N(2)=1/4,ans),k);
evalf(get_k);
```

The *half-life* τ of the radioactive nuclei can be used to determine k, where τ is the time required for half of the nuclei to decay. The half-life τ is more commonly known than the value of the rate constant k for radioactive elements.

Example 2.3: *If the half-life is τ, then find k in terms of τ.*

Solution: *Using solution (2.5) and setting $N(t + \tau) = N(t)/2$, we have*

$$\frac{N(t + \tau)}{N(t)} = \frac{1}{2}, \tag{2.6}$$

which gives

$$\frac{n_0 e^{-k(t+\tau-t_0)}}{n_0 e^{-k(t-t_0)}} = \frac{1}{2}.$$

This simplifies to

$$e^{-k\tau} = \frac{1}{2}.$$

Taking logarithms of both sides gives $\ell n(1/2) = -k\tau$. Hence

$$k = \frac{\ell n(2)}{\tau}. \tag{2.7}$$

Note that both τ and k are independent of n_0 and t_0.

Residence time

Similar to the idea of half-life is the concept of residence time. This is defined as the mean time that an individual particle is in the compartment. A nice interpretation is that the residence time for a single compartment is k^{-1}, where k is the rate constant in the differential equation

$$\frac{dN}{dt} = -kN.$$

To see how to deduce this result, we think of particles leaving the compartment randomly and find the probability density function for the random variable T, the time each particle spends inside the compartment before leaving.

If we start with n_0 particles, then it is easy to compute the fraction of particles remaining in the compartment at time t. Since $N(t) = n_0 e^{-kt}$ then this fraction is given by e^{-kt}. This can be given a probabilistic interpretation. Since e^{-kt} is the fraction of particles left in the compartment at time t, quantity $F(t) = 1 - e^{-kt}$ represents the probability an individual particle has left the compartment by time t. This is the cumulative probability function for the time each particle was in the compartment, that is, T is the random variable representing the time for each particle. It follows that

$$F(t) = \Pr\{T < t\} = 1 - e^{-kt}$$

is the cumulative probability of the time of each particle in the compartment. The probability density function (pdf), $f(t)$, is given as the derivative of the cumulative probability function, so the random variable corresponding to the time a particle stays in the compartment is

$$f(t) = k e^{-kt},$$

which shows that the time in the compartment T has an exponential distribution. We can calculate the mean of this distribution as the integral

$$\int_0^\infty t \times k e^{-kt} \, dt = \frac{1}{k}.$$

Hence we have the interpretation that *the reciprocal of the rate constant, k^{-1}, is the mean time that an individual particle spends in the compartment.*

Radiocarbon dating

We can apply the above theory to the problem of dating paintings by considering the decay process of certain radioactive elements in each.

The following description and example are adapted from Borelli and Coleman (1996). All living organisms absorb carbon from carbon dioxide (CO_2) in the air, and thus all contain some radioactive carbon nuclei. This follows since CO_2 is composed of a radioactive form of carbon ^{14}C, as well as the common ^{12}C. (^{14}C is produced by the collisions of cosmic rays (neutrons) with nitrogen in the atmosphere, and the ^{14}C nuclei decay back to nitrogen atoms by emitting β particles.) Nobel Prize winner Willard Libby, during the late 1940s, established how the known decay rate and half-life of ^{14}C, together with the carbon remaining in fragments of bones or other dead tissue, could be used to determine the year of death. Because of the particular half-life of carbon, internationally agreed upon as $5{,}568 \pm 30$ years for ^{14}C, this process is most effective with material between 200 and 70,000 years old.

Carbon dating depends on the fact that for any living organism the ratio of the amount of ^{14}C to the total amount of carbon in the cells is the same as that ratio in the surroundings. Assuming the ratio in air is constant, then so is the ratio in living organisms. However, when an organism dies, CO_2 from the air is no longer absorbed although ^{14}C within the organism continues to undergo radioactive decay.

In the Cave of Lascaux in France there are some ancient wall paintings, believed to be prehistoric. Using a Geiger counter, the current decay rate of ^{14}C in charcoal fragments collected from the cave was measured as approximately 1.69 disintegrations per minute per gram of carbon. In comparison, for living tissue in 1950 the measurement was 13.5 disintegrations per minute per gram of carbon.

Example 2.4: *How long ago was the radioactive carbon formed and, within an error margin, the Lascaux Cave paintings painted?*

Solution: *Let $N(t)$ be the amount of ^{14}C per gram in the charcoal at time t. We apply the model of exponential decay, $N' = -kN$. We know $\tau = 5{,}568 \pm 30$ years (the half-life of ^{14}C). From equation (2.7), $k \simeq 0.0001245$ per year.*

Let $t = t_0 = 0$ be the current time. Let T be the time that the charcoal was formed, and thus $T < 0$. For $t > T$, ^{14}C decays at the rate

$$\frac{dN}{dt} = -kN \quad \text{with} \quad N(0) = n_0$$

and

$$N(T) = n_0 e^{-kT}$$

or

$$T = -\frac{1}{k} \ell n \left(\frac{N(T)}{n_0} \right).$$

But we do not know $N(T)$ or n_0. However,

$$\frac{N'(T)}{N'(0)} = \frac{-kN(T)}{-kN(0)} = \frac{N(T)}{n_0},$$

and we do have $N'(T) = 1.69$ and $N'(0) = 13.5$. Thus

$$T = -\frac{1}{k} \ell n \left(\frac{N(T)}{n_0} \right) = 16{,}692 \pm 90 \, years.$$

The accuracy of this process depends on having prior knowledge of the exact ratio of ^{14}C to the total C in the atmosphere, but this has changed over the years. There is a basic sinusoidal variation with an 8,000-year period (approximately) to consider. Further, volcanic eruptions and industrial smoke emit only ^{12}C (from tissue older than 100,000 years) into the atmosphere and decrease the ratio, while nuclear testing has resulted in a 100% increase in the ratio in certain parts of the northern hemisphere! These are now factored into dating calculations (see Borelli and Coleman, 1996).

As previously mentioned, radiocarbon dating is inaccurate for the recent past and for very long time periods. Thus the lead isotope with a relatively short half-life ($\simeq$ 22 years) is used for the dating of paintings in our recent history, while at the other extreme, radioactive substances (such as Uranium-238) with half-lives of billions of years are used to date the Earth.

Summary of skills developed here:

- *Understand the concept of modelling with compartments and the balance law.*
- *Find the constant of proportionality from the half-life for radioactive elements.*
- *Generate a solution, using* `Maple` *or analytical methods, to simple separable differential equations.*

2.3   Case Study: Detecting art forgeries

This case study is based on an article in Braun (1979).

After World War II, Europe was in turmoil. Some well-known artworks had disappeared, others had turned up miles from the galleries to which they belonged, and some new ones had appeared! The Dutch Field Security uncovered a firm that, acting through a banker, had been involved in selling artwork to the Nazi leader, Goering. In particular in the case of a painting by the seventeenth century artist Jan Vermeer, "Woman Taken in Adultery," the banker claimed that he had been acting on behalf of Van Meegeren, a third-rate painter, who was duly imprisoned for the crime of collaboration. However, shortly after his arrest, Van Meegeren announced that he had never sold this painting to Goering and that it was in fact a forgery. He claimed that, together with "Disciples at Emmaus" and several others, he himself had painted them. In order to prove this claim, he began to forge a further work by Vermeer while in prison. When he learned that his charge had been changed from that of collaboration to one of forgery, he refused to finish the work as it would provide evidence against himself and reveal the means by which he had 'aged' the paintings.

A team of art historians, chemists and physicists was employed to settle the dispute. They took x-rays to establish what paintings might lie underneath, analysed pigments in the paints and looked for chemical and physical signs of old age authenticity. They found traces of modern pigment in the cobalt blue, although Van Meegeren had been very careful, in all other cases, to use pigments that the artists themselves might have used. Further examination indicated Van Meegeren had scraped old canvases that he had used as a base for the forgeries. The discovery of phenoformaldehyde, which he had cleverly mixed into the paints and then baked in the oven so that it would harden to bakelite as was typical of old paint, finally sealed his fate and he was convicted in October 1947 and sentenced to a year in prison. (He died of a heart attack while in prison in December of that year.)

But the controversy was not over. Art lovers claimed that "Disciples at Emmaus" was far too beautifully painted to be a forgery and was in quite a different class from the other forgeries. They demanded more scientific evidence to prove it a fake. Besides, a noted art historian A. Bredius had previously certified it as authentic, and it had been bought by the Rembrandt Society for a sizeable sum of money. Reputations and money were at stake here.

A scientific study was undertaken with the results based on the notion of radioactive decay. The atoms of certain elements are unstable, or radioactive, and within a given time (which depends on the element involved) a number of the atoms disintegrate. This rate of disintegration has been shown to be proportional to the number of atoms present at that time, $N(t)$. So, starting from $N(t_0) = n_0$, where n_0 is the original amount of material,

$$\frac{dN}{dt} = -\lambda N, \qquad \text{with} \qquad N = n_0 e^{-\lambda(t-t_0)},$$

where λ, the decay constant, is chosen positive.

From here, if the initial formation of the substance was at time t_0, then its age is $(1/\lambda)\ln(n_0/N)$. N can be measured, but n_0, the original amount of the material, is harder to find.

A useful quantity to know is the half-life of an element, the time taken for the number of atoms present at a particular time to halve. Now if $N = \frac{1}{2}n_0$, then rearranging the above gives

$$\text{half-life} = \frac{\ln 2}{\lambda} \simeq \frac{0.69}{\lambda}.$$

The half-lives of many radioactive substances are well known: for Uranium-238 it is 4.5 billion years, for Carbon-14 is 5,568 years, for Lead-210 it is 22 years and for Polonium-214 it is less than 1 second.

All paintings contain small amounts of radioactive Lead-210, which we know to decay with a half-life of 22 years, as this is contained in lead white which has been used by artists as a pigment for over 2,000 years. Lead white contains lead metal (which is smelted from rocks), which in turn contains small amounts of Lead-210 and extremely small amounts of Radium-226 (mostly removed in the smelting process). There is no further supply of lead due to the absence of Radium-226, which decays to Lead-210 in a series of steps, and so Lead-210 initially decays rapidly. Radium-226 has a half-life of 1,600 years and since it exists in very small amounts in lead white, eventually the process of lead decay stabilises when the amount disintegrating balances with the amount produced by radium disintegration.

Let $N(t)$ be the amount of Lead-210, and then, as before,

$$\frac{dN}{dt} = -\lambda N + R(t), \qquad n_0 = N(t_0),$$

where $R(t)$ is the rate of disintegration of Radium-226 per minute per gram of lead white. Since we are interested in a time period of 300 years at the most, and the half-life of Radium-226 is 1,600 years, and the amount present is so small, we can consider $R(t) = R$ to be a constant. Solving the equation, using an integrating factor of $e^{\lambda t}$, we conclude that

$$N(t) = \frac{R}{\lambda}\left(1 - e^{-\lambda(t-t_0)}\right) + n_0 e^{-\lambda(t-t_0)}.$$

R and N can be measured easily and λ is known; however, n_0 is more difficult and is dealt with in the following way. The above equation can be rearranged to give

$$\lambda n_0 = \lambda N e^{\lambda(t-t_0)} - R(e^{\lambda(t-t_0)} - 1),$$

where λn_0 is the disintegration rate when the white lead was extracted from the ore at time t_0. A good estimate of a reasonable size for this disintegration rate can be obtained by taking current measurements for such ore, and they provide a range from 0.18 to 140 disintegrations per minute per gram. In exceptional circumstances, the rate has been measured as 30,000 disintegrations (per minute per gram). This gives a very wide range and thus excludes any possibility of predicting the age of the painting with much accuracy. However, an exact date is not required. What is needed is to establish if the painting is old (about 300 years old) or modern (painted within the last 30 to 50 years).

Suppose it were 300 years old, then λn_0 as expressed above should provide an original (at t_0) disintegration rate of much less than 30,000. Alternatively, if it is a modern forgery then λn_0 should be ridiculously large, or far greater than 30,000 disintegrations per minute per gram.

Using the half-life formula to find λ ($\lambda = (\ln 2)/22$) it follows that $e^{300\lambda} = 2^{150/11}$. The current disintegration rate can be measured, and since it is the same as for Polonium-210 after a few years, and as the latter is easier to measure in white lead, it is taken as 8.5 disintegrations $(\text{min}^{-1}\text{g}^{-1})$ (measured from ^{210}Po) for the painting "Disciples of Emmaus." Now, with $R = 0.8$, the equation gives

$$\lambda n_0 = 8.5 \times 2^{150/11} - 0.8(2^{150/11} - 1) > 98,000.$$

This painting was definitely not painted by Vermeer in the seventeenth century.

2.4 Scenario: Pacific rats colonise New Zealand

The scenario that follows emphasises the care with which such models should be adopted when applying this dating procedure as 'proof' of a proposed theory. This scenario is based on an article by Anderson (1996).

Opinions are divided as to when humans first occupied New Zealand. Some, due to changes in the vegetation, favour an arrival date of 1400BP (where BP means before present) or earlier, while others argue, based on archaeological evidence, for a date of 900BP or later. Anderson (Anderson, 1996) discusses the dating of Pacific rats (*Rattus exulans*) as an indicator of human arrival as they would have arrived as 'stowaways' with the first Polynesians. The advantage of this means of establishing an arrival time is that the rat population would have expanded rapidly and thus evidence of them should be easy to find.

To ensure the least contamination of the material by other carbon sources, clean hard bones were carefully selected from dry shelters and dunes in the Shag River Mouth. The idea was to use the bone protein (collagen) for obtaining the age as it would hold no inbuilt age, a potential problem associated with marine reservoir effects. (Some shell exposed only to water from deep in the ocean is exposed to a very different ratio, a lower ratio, of carbon to ^{14}C than that in air. Thus different dates can result. These are marine reservoir effects.)

The AMS (accelerator mass spectroscopy) radiocarbon dating method was used since it only requires very tiny amounts of carbon and the Institute of Geological and Nuclear Sciences, Ltd. (IGNS) produced a range with dates as early as 2000BP. This suggested that the rats were in existence in New Zealand at least a millennium earlier than previously thought. By 1996 this date had been accepted as 'fact'.

Anderson proposes that these 'facts' and dates are questionable. Such dates are not consistent with the carbon chronologies associated with moa eggshell, charcoal and bones using non-protein (non-collagen) samples. He suggests that there are further factors that contribute to carbon levels. There is a considerable likelihood and evidence that the rats ate insects living in rotting wood, in which case the radiocarbon activity in the rats may have been depleted to levels approaching the age of the old logs — possibly a thousand or thousands of years older than the rats.

Thus, in the case of the New Zealand rat bones, the radiocarbon dating procedure was in fact dating these ancient New Zealand logs and not the rats which, 1,000 or so years later, arrived with the first New Zealanders and set about eating the insects that were eating the rotten old logs! All this in spite of the extreme care taken in avoiding any possible contamination of the samples.

2.5   Lake pollution models

First we consider the simple example of the concentration of salt in a tank of water. We then apply this method to model the changing concentration of pollution in a lake system.

Background

Pollution in our lakes and rivers has become a major problem, particularly over the past 50 years. In order to improve this situation in the future, it is necessary to gain a good understanding of the processes involved. Some way of predicting how the situation might improve (or decline) as a result of current management practices is vital. To this end we need to be able to predict how pollutant amounts or concentrations vary over time and under different management strategies.

General compartmental model

This problem can be considered as a compartmental model with a single compartment, the lake, as is illustrated in Figure 2.5. Applying the balance law there is an input of polluted water from the river(s) flowing into the lake, or due to a pollution dump into the lake, and an output as water flows from the lake carrying some pollution with it.

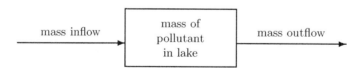

Figure 2.5: Input-output compartmental diagram for pollution in a lake.

This leads us to the word equation, for the mass of pollutant in the lake,

$$\left\{ \begin{array}{c} rate\ of\ change \\ of\ mass \\ in\ lake \end{array} \right\} = \left\{ \begin{array}{c} rate \\ mass \\ enters\ lake \end{array} \right\} - \left\{ \begin{array}{c} rate \\ mass \\ leaves\ lake \end{array} \right\}. \qquad (2.8)$$

Before developing the differential equations describing this process, we examine the simple model of a salt solution in a tank.

An example problem of salt dissolved in a tank

(The following example is adapted from Borelli and Coleman (1996).) As an example problem involving concentrations, let us consider how to model the concentration of salt dissolved in a tank of water. The solution of water and salt is sometimes called brine. A large tank contains 100 litres of salt water. Initially s_0 kg of salt is dissolved. Salt water flows into the tank at the rate of 10 litres per minute, and the concentration $c_{in}(t)$ (kg of salt/litre) of this incoming water-salt mixture varies with time. We assume that the solution in the tank is thoroughly mixed and that the salt solution flows out at the same rate at which it flows in: that is, the volume of water-salt mixture in the tank remains constant.

Example 2.5: *Find a differential equation for the amount of salt in the tank at any time t. (Note that concentration can be defined as the mass of salt per unit volume of mixture.)*

Solution: *The tank is the 'compartment'. Let $S(t)$ be the amount (mass) of salt in the tank at time t. $S(t)$ is what we want to find. First we draw a diagram of the situation as in Figure 2.6.*

Figure 2.6: Input-output compartmental diagram for salt solution in tank.

Then, applying the balance law to the mass of salt, this can be described by the word equation

$$\left\{ \begin{array}{c} \text{rate of change} \\ \text{of mass of salt} \\ \text{in tank} \end{array} \right\} = \left\{ \begin{array}{c} \text{rate} \\ \text{mass of salt} \\ \text{enters tank} \end{array} \right\} - \left\{ \begin{array}{c} \text{rate} \\ \text{mass of salt} \\ \text{leaves tank} \end{array} \right\}. \tag{2.9}$$

The rate at which the mass of salt is added is the flow rate 10 (litres) multiplied by the concentration of the incoming mixture $c_{in}(t)$ (in kg/litre),

$$\left\{ \begin{array}{c} \text{rate} \\ \text{salt} \\ \text{enters tank} \end{array} \right\} = 10 c_{in}(t).$$

The rate at which the salt is leaving the tank is the flow rate out multiplied by the concentration of salt in the tank, $S(t)/100$, so

$$\left\{ \begin{array}{c} \text{rate} \\ \text{salt} \\ \text{leaves tank} \end{array} \right\} = 10 \times \frac{S(t)}{100} = \frac{S(t)}{10},$$

again measured in kg/min.

So the differential equation describing the rate of change in the mass of salt in the tank is given by

$$\frac{dS}{dt} = 10 c_{in}(t) - \frac{S(t)}{10}.$$

To solve the differential equation we also need an initial condition. At time $t = 0$ we assume the amount of salt dissolved in the tank is s_0 (measured in kg). Hence we are required to solve the IVP

$$\frac{dS}{dt} = 10 c_{in}(t) - \frac{1}{10} S(t), \qquad S(0) = s_0. \tag{2.10}$$

This is a linear, first-order ODE (ordinary differential equation). It is not a separable equation, but we can solve it using the technique of integrating factors (see Appendix A.4 on integrating factors). We obtain a solution for an arbitrary function for input concentration $c_{\text{in}}(t)$ and then later investigate different functional forms of $c_{\text{in}}(t)$.

Example 2.6: *Using the technique of integrating factors, solve the initial value problem (2.10) on the interval $[0, t]$ in terms of an arbitrary function $c_{\text{in}}(t)$.*

Solution: *Multiplying equation (2.10) by the integrating factor (IF)* $e^{\int_0^t \frac{1}{10}\, ds} = e^{(t-0)/10} = e^{t/10}$ *we get*

$$\frac{d}{dt}\left(S(t) \times e^{t/10}\right) = 10 c_{\text{in}}(t) e^{t/10},$$

whence, integrating over $[0, t]$, gives

$$S(t)e^{t/10} - s_0 e^{0/10} = 10 \int_0^t c_{\text{in}}(s) e^{s/10}\, ds,$$

where s is a dummy variable of integration. Hence,

$$S(t) = s_0 e^{-t/10} + 10 e^{-t/10} \int_0^t c_{\text{in}}(s) e^{s/10}\, ds. \tag{2.11}$$

Note that the initial condition $S(0) = s_0$ was included when applying the boundaries of the given interval of integration. Alternatively, for the more general case of indefinite integrals, see Appendix A.4.

The original question was: How much salt is in the tank at time t? To answer this we need to know $c_{\text{in}}(t)$, the concentration of the incoming salt solution. Suppose $c_{\text{in}}(t) = c_1$ is a constant; then (2.11) gives

$$S(t) = s_0 e^{-t/10} + 100 c_1 (1 - e^{-t/10}). \tag{2.12}$$

Suppose, instead, c_{in} is a sinusoidal function, say $c_{\text{in}}(t) = 0.2 + 0.1 \sin t$; then evaluating the integral (by integration tables or by using `Maple` or `MATLAB`) gives

$$S(t) = s_0 e^{-t/10} + 20 + \frac{10}{101}(\sin t - 10 \cos t - 192 e^{-t/10}). \tag{2.13}$$

Note that in equations (2.11), (2.12) and (2.13), as $t \to \infty$ the effect of the initial condition on the solution decreases and becomes negligible. Also, these solutions have two parts: one that is the response to the initial data, and one that is the response to the input. We discuss this further in Section 2.10 where we cover some general theory about first-order linear differential equations.

A lake pollution model

We return now to the problem of pollution in a lake. We apply the above theory to investigate the changing concentration of a pollutant.

As usual, we need to make some assumptions while developing the model and to this end we assume that the lake has a constant volume V, and that it is continuously well mixed so that the pollution is uniform throughout.

Let $C(t)$ be the concentration of the pollutant in the lake at time t. Let F be the rate at which water flows out of the lake in m^3/day. Since the volume is constant, we have

$$\left\{\begin{array}{c} \text{flow of} \\ \text{mixture} \\ \text{into lake} \end{array}\right\} = \left\{\begin{array}{c} \text{flow of} \\ \text{mixture} \\ \text{out of lake} \end{array}\right\} = F.$$

Applying the balance law to the mass of the pollutant $M(t)$ we can describe the process in words,

$$\left\{\begin{array}{c} \text{rate of} \\ \text{change of} \\ \text{mass of} \\ \text{pollutant} \\ \text{in lake} \end{array}\right\} = \left\{\begin{array}{c} \text{rate at} \\ \text{which the} \\ \text{pollutant enters} \\ \text{the lake} \end{array}\right\} - \left\{\begin{array}{c} \text{rate at} \\ \text{which the} \\ \text{pollutant leaves} \\ \text{the lake} \end{array}\right\}.$$

This translates into the differential equation for the changing mass

$$M'(t) = Fc_{\text{in}} - F\frac{M(t)}{V}, \tag{2.14}$$

where c_{in} is the concentration (in units of mass per unit of volume, such as g/m^3) of the pollutant in the flow entering the lake.

Now, since $M(t) = C(t)V$ we have that $M'(t) = C'(t)V$ (since V is constant) and hence that $C'(t) = M'(t)/V$. With this change of variable the differential equation (2.14) for the mass is transformed to a differential for the concentration of the pollutant in the lake,

$$\frac{dC}{dt} = \frac{F}{V}c_{\text{in}} - \frac{F}{V}C. \tag{2.15}$$

If the flow rate F is constant with time, then we can use the technique of separating the variables to solve the equation, as in the following example.

Example 2.7: *Solve the differential equation (2.15) with the initial condition $C(0) = c_0$.*

Solution: *Separating the variables,*

$$\int \frac{1}{c_{\text{in}} - C}\, dC = \int \frac{F}{V}\, dt.$$

Integrating gives

$$-\ell n\, |c_{\text{in}} - C| = \frac{F}{V}t + K \quad (K \text{ an arbitrary constant}),$$

and thus

$$C(t) = c_{\text{in}} - e^{-K}e^{-Ft/V}.$$

Using the initial condition $C(0) = c_0$, we obtain

$$C(t) = c_{\text{in}} - (c_{\text{in}} - c_0)e^{-Ft/V}. \tag{2.16}$$

Note again that the solution can be divided into two parts: $c_0 e^{-Ft/V}$ is the contribution from the initial data, and $(c_{\text{in}} - c_{\text{in}}e^{-Ft/V})$ is the contribution from the pollution inflow to the system. Also note that as $t \to \infty$, $C(t) \to c_{\text{in}}$. That is,

$$\lim_{t \to \infty} (c_{\text{in}} - (c_{\text{in}} - c_0)e^{-Ft/V}) = c_{\text{in}},$$

and the concentration in the lake increases/decreases steadily to this limit.

`Maple` or `MATLAB` could have been used to establish the solution and plot the results. The following code (see Listing 2.7 and Listing 2.8) provides the time-dependent plot for a variety of initial pollution concentrations where the safety threshold for the level of pollution is illustrated with a line.

Listing 2.7: Maple code: c_cm_lake.mpl

```
restart:with(plots):
cin:=3;V:=28;F:=4*12;threshold:=4;init_c:=10;
de1:=diff(C(t),t)=F/V*(cin-C(t));
soln:=c0->dsolve({de1,C(0)=c0},C(t),numeric):
plot1:=c0->odeplot(soln(c0),[t,C(t)],0..8):
list1:=seq(plot1(i/2),i=1..12):
line1:=plot([[0,threshold],[8,threshold]]):
display(list1,line1);
```

Listing 2.8: MATLAB code: c_cm_lake.m

```
function c_cm_lake
 global F V cin %set global variables
 N = 100;
 cin=3; V=28; F=4*12; c0=10; tend=4;
 t = linspace(0,1,N)
 [tsol, ysol] = ode45( @derhs, [0, tend], c0 );
 plot(tsol, ysol)

function ydot = derhs(t, c)
 global F V cin %set global variables
 ydot = F/V*(cin - c);
```

Suppose we start with a polluted lake, with initial concentration c_0, but subsequently only fresh water enters the lake. We are interested in calculating the time for the pollution level in the lake to reduce to a specified value.

Example 2.8: *How long will it take for the lake's pollution level to reach 5% of its initial level if only fresh water flows into the lake?*

Solution: *From the solution (2.16), and setting $c_{in} = 0$, we have*

$$C(t) = c_0 e^{-Ft/V}.$$ (2.17)

Rearranging to solve for t,

$$t = -\frac{V}{F} \ell n \left(\frac{C}{c_0} \right).$$

Since $C = 0.05 c_0$, we find t given by

$$t = -\ell n(0.05) \times \frac{V}{F} \simeq \frac{3V}{F}.$$

Application to American lakes

Consider two American lakes where Lake Erie flows into Lake Ontario (Mesterton-Gibbons (1989)). Lake Erie has $V = 458 \times 10^9 \, \text{m}^3$ and $F = 480 \times 10^6 \, \text{m}^3 \text{day}^{-1} = 1.75 \times 10^{11} \, \text{m}^3 \text{year}^{-1}$. Thus if $t_{0.05}$ is the time it takes for the pollution level to reach 5% of its current level, then from above $t_{0.05} \approx 7.8$ years.

Similarly, Lake Ontario has $V = 1{,}636 \times 10^9 \, \text{m}^3$, $F = 572 \times 10^6 \, \text{m}^3 \text{day}^{-1} = 2.089 \times 10^{11} \, \text{m}^3 \text{year}^{-1}$ and we calculate $t_{0.05} \approx 23.5$ years if only fresh water flows into Lake Ontario. Although the flow rate in and out of Lake Ontario is similar to Lake Erie, it takes significantly longer to clear the pollution from Lake Ontario due to the larger volume water in Lake Ontario.

Maple provides a method of substitution that can be used to find this time. The Maple code, in Listing 2.9, with $c_0 = 10$ in the above example, shows how to do this.

Listing 2.9: Maple code: c_cm_lake_subs.mpl

```
restart:
de1:=diff(C(t),t)=(F/V)*(cin-C(t));
soln:=c0->dsolve({de1,C(0)=c0},C(t)):
cin:=0;c0:=10:
ans:=soln(c0):
get_t:=solve(subs(C(t)=(0.05*c0),ans),t):
evalf(get_t);
```

We have made some assumptions in this model. First, we assumed the lake to be well mixed. However, if the lake were not well mixed, then the time taken to flush the lake is likely to be longer since poor mixing will prolong the process. We have also assumed the flow rate F to be constant.

Seasonal flow rate

Suppose that the flow rate into and out of the lake, $F(t)$, varied seasonally. We could let $F(t) = F_0(1 + \epsilon \sin(2\pi t))$, where t is now measured in years, F_0 is the mean in/out flow (with units of litres per year) and $0 < epsilon < 1$ so that $F \geq 0$. Here $2F_0\epsilon$ is the maximum variation (maximum-minimum) in the flow rate, and we assume this to be small compared to the mean flow rate F_0. The period of the sin function is 1 year.

Let us look at the modified equation for the concentration:

$$\frac{dC}{dt} = -\frac{F_0}{V}\left(1 + \epsilon \cos(2\pi t)\right)C.$$

Separating the variables and integrating gives

$$\ell n\,|C(t)| = \frac{-F_0}{V}t - \frac{F_0}{V}\frac{\epsilon}{2\pi}\sin(2\pi t) + K,$$

with K some arbitrary constant, and applying the initial condition and solving for $C(t)$,

$$C(t) = c_0 e^{-F_0(t + \frac{\epsilon}{2\pi}\sin(2\pi t))/V}.$$

Since $\sin(2\pi t) \leq 1$ and ϵ is small, we have

$$C(t) \leq c_0 e^{-F_0(t + \frac{\epsilon}{2\pi})/V}$$
$$\leq c_0 e^{-F_0(t + \frac{1}{2\pi})/V} \qquad (2.18)$$

since $\epsilon < 1$. For large t and small ϵ, there is little difference between the predictions from this solution and that of the simpler model as given in equation (2.17), concerning the time taken for the lake to purify. Thus, from (2.18) and (2.17), we can be reasonably sure that the times we estimated applying the previous very simple model will provide a lower bound for the purification time.

Further extensions

This model can easily be extended to encompass a system of lakes (one lake flowing into another) by setting the inflow in the model of the downstream lake equal to the outflow of the upstream lake. Clearly, there may be several lakes/rivers feeding into the downstream lake at any time, and a single upstream lake may only be a fraction of the flow feeding the lower lake.

> **Summary of skills developed here:**
> - *Apply the concept of modelling with compartments and the balance law.*
> - *Extend the model developed in this section to a system of lakes, or the addition of further pollution sources into a single lake.*
> - *Sketch graphs of pollution levels over time.*
> - *Generate solutions, using* Maple *or analytical methods, to the model and its extensions.*

2.6   Case Study: Lake Burley Griffin

The information for the case study is adapted from the research paper Burgess and Olive (1975).

Lake Burley Griffin in Canberra, the capital city of Australia, was created artificially in 1962 for both recreational and aesthetic purposes. In 1974 the public health authorities indicated that pollution standards set down for safe recreational use were being violated and that this was attributable to the sewage works in Queanbeyan upstream (or rather the discharge of untreated sewage into the lake's feeder river).

After extensive measurements of pollution levels taken in the 1970s it was established that, while the sewage plants (of which there are three above the lake) certainly exacerbated the problem, there were significant contributions from rural and urban runoff as well, particularly during summer rainstorms. These contributed to dramatic increases in pollution levels and at times were totally responsible for lifting the concentration levels above the safety limits. As a point of interest, Queanbeyan (where the sewage plants are situated) is in the state of New South Wales (NSW) while the lake is in the Australian Capital Territory, and, although they are a ten-minute drive apart, the safety levels/standards for those who swim in NSW are different from the standards for those who swim in the Capital Territory!

In 1974 the mean concentration of the bacteria faecal coliform count was approximately 10^7 bacteria per m^3 at the point where the river feeds into the lake. The safety threshold for this faecal coliform count in the water is such that for contact recreational sports no more than 10% of total samples over a 30-day period should exceed 4×10^6 bacteria per m^3.

Given that the lake was polluted, it is of interest to examine how, if sewage management were improved, the lake would flush out and if and when the pollution levels would drop below the safety threshold.

The system can be modelled, very simply, under a few assumptions. Flow (F) into the lake is assumed equal to flow out of the lake, and the volume (V) of the lake will be considered constant and is approximately $28 \times 10^6 \, m^3$. Further, the lake can be considered as well mixed in the sense that the pollution concentration throughout will be taken as constant. Under these assumptions a suitable differential equation model for the pollutant concentration is

$$\frac{dC}{dt} = \frac{F}{V}c_{\text{in}} - \frac{F}{V}C, \tag{2.19}$$

where c_{in} is the concentration of the pollutant entering the lake. With the initial concentration taken as c_0, the solution is

$$C(t) = c_{\text{in}} - (c_{\text{in}} - c_0)e^{-Ft/V}. \tag{2.20}$$

With only fresh water entering the lake ($c_{\text{in}} = 0$), with a mean monthly flow of $4 \times 10^6 \text{m}^3 month^{-1}$ and with the initial faecal coliform count of 10^7 bacteria per m^3 (as was measured in 1974), the lake will take approximately 6 months for the pollution level to drop below the safety threshold. However, pure water entering the lake is not a very realistic scenario with three sewage plants and much farmland upstream, and so including the entrance of polluted river water into the lake model is essential.

From the solution (2.20), as time increases so the concentration of a pollutant in the lake will approach the concentration of the polluted water entering the lake. This level is independent of the initial pollution level in the lake and if $c_0 > c_{\text{in}}$ then the level of pollution in the lake decreases monotonically to c_{in}, while if $c_0 < c_{\text{in}}$ then the level increases steadily until it reaches c_{in}. Thus with the faecal coliform entering the lake at a count of 3×10^6 bacteria per m^3 the concentration of the pollutant in the lake will approach this level with time. This is evident in Figure 2.7, where `Maple` has been used to solve the differential equation (2.19) and plot the results, for a number of different initial concentrations.

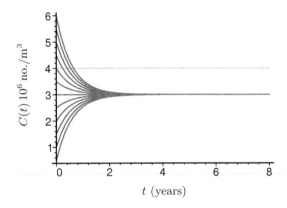

Figure 2.7: Pollution levels in Lake Burley Griffin over time with constant rates of flow and pollution. A range of initial concentrations has been used. The grey line is the pollution threshold level.

However, the model is still simplistic. Flow rates change over the year with on average a seasonal pattern, while the amount of pollution reaching the lake will itself be seasonal. Assuming a sinusoidal pattern over the year, a rough estimate of the inflow concentration from the available data in the 1970s is $c_{\text{in}}(t) = 10^6(10 + 10\cos(2\pi t))$ bacteria m^{-3} and for the flow rate, $F(t) = 10^6(1 + 6\sin(2\pi t))\,\text{m}^3 year^{-1}$. Note here that the concentration of the pollutant increases when the flow is low. We need to re-solve the differential equation (2.19) with F no longer constant. Using `Maple` to determine the solution and plot the concentration of pollution in the lake over time, the results are displayed in Figure 2.8.

It should be noted that this model for Lake Burley Griffin is still simplistic in its assumption of a well-mixed body of water. If the concentration decreased from the point of river entry to the point of outflow, then the flushing time could take considerably longer. Further, in most lakes there is a main channel of water flow that flushes regularly, and adjacent to this channel are areas of trapped water that flush less frequently and through a very different process. The process is that of diffusion, which operates at a microscopic level and

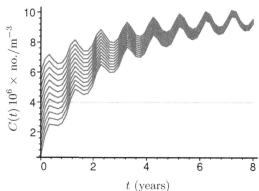

Figure 2.8: Pollution levels in Lake Burley Griffin over time with seasonal incoming flows and pollutant concentrations. A range of initial concentrations has been used. The grey line is the pollution threshold level.

is extremely gradual. Thus pockets of the lake may have a much higher (or lower) pollution concentration than others, and these may also be the protected bays where swimming is most likely to take place.

2.7 Drug assimilation into the blood

We investigate two simple models of cold pill assimilation into the bloodstream that are adapted from Borelli and Coleman (1996). In the first model we consider a single cold pill and in the second a course of cold pills. The basic idea of using a compartmental diagram to build the models is used.

Background

We readily take pills or drink alcohol without necessarily having a good understanding of how these drugs are absorbed into the bloodstream or for how long they have an effect on us. The warnings on the packaging list some of the effects and are intended to ensure safety for all users. In the following example and case study, we see that different drugs are absorbed into, and extracted from, the blood at very different rates. Some may affect us for many hours after the medication has ceased. Further, in the case study on alcohol absorption, we see how different body masses and the sex of an individual can radically modify the effects of alcohol.

The drug dissolves in the gastrointestinal tract (GI-tract) and each ingredient is diffused into the bloodstream. They are carried to the locations in which they act and are removed from the blood by the kidneys and the liver. The assimilation and removal may occur at different rates for the different ingredients of the same pill.

General compartmental model

We can consider this problem as a compartmental model with two compartments, corresponding to the GI-tract and the bloodstream. This is illustrated in Figure 2.9. The GI-tract compartment has a single input and output and the bloodstream compartment has a single input and output.

Figure 2.9: Input-output compartmental diagram for drug assimilation.

This application of the balance law gives us two word equations, one for each compartment. We have

$$\left\{\begin{array}{c} rate\ of\ change \\ of\ drug \\ in\ GI\ tract \end{array}\right\} = \left\{\begin{array}{c} rate\ of \\ drug \\ intake \end{array}\right\} - \left\{\begin{array}{c} rate \\ drug \\ leaves\ GI\text{-}tract \end{array}\right\},$$

$$\left\{\begin{array}{c} rate\ of\ change \\ of\ drug \\ in\ blood \end{array}\right\} = \left\{\begin{array}{c} rate \\ drug \\ enters\ blood \end{array}\right\} - \left\{\begin{array}{c} rate \\ drug \\ leaves\ blood \end{array}\right\}. \tag{2.21}$$

We let $x(t)$ be the amount of a drug in the GI-tract at time t and $y(t)$ the amount in the bloodstream at time t. We consider two models: a single cold pill where there is no ingestion of the drug except that which occurs initially, and a course of cold pills where the drug intake is assumed to occur continuously.

The common cold remains without a cure. However, there are pills that can be taken to relieve some of the congestion and symptoms, such as watering eyes and a running nose, through the action of a decongestant and an antihistamine. The cold pills we consider in the following two models consist of these two drugs.

Model I: A single fast-dissolving cold pill

In the GI-tract we consider the pill to have been swallowed, and so after this event (over subsequent time) we have nothing more entering the GI-tract. The pill quickly dissolves and the drug begins to enter the bloodstream from the GI-tract. So, for the GI-tract there is only an output term.

Assuming the output rate is proportional to the GI-tract drug concentration, which is therefore proportional to the amount of drug in the bloodstream, then

$$\frac{dx}{dt} = -k_1 x, \qquad x(0) = x_0, \tag{2.22}$$

where x_0 is the amount of a drug in the pill, our initial condition, and k_1 is a positive coefficient of proportionality (the rate constant). We assume that the instant the pill enters the GI-tract at $t = 0$ it dissolves instantaneously so $x(0) = x_0$.

In the bloodstream the initial amount of the drug is zero, so $y(0) = 0$. The level in the bloodstream increases as the drug diffuses from the GI-tract and decreases as the kidneys and liver remove it. Thus,

$$\frac{dy}{dt} = k_1 x - k_2 y, \qquad y(0) = 0, \tag{2.23}$$

with k_2 another positive constant of proportionality. The cold pill is made up of a decongestant and an antihistamine, and the coefficients of proportionality, k_1 and k_2, are different for the different component drugs in the pill.

As expected, as t increases both x and y approach zero although the rate at which this occurs depends on the coefficients k_1 and k_2 associated with each drug. Some values are given in Table 2.1.

Table 2.1: Values of constants for decongestant and antihistamine in cold pills. From Borelli and Coleman (1996).

	Decongestant	Antihistamine
k_1	1.3860/hr	0.6931/hr
k_2	0.1386/hr	0.0231/hr

Using these values we easily solve the differential equations numerically, with MATLAB or Maple, and we can graph the changing values for x and y in each of the two 'compartments'. Some sample MATLAB code is given in Listing 2.10 to graph the amounts of decongestant as a function of time. In Figure 2.10 the solutions are plotted for both decongestant and anti-histimine. The Maple code used to produce this is given in Listing 2.11.

Listing 2.10: MATLAB code: c_cm_coldpills.m

```
function c_cm_coldpills1
global k1 k2;
k1=1.386; k2=0.1386;
tend=15; %end time in hours
x0=1; y0=0; u0 = [x0; y0];
[tsol, usol] = ode45(@rhs, [0, tend], u0);
xsol = usol(:,1);
ysol = usol(:,2);
plot(tsol, xsol,'k'); hold on
plot(tsol, ysol,'r:'); hold off

function udot = rhs(t, u)
global k1 k2
x = u(1); y=u(2);
xdot = -k1*x;
ydot = k1*x - k2*y;
udot = [xdot; ydot];
```

Listing 2.11: Maple code: c_cm_coldpills.mpl

```
restart:with(plots):with(DEtools):
k1:=1.386;k2:=0.1386;k3:=0.6931;k4:=0.0231;
I1:=0;I2:=0;hours:=15;initx:=1;inity:=0;initw:=1;initz:=0;
de1:=diff(x(t),t)=I1-k1*x(t);
de2:=diff(y(t),t)=k1*x(t)-k2*y(t);
pair1:=[de1,de2]:
plot1:=DEplot(pair1,[x,y],t=0..hours,{[0,initx,inity]},
    stepsize=0.1,scene=[t,x],linecolour=gray,arrows=NONE):
plot2:=DEplot(pair1,[x,y],t=0..hours,{[0,initx,inity]},
    stepsize=0.1,scene=[t,y],linecolour=gray,arrows=NONE):
de3:=diff(w(t),t)=I1-k3*w(t);
de4:=diff(z(t),t)=k3*w(t)-k4*z(t);
pair2:=[de3,de4]:
plot3:=DEplot(pair2,[w,z],t=0..hours,{[0,initw,initz]},
    stepsize=0.1,scene=[t,w],linecolour=black,arrows=NONE):
plot4:=DEplot(pair2,[w,z],t=0..hours,{[0,initw,initz]},
    stepsize=0.1,scene=[t,z],linecolour=black,arrows=NONE):
display(plot1,plot3);display(plot2,plot4);
```

The coefficients also depend on the age and health of the person involved, and the concentration of a drug may also depend on the person's body mass, which means that for some people the doses may peak higher and/or faster than for the average person, and of course this can be potentially dangerous.

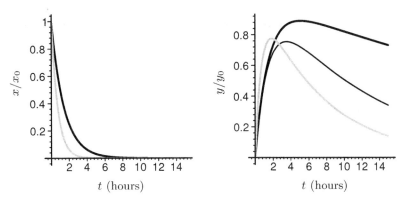

Figure 2.10: Single cold pill. The graph on the left represents the GI-tract and that on the right the bloodstream. The levels of antihistamine are illustrated with black lines ($k_2 = 0.0231$ thick line, $k_2 = 0.08$ thin line) and the decongestant with a grey line. The other parameter values are given in Table 2.1, with typical initial values for the antihistamine and decongestant of 4 mg and 60 mg.

Exact solution for Model I

The differential equations (2.22) and (2.23) are not coupled because the first equation is totally independent of y. So we can solve it independently of the second equation. The solution to this first differential equation is (see Exercise 2.13)

$$x = x_0 e^{-k_1 t}.$$

Using this solution, the second differential equation (2.23) becomes

$$\frac{dy}{dt} = k_1 x_0 e^{-k_1 t} - k_2 y.$$

This equation is a first-order linear differential equation whose solution when $k_1 \neq k_2$ is (see Exercise 2.13)

$$y = \frac{k_1 x_0}{k_1 - k_2} \left(e^{-k_2 t} - e^{-k_1 t} \right).$$

(Note that a different solution is valid when $k_1 = k_2$.)

Model II: A course of slowly dissolving cold pills

In reality, particularly for a cold, we take a course of pills rather than just one. To model this let us assume that the drug is delivered to the GI-tract continuously, which is reasonable for pills that dissolve slowly in the GI-tract. Thus we assume a constant rate of drug input, I (e.g. ml of drug per hour). Also, since the pill dissolves slowly, we assume that initially there is no drug in the GI-tract.

We adjust our model from the previous section to give

$$\frac{dx}{dt} = I - k_1 x, \qquad x(0) = 0,$$
$$\frac{dy}{dt} = k_1 x - k_2 y, \qquad y(0) = 0, \tag{2.24}$$

where I is a positive constant representing the rate of ingestion of the drug (in grams per unit of time). Note that here $x(0) = 0$, whereas in the previous model $x(0) = x_0$.

Once again we have a sequence (cascade) of linear equations that are not coupled and thus can be exactly solved sequentially. For

$$\frac{dx}{dt} = I - k_1 x$$

the solution is (see Exercises, Exercise 2.14)

$$x(t) = \frac{I}{k_1}(1 - e^{-k_1 t}).$$

Then solving for $y(t)$ in the second equation gives (see Exercises, Exercise 2.14)

$$y(t) = \frac{I}{k_2} \left[1 - \frac{1}{k_2 - k_1} \left(k_2 e^{-k_1 t} - k_1 e^{-k_2 t} \right) \right].$$

(Note again that this solution is valid only if $k_1 \neq k_2$.)

As usual we are interested in what happens over time; in this case, as $t \to \infty$, then $x \to I/k_1$ and $y \to I/k_2$. Now we draw the graphs of these functions to establish the accumulating levels of the decongestant and antihistamine in the GI-tract and the bloodstream over a time period where pills are taken each hour and the value of I is constant.

With `Maple` or `MATLAB`, we can easily explore the effects of the antihistamine in the bloodstream as k_2 increases/decreases and determine how sensitive the system is to changes in the parameter k_2; see Figure 2.10.

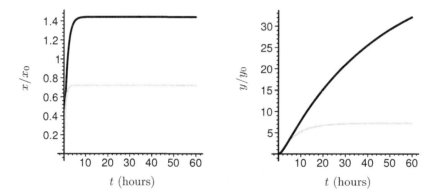

Figure 2.11: Course of cold pills. The graph on the left represents the GI-tract and that on the right the bloodstream. The levels of antihistamine are illustrated with black lines and the decongestant with a grey line. The parameter values are given in Table 2.1, with typical initial values for the antihistamine and decongestant of 0.5 mg and 15 mg.

Note that the level of the antihistamine (which makes you sleepy) is slow to rise, but settles to a high level, while that of the decongestant is faster to rise but settles to a lesser level with time. This is illustrated in Figure 2.11.

Our assumption of I constant holds in the case where drugs are embedded in resins and dissolve at constant rates allowing the drug to be released slowly and evenly over a period of hours. In reality some pills dissolve quickly and thus $I(t)$ should be a pulsing function, possibly a sinusoidal function, representing repeated doses, or some other function of t providing an initial and substantial boost to the drug level and then very little during the remaining time period before the next dose is taken. We do not consider those cases here, but what is required in the model is the replacement of I with an appropriate time dependent function.

> ### Summary of skills developed here:
> - *Be able to model a sequence of processes using the compartmental technique.*
> - *Understand how to find numerical solutions, with* `Maple` *or* `MATLAB`, *and analytically, in each of the sequence of compartments.*
> - *Sketch graphs of solutions in each compartment over time.*
> - *Establish how changes in parameters impact on these solutions.*

2.8 Case Study: Dull, dizzy or dead?

This case study has been adapted from Oakley and Ksir (1993).

Alcohol is unusual in that it requires no digestion and can be absorbed extremely rapidly from the stomach into the bloodstream (particularly if the stomach is empty of food or other liquids), and is absorbed even more rapidly from the intestines. The vapour can also be absorbed through the lungs. Thus, unless alcohol is heavily diluted or taken with food, very little metabolism occurs in the GI-tract (gastrointestinal tract) and all the alcohol is absorbed into the bloodstream. Alcohol is distributed freely to all body fluids and the concentration of alcohol in the brain rapidly approaches that in the blood. Most of the alcohol (90–98%) is oxidised through the liver and excreted, while the remainder leaves the body through the lungs, urine, saliva and sweat. However, the liver can only metabolise alcohol at a constant rate if the concentrations are not small.

The state of drunkenness (or 'happy feeling') experienced is a measure of the blood alcohol level (BAL) or the alcohol concentration in the blood. This concentration is a measure of the total mass of alcohol in grams divided by the total fluid volume in the body. It is measured in grams/100 ml of blood. Thus 100 g/100 ml corresponds to a BAL of 100 (grams/100 ml), and 100 mg/100 ml measures a BAL of 0.1 (grams/100 ml). As is common practice, we will not stipulate the units of BAL on each occurrence, but assume that grams/(100 ml blood) is understood.

Australian law prohibits the driving of vehicles (including boats and horse- or camel-drawn vehicles) for those with a BAL above 0.05. This then relates to 50 mg/100 ml alcohol in the bloodstream. The restriction is a result of U.S. statistics which indicate that a person with a BAL of 0.15 is 25 times more likely to have a fatal accident than one with no alcohol. Furthermore, for 41% of Australian men, excessive alcohol leads to confrontational behaviour. (For women the figure is 12%.)

The alcohol intake into the GI-tract is 'controlled' by the drinker. The amount of alcohol subsequently absorbed into the bloodstream depends on the concentration of alcohol, other liquids and food in the GI-tract, as well as on the weight and sex of the individual. Alcohol is removed from the bloodstream at a constant rate by the liver. This is independent of the body weight, sex of the individual and concentration of alcohol in the bloodstream, and assumes that the liver has not been damaged by large doses of alcohol. (As mentioned above, a small percentage leaves through sweat, saliva, breath and urine. Ignoring this could mean the BAL estimate may be slightly above the true value.)

Table 2.2: Blood alcohol level (BAL in g/(100 ml blood)) and behavioural effects.

BAL	Behavioural effects	
0.05	Lowered alertness, usually good feeling, release of inhibitions, impaired judgement	Dull and dignified
0.10	Slowed reaction times and impaired motor function, less caution	Dangerous and devilish
0.15	Large consistent increases in reaction time	Dizzy
0.20	Marked depression in sensory and motor capability, decidedly intoxicated	Disturbing
0.25	Severe motor disturbance, staggering, sensory perceptions greatly impaired, smashed!	Disgusting and dishevelled
0.30	Stuporous but conscious, no comprehension of what's going on	Delirious and disoriented
0.35	Surgical anaesthesia; minimal level causing death	Dead drunk
0.40	50 times the minimal level; causing death	Dead!

Table 2.3: Effective alcohol intake during drinking.

Absolute alcohol content	Beverage intake	Number of drinks
14 g	1 oz spirits, or 1 glass wine, or 1 can beer	$n = 1$
28 g	2 oz spirits, or 2 glasses of wine, or 2 cans of beer	$n = 2$
$(n \times 14)$ g	n oz spirits, or n glasses of wine, or n cans of beer	n

Let $C_1(t)$ be the concentration of alcohol (effective BAL) in the GI-tract at time t and let $C_2(t)$ be the concentration of alcohol (in BAL) in the bloodstream at time t. Applying the balance law, the rates of change of alcohol in the GI-tract and the bloodstream can be modelled as

$$\frac{dC_1}{dt} = I - k_1 C_2,$$
$$\frac{dC_2}{dt} = k_2 C_1 - \frac{k_3 C_2}{C_2 + M},$$

(2.25)

with constants k_1, k_2 and M positive. Here I is related to the amount of alcohol consumed per unit of time (here taken as one hour). Since C_1 and C_2 are concentrations measured in units of BAL, we can show that $I = i/V_b$, where i is the ingestion rate of alcohol measured in units of BAL/hour and V_b is the volume of fluid in the blood, measured in 100 ml (see Exercise 2.19).

In the case of drinking on an empty stomach, $k_1 = k_2$; if drinking occurs together with a meal (or is diluted), then $k_1 > k_2$. Note that the amount of alcohol leaving the GI-tract and the amount entering the bloodstream is directly proportional to the amount present in the GI-tract.

On the other hand the amount of alcohol metabolised through the liver is represented as a more complicated function, a Michaelis–Menten type function (Michaelis and Menten (1913)). As mentioned above, the amount removed by the liver is constant regardless of the amount of alcohol or its concentration.

This has been modelled with the function $-k_3C_2/(C_2+M)$ for the following reason. If C_2 is large compared with M, then $C_2' \simeq -k_3$ and $C_2(t)$ is decreasing at the constant rate k_3. As C_2 decreases and becomes small compared with M, then $C_2' \simeq -(k_3/M)C_2$ and $C_2(t)$ is an exponentially decaying function ($C_2 \simeq c_0 e^{-(k_3/M)t}$) that ensures that $C_2(t)$ remains positive. An appropriate value for M in this case is $M = 0.005$ in units of BAL.

What follows are model runs for cases where drinking occurred on an empty stomach or with a meal, and where drinking occurred within a short period or continued over several hours. In each case, `Maple` or `MATLAB` can be used to solve the equations and display the graphs.

Drinking on an empty stomach: In this case the alcohol is modelled as being very rapidly absorbed into the bloodstream in the first hour, that is, $k_1 = k_2 \approx 6$, and the intent is to model the increasing/decreasing BAL in subsequent hours. In the case where n initial drinks are consumed and no more alcohol is taken, we have $C_1(0) = c_0$, where c_0 is the effective BAL from the initial drink(s) consumed, and $I = 0$ as there is no subsequent drinking in this case. In the bloodstream, the initial amount of alcohol is 0 (that is, there is no alcohol in the bloodstream prior to drinking) and so $C_2(0) = 0$. The system above becomes

$$\frac{dC_1}{dt} = -k_1C_1, \qquad C_1(0) = c_0,$$
$$\frac{dC_2}{dt} = k_2C_1 - \frac{k_3C_2}{C_2 + M}, \qquad C_2(0) = 0. \tag{2.26}$$

The initial value c_0 is given by $c_0 = nc_s$, where n is the number of drinks as listed in Table 2.3 and c_s is the effective BAL produced by a single standard drink. The value of c_s depends on the sex and weight of the individual, since they dictate the total amount of body fluid for that person. One drink produces 14 g effective alcohol. The total volume of blood fluids in a woman is approximately $0.67 \times W$ litres, where W is her body weight in kg. (For a man the volume of blood fluids is approximately $0.82 \times W$ litres.) The concentration of alcohol will be (amount of alcohol/total fluids in grams per litre) and this can be expressed in g/100 ml, that is, BAL.

So for a male of 68 kg who has three glasses of wine rapidly and then stops drinking, $c_0 = 3 \times c_s$. His total body fluids are $\approx 0.82 \times 68 = 55.76$ litres. Then the concentration of alcohol per drink is given as $c_s = 14/55.76$ g/litre which converts to an effective BAL of 0.025. Thus $c_0 = 3c_s = 0.075$.

Further, k_3 is a measure of the rate at which the liver removes alcohol from the blood, and this value is the same for all persons and is approximately 8g/hr. The associated reduction in BAL depends on the total body fluids of the individual, and for the 68 kg man can be calculated as $8/(55 \times 10) = 0.014$ g/100 ml. Then $k_3 = 0.014$ BAL.

For the case of drinking continuously over time we take $I = (n/T)c_s$ (BAL/hr) and $c_0 = 0$ in the differential equations (2.25), where n is the average number of drinks consumed in T hours. The results are illustrated in Figure 2.12 for the 68 kg male. (The constants are as described above.)

Drinking with a substantial meal: The rate at which alcohol is absorbed into the bloodstream from the GI-tract is substantially reduced when alcohol is diluted or taken with a meal. The presence of fats in particular reduces the resulting BAL. Returning to equation system (2.26) above, we now have that $k_1 > k_2$, that is, the rate at which alcohol leaves

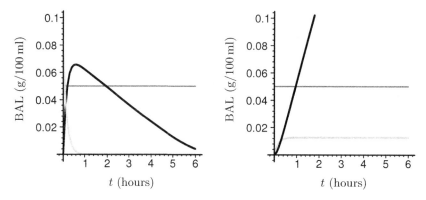

Figure 2.12: The BAL of a 68 kg man drinking on an empty stomach, with $n = 3$ drinks per hour. The first diagram is for a single drinking bout, and the second for a continuous drinking binge. The black line illustrates the BAL in the bloodstream, $C_2(t)$, and the grey line the level in the GI-tract, $C_1(t)$. The legal limit BAL 0.05 is also indicated by the horizontal line.

the GI-tract is greater than the rate at which alcohol enters the bloodstream. In fact, after a substantial meal, the rate of absorption into the bloodstream is approximately halved so that we can replace k_2 with $k_1/2$. (This is a very simplistic approximation and the model would be improved by including a more realistic mathematical relationship between food intake and alcohol absorption into the bloodstream.) The graphs of Figure 2.13 illustrate the effect of a meal on the BAL for the 68 kg man in the case of an initial alcohol intake without further drinking, and in the case of continuous drinking.

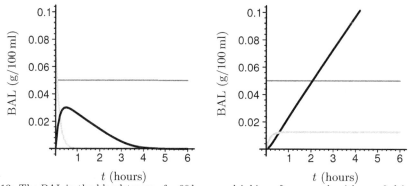

Figure 2.13: The BAL in the bloodstream of a 68 kg man drinking after a meal, with $n = 3$ drinks per hour. The first diagram is for a single drinking bout, and the second for a continuous drinking binge. The black line illustrates the BAL in the bloodstream, $C_2(t)$, and the grey line the level in the GI-tract, $C_1(t)$. The legal limit BAL 0.05 is also indicated by the horizontal line.

As for the aftermath of all this? Well the Germans hear the 'wailing of cats', the Italians call it 'out of tune', the French refer to it as 'woody mouth', the Norwegians say there are 'workmen in my head', the Swedes have 'pain in the roots of my hair' and the English suffer the hangover.

2.9   Cascades of compartments

As exemplified in the previous section(s), linear cascades are models with a series of compartments, as in Figure 2.14. Here each compartments feeds into the next one without any loops. An example was the case of drug assimilation or a series of lakes in Section 2.5.

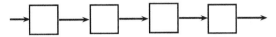

Figure 2.14: Linear cascade of compartments.

In other cases, more than one source can feed into one compartment, or a compartment may output into many compartments, as in Figure 2.15. Such an arrangement is called a branching linear cascade.

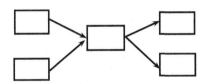

Figure 2.15: Branching cascade of compartments.

Usually in these cases the equations can be solved using the top-down approach that we followed in Section 2.7. That is, solving the first equation and using this solution to solve the second, and so on. Such models are also applicable to the decay of radioactive elements such as Uranium-238, which decays to Lead-206 in a decay cascade involving 16 intermediate elements and isotopes; see Borelli and Coleman (1996):

$$x'_1 = -k_1 x_1$$
$$x'_2 = k_1 x_1 - k_2 x_2$$
$$\vdots$$
$$x'_{17} = k_{18} x_{15} - k_{19} x_{16}$$
$$x'_{18} = k_{19} x_{16} + k_{20} x_{17}.$$

However, situations leading to simultaneous differential equations, which cannot be solved this way, form the basis of Chapter 5.

As with the section on exponential decay, these constants k_i are expressed in terms of the half-lives of the radioactive elements using the relation $\ell n\, 2 = \tau_i k_i$ (see equation (2.7) in Section 2.2). This particular problem is difficult to solve using a computer because of the very different time lengths involved. The half-life of Uranium-238 is approximately 4.5 billion years, while that for the next element, Thorium-234, is 24 days. This problem of very different time lengths in two parts of the same problem, a *stiff problem*, is discussed in Chapter 4.

2.10   First-order linear DEs

When we examined the solutions to the linear differential equations above more closely, there were some interesting features. For the lake problems, we had

$$\frac{dC}{dt} = \frac{F}{V}c_{\text{in}} - \frac{F}{V}C,$$
$$C = c_0 e^{-Ft/V} + c_{\text{in}}(1 - e^{-Ft/V}).$$

Notice that this solution has two parts: one dependent on the initial value of the system c_0 and the other dependent on the input to the system c_{in}. Further, the effect of c_0 decreases, approaching 0 with time while that of c_{in} grows with time.

This phenomenon is a general feature of such linear differential equations and there are some elegant general results that follow. (They are included here to provide a better 'feel' for expected results and the reader is referred to other texts for proofs and more details if they are required.)

Consider the *normal form* of a first-order linear ordinary differential equation (ODE)

$$y' + p(t)y = q(t) \tag{2.27}$$

(normal form just means that the coefficient of y' is one) where the coefficient $p(t)$ and the driving term (or input) $q(t)$ are continuous over a finite time interval. If there is no input ($q(t) = 0$), the equation is said to be *homogeneous*. If $q(t) \neq 0$, then the equation is *inhomogeneous*, that is, it is *driven* or *forced*.

Such an ODE has infinitely many solutions, each associated with a distinct constant of integration. Given an initial condition for the above equation, $y(t_0) = y_0$, the solution will be unique. Further, as we noticed above, the solution to the system can be divided into two parts:

- The response to the initial conditions

- The response to the forcing term

We state this in some theorems:

Theorem 1 (General solution theorem) *Consider the linear ODE* (2.27) *with* $p(t)$ *and* $q(t)$ *continuous on some interval containing* t_0. *Let*

$$P(t) = \int_{t_0}^{t} p(r)\,dr, \quad R(t) = \int_{t_0}^{t} e^{P(s)}q(s)\,ds$$

on that interval. Then every solution $y(t)$ *of the ODE has the form*

$$y(t) = e^{-P(t)}K + e^{-P(t)}R(t), \qquad with\ K\ an\ arbitrary\ constant. \tag{2.28}$$

Conversely, any function $y(t)$ *of the form (2.28) is a solution of the differential equation on the interval.*

Note that one important consequence of this theorem is that every solution of the ODE exists at least in the interval in which $p(t)$ and $q(t)$ are continuous. Another important consequence is that no two solutions corresponding to different values of K can ever intersect.

Why? Because suppose they did for $t = T$, then rearranging $y_1(T) = y_2(T)$ gives

$$
\begin{aligned}
0 &= y_1(T) - y_2(T) \\
&= \left[e^{-P(T)} K_1 + e^{-P(T)} R(T) \right] - \left[e^{-P(T)} K_2 + e^{-P(T)} R(T) \right] \\
&= e^{-P(T)} (K_1 - K_2) \\
&\neq 0 \qquad \text{since } K_1 \neq K_2.
\end{aligned}
$$

Since T is arbitrary, no two solutions corresponding to different initial values can intersect.

Theorem 2 (Solution of initial value problem (IVP)) *Let $p(t)$ and $q(t)$ be continuous in some interval containing t_0, and let y_0 be any constant. Then the IVP*

$$
y' + p(t)y = q(t), \qquad y(t_0) = y_0
$$

has exactly one solution.

Writing $P_0(t) = \int_{t_0}^{t} p(r)\,dr$, this solution is given by

$$
\underbrace{y(t)}_{\substack{total \\ response}} = \underbrace{e^{-P_0(t)} y_0}_{\substack{response\ to \\ initial\ data}} + \underbrace{e^{-P_0(t)} \int_{t_0}^{t} e^{P_0(s)} q(s)\,ds}_{\substack{response\ to \\ input}}.
$$

There is no need to memorise these theorems as they are just the application of the integrating factor technique.

Rewriting the IVP above as $y' = f(t, y) = q(t) - p(t)y$, $y(t_0) = y_0$, we have

Theorem 3 (Existence theorem) *If f and the partial derivative $\partial f / \partial y$ are continuous on a region R in the (t, y)-plane, and (t_0, y_0) is a point in R, then the IVP above has a unique solution $y(t)$ on an interval containing t_0.*

Summary of skills developed here:
- *Identify a linear differential equation for which there exists a unique solution.*
- *Understand the significance of the two separate parts to the solution.*

2.11 Equilibrium points and stability

We have seen, in Section 2.5, that the solutions of differential equations can tend to a steady state, that is, a constant value, as the time becomes very large. It can be useful to determine the value of the steady state, when it exists, directly from the differential equation, in analytic form. This can then allow us to determine how a steady-state solution (also called an equilibrium solution) depends on the various parameters in the problem.

Finding equilibrium solutions

For a differential equation

$$\frac{dx}{dt} = f(x), \tag{2.29}$$

the equilibrium solutions are the solutions $x = x_e$ such that $f(x) = 0$. The following example demonstrates how to find an equilibrium solution.

Example 2.9: *Find all equilbrium points for the differential equation for the concentration of pollutant in a lake, from Section 2.5,*

$$\frac{dC}{dt} = \frac{F}{V}c_{in} - \frac{F}{V}C,$$

where F and V are positive constants.

Solution: *Setting $dC/dt = 0$ we obtain*

$$\frac{F}{V}(c_{in} - C) = 0 \qquad \Rightarrow \qquad C = c_{in}.$$

Stability

An equilibrium solution, x_e, can only be a steady-state solution if the solution tends towards it as the time increases without bound. This is associated with the equilibrium solution being stable. By stable we mean that any solution close to the equilibrium solution will tend towards the equilibrium solution and by unstable the solution will not get closer to the equilibrium point.

We can determine this using Taylor series (see Appendix B.2). We let $x(t) = x_e + \xi(t)$, with $\xi(t) \ll 1$, where the new variable ξ represents the small perturbation from the equilibrium solution. For the differential equation (2.29), if we expand the RHS about the equilibrium solution by letting $x = x_e + \xi$, then the differential equation for the variable ξ is

$$\frac{dx}{dt} = \frac{d(x_e + \xi)}{dt} = f(x) = f(x_e + \xi) \simeq f(x_e) + \xi f'(x_e).$$

Since $f(x_e) = 0$, by the definition of an equilibrium point, then the original differential equation is approximated, close to the equilibrium solution, by

$$\frac{d\xi}{dt} \simeq \xi f'(x_e)$$

for small values of ξ. We can now interpret what happens without actually solving this differential equation. For $\xi > 0$, if $f'(x_e) < 0$ then $d\xi/dt < 0$, so $x(t) = x_e + \xi(t)$ approaches x_e, the equilibrium point. Similarly, for $\xi(t) < 0$, then $x(t) = x_e + \xi(t)$ increases towards the equilibrium solution. Thus the solution is attracted to the equilibrium solution. By a similar argument, when $f'(x_e) > 0$ the solution is repelled from the equilibrium solution. Thus we have

$$\text{equilibrium solution is stable if} \quad f'(x_e) < 0$$

and unstable otherwise.

Example 2.10: Determine if the equilibrium solution $C_e = c_{in}$ is stable or unstable.

Solution: Here

$$f(C) = \frac{F}{V}(c_{in} - C), \qquad C_e = c_{in}, \qquad \Rightarrow f'(C_e) = -\frac{F}{V}.$$

Since F and V are positive parameters, this means that the equilibrium solution $C_e = c_{in}$ is always stable.

We have introduced the idea of equilibrium solutions and stability. We applied these to a linear differential equation where we already knew the solution, so we did not get any great benefit from this analysis. However, the technique can be applied to nonlinear differential equations, some of which we cannot solve easily. In such cases we often end up with more than one equilibrium solution, some of which may be stable and some unstable. More importantly, the concept of equilibrium solutions and stability can also be applied to systems of differential equations where it can provide valuable information about the nature of solutions. We postpone a discussion of this until Chapter 7.

Summary of skills developed here:

- *Know the definition for an equilibrium solution and be able to find it for a single differential equation.*
- *Apply the criteria for stability of an equilibrium solution.*

2.12 Case Study: Money makes the world go around

We have not considered any economic theories in this book, but differential equation dynamics and equilibrium solutions have been applied in this field also. In the following case study we introduce a simple model for economic growth and look at some of the basic concepts underlying such models. This particular model is one of production, or units of 'output', based on investment capital and available labour. As an example of this structure, we introduce the Cobb–Douglas model. This case study is adapted from Solow (1956).

Economic growth is currently seen in Western cultures as the production of a greater economic surplus, which can be used to advance society. As early as 1377, an Arabian economic thinker Ibn Khaldun recognised the concept of economic growth. He noted that, with an increase in population, the available labour increases and production thrives, and that excess wealth so generated can be used to support luxuries, distinguishing labour for the necessities of life from surplus labour for serving luxury. However, Gross Domestic Product (GDP) per capita changed little for most of human history prior to the industrial revolution and mass education.

The neo-classical economic growth model, the type of model we consider here, deals with economic growth considered as increased stocks of capital goods (production or output) that are dependent on available labour and investment capital. Solow and Swan introduced the first attempt at long-term predictions for economic growth in the 1950s (Solow (1956); Swan (1956)). One of the consequent predictions was that economies will attain an equilibrium, in the sense that further capital investment and/or labour will not increase growth. Advances in technology, however, have the potential to alter the equilibrium, as does education and other changes in social structure, and data suggest that the world has, slowly, continued to improve its rate of growth, not settling to a fixed equilibrium.

The model presented here (Solow (1956)) is one of production, that is, units of output, based on investment or capital and available labour. The *warranted rate of growth* is the terminology used for the profitable rate of investment, that is, the growth of capital, and the *natural rate of growth* is that used for the rate of growth of the available labour force. A further term applied in economics is constant returns to scale, which implies that a function F of two variables K and L has the property $F(aK, aL) = aF(K, L)$.

We now introduce the model. We define output (or production) $Y(t)$ as a function of the two factors of production: the stock of capital $K(t)$ and the available labour force $L(t)$, all being functions of time t. Suppose the rate of output savings is a fraction s of the total output so that

$$\frac{dK}{dt} = sY \tag{2.30}$$

at any instant in time, with the production function $Y = F(K, L)$ and F is a function with constant returns to scale. We now make an assumption concerning the labour force: we assume Harrod's natural growth rate (constant n), in the absence of technological advancement, so that

$$L(t) = L_0 e^{nt}, \tag{2.31}$$

which provides a curve describing labour. The basic equation for capital accumulation over time is then

$$\frac{dK}{dt} = sF(K, L_0 e^{nt}). \tag{2.32}$$

Many alternatives are possible for F, and thus many solutions to equation (2.32); however, we are interested in the qualitative nature of these solutions and the possible economic consequences. To investigate this aspect, we combine the two variables K and L, to reduce the number of variables, and carry out some simple analysis.

Let $r = K/L$ be the ratio of capital to labour, so that $K = rL_0 e^{nt}$ and thus

$$\frac{dK}{dt} = L_0 e^{nt} \frac{dr}{dt} + nr L_0 e^{nt}.$$

Combining this equation with equation 2.32, and applying the property of constant returns to scale,

$$\left(\frac{dr}{dt} + nr \right) L_0 e^{nt} = sF\left(K, L_0 e^{nt}\right)$$

$$\Rightarrow \quad \left(\frac{dr}{dt} + nr \right) L_0 e^{nt} = sL_0 e^{nt} F\left(\frac{K}{L_0 e^{nt}}, 1 \right).$$

Defining $f(r) = F(K/L, 1)$ we obtain

$$\frac{dr}{dt} = sf(r) - nr. \tag{2.33}$$

The function f has a simple interpretation: the total product (output) as capital (per single unit of labour) varies. Equivalently, f is the output per worker as a function of capital per worker. Thus, the rate of change in the capital-labour ratio, r, is the difference between the increment of capital and the increment of labour.

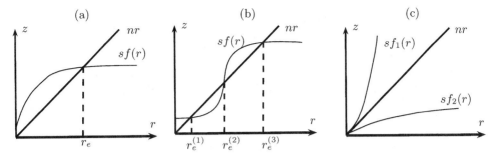

Figure 2.16: Illustration of graphical determination of equilibrium points for the differential equation (2.33). In each graph the terms $z = nr$ and $z = sf(r)$ are plotted and the equilibrium points are where these two intersect. In (a), the function f is such that there is only one equilibrium point. In (b), there are three equilibrium points, denoted by $r_e^{(1)}$, $r_e^{(2)}$ and $r_e^{(3)}$. In (c), there are no equilibrium points for either of f_1 or f_2.

This equation has an equilibrium point r_e, where the capital is expanding at the same rate as the labour force, that is, where $dr/dt = 0$. Figure 2.16 illustrates three distinct scenarios that occur for different practical definitions of the production function f.

In Figure 2.16(a), r_e is stable since for $sf(r) > nr$, dr/dt is positive and r increasing, and for $sf(r) < nr$, r is decreasing. Thus, no matter where one starts with labour and capital, this model predicts that r approaches r_e. In Figure 2.16(b), for similar reasons, both $r_e^{(1)}$ and $r_e^{(3)}$ are stable, while $r_e^{(2)}$ is unstable; thus, depending on the initial conditions for $r(t)$, the system tracks towards one of the stable points. Note that the figure has been drawn such that production is possible without capital, and hence the curve does not pass through the origin. Figure 2.16(c) illustrates the case where no balanced growth equilibrium exists. For the upper curve, $sf(r) > nr$, and $r(t)$ continues to increase with full employment and capital and income increasing faster than the supply of labour. Alternatively, for the lower case where $sf(r) < nr$, the ratio continues to decrease, approaching 0. For this case, ongoing full employment leads to decreasing output. Many different curves for $F(K, L)$ are possible; however, the cases in Figure 2.16 provide a framework through which to interpret them.

One simple example of a production function F is the Cobb–Douglas model, where $Y = F(K, L) = K^a L^{1-a}$ with $a < 1$. The function F in this case has the property of constant returns to scale and thus, in terms of the ratio $r = K/L$, $Y = sf(r) = sr^a$ and equation (2.33) becomes

$$\frac{dr}{dt} = sr^a - nr. \qquad (2.34)$$

Note that f is a monotonically increasing function of r. For small r and with $a < 1$, the initial slope of sr^a is close to vertical, and thus $sr^a > nr$. Alternatively, as r increases, and since $a < 1$, the slope of sr^a decreases with sr^a, for large values of r, such that $sr^a < nr$. This corresponds to Figure 2.16(a) regardless of the parameter values for n and a, and thus implies the existence of a single stable equilibrium point located at $r_e = (s/n)^{1/(1-a)}$ (from equation (2.34)). We note that this value increases with increased savings, s, as would be expected.

Further, it is not straightforward to solve equation (2.34) analytically (although straightforward numerically, if the parameter values were known accurately). However, we can establish the dynamics of K and L directly. Since

$$\frac{dK}{dt} = sK^a (L_0 e^{nt})^{1-a}, \qquad (2.35)$$

from equation (2.35), we can integrate to get

$$K(t) = \left[K_0^{1-a} - \frac{s}{n} L_0^{1-a} + \frac{s}{n} L_0^{1-a} e^{(1-a)nt} \right]^{1/(1-a)}$$

where K_0 and L_0 are the initial values. Thus, according to this model, capital grows as $(s/n)^{1/(1-a)} L_0 e^{nt}$, namely at the same rate as labour. We also get an equilibrium value for K/Y, that is, capital per unit of production. Income per labour unit, is

$$Y = \left(\frac{K}{L} \right)^a L \quad \Rightarrow \quad \frac{Y}{L} = \left(\frac{K}{L} \right)^a = r^a.$$

Clearly, many factors that are not considered in this model have a direct impact on production or output. Technological advances, for example, may reduce the amount of labour and/or capital required for the same output. Work by Romer (1986), Lucas (1988) and Barro (1997) recognise the impact of technology (e.g., innovation) as well as worker productivity (e.g., education) on economic growth predictions. Moreover, recent work indicates a correlation with climate (but perhaps that works both ways?).

Whatever model is adopted, it should be kept in mind that "All theory depends on assumptions that are not quite true", (Solow, 1956, page 65), and economic models are no exception. Debate continues as to what a 'best' model might look like, and is likely to continue indefinitely. However, as Robert Lucas commented on growth theory: "The consequences for human welfare are simply staggering."

2.13   Exercises for Chapter 2

2.1. Solving differential equations. *For the differential equation*

$$\frac{dy}{dt} = 2y,$$

(a) *Show that $y(t) = ce^{2t}$, where c is any real valued constant, satisfies the differential equation by substituting into both sides of the equation. Is there any value of c that isn't a solution?*

(b) *Find the one solution that corresponds to the initial condition $y(0) = 5$.*

See Appendix A.1 if you need any revision of the meaning of differential equations.

2.2. Solving first-order DEs. *For the following first-order differential equations, find the general solution, solving for the dependent variable. See Appendix A.3 if you need to revise how to solve first-order differential equations.*

(a) $\dfrac{dy}{dt} = -3y$,   (b) $\dfrac{dC}{dt} = 3C - 1$,   (c) $\dfrac{dy}{dt} = 3yt^{-1}$.

2.3. Atmospheric pressure. *The Earth's atmospheric pressure p is often modelled by assuming that dp/dh (the rate at which pressure p changes with altitude h above sea level) is proportional to p. Suppose that the pressure at sea level is 1,013 millibars and that the pressure at an altitude of 20 km is 50 millibars.*

Answer the following questions and then check your calculations with `Maple` *or* `MATLAB`.

(a) *Use an exponential decay model*

$$\frac{dp}{dh} = -kp$$

to describe the system, and then by solving the equation find an expression for p in terms of h. Determine k and the constant of integration from the initial conditions.

(b) *What is the atmospheric pressure at an altitude of 50 km?*

(c) *At what altitude is the pressure equal to 900 millibars?*

2.4. The Rule of 72. *Continuous compounding for invested money can be described by a simple exponential model, $M'(t) = 0.01rM(t)$, where $M(t)$ is the amount of money at time t and r is the percent interest compounding. Business managers commonly apply the Rule of 72, which says that the number of years it takes for a sum of money invested at r% interest to double, can be approximated by 72/r. Show that this rule always overestimates the time required for the investment to double.*

2.5. Dating a seashell. *If an archaeologist uncovers a seashell which contains 60% of the ^{14}C of a living shell, how old do you estimate that shell, and thus that site, to be? (You may assume the half-life of ^{14}C to be 5,568 years.)*

2.6. Olduvai Gorge. *(From Borelli and Coleman (1996).) Olduvai Gorge, in Kenya, cuts through volcanic flows, tuff (volcanic ash), and sedimentary deposits. It is the site of bones and artefacts of early hominids, considered by some to be precursors of man. In 1959, Mary and Louis Leakey uncovered a fossil hominid skull and primitive stone tools of obviously great age, older by far than any hominid remains found up to that time. Carbon-14 dating methods being inappropriate for a specimen of that age and nature, dating had to be based on the ages of the underlying and overlying volcanic strata.*

The method used was that of potassium-argon decay. The potassium-argon clock is an accumulation clock, in contrast to the ^{14}C dating method. The potassium-argon method depends on measuring the accumulation of 'daughter' argon atoms, which are decay products of radioactive potassium atoms. Specifically, potassium-40 (^{40}K) decays to argon (^{40}Ar) and to Calcium-40 (^{40}Ca) by the branching cascade illustrated below in Figure 2.17. Potassium decays to calcium by emitting a β particle (i.e. an electron). Some of the potassium atoms, however, decay to argon by capturing an extra-nuclear electron and emitting a γ particle.

Figure 2.17: Compartment diagram for Exercise 2.6.

The rate equations for this decay process may be written in terms of $K(t)$, $A(t)$ and $C(t)$, the potassium, argon and calcium in the sample of rock:

$$K' = -(k_1 + k_2)K,$$
$$A' = k_1 K,$$
$$C' = k_2 K,$$

where

$$k_1 = 5.76 \times 10^{-11} \text{ year}^{-1}, \qquad k_2 = 4.85 \times 10^{-10} \text{ year}^{-1}.$$

(a) Solve the system to find $K(t)$, $A(t)$ and $C(t)$ in terms of k_1, k_2, and $k_3 = k_1 + k_2$, using the initial conditions $K(0) = k_0$, $A(0) = C(0) = 0$.

(b) Show that $K(t) + A(t) + C(t) = k_0$ for all $t \geq 0$. Why would this be the case?

(c) Show that $K(t) \to 0$, $A(t) \to k_1 k_0 / k_3$ and $C(t) \to k_2 k_0 / k_3$ as $t \to \infty$.

(d) The age of the volcanic strata is the current value of the time variable t because the potassium-argon clock started when the volcanic material was laid down. This age is estimated by measuring the ratio of argon to potassium in a sample. Show that this ratio is

$$\frac{A}{K} = \frac{k_1}{k_3} \left(e^{k_3 t} - 1 \right).$$

(e) Now show that the age of the sample in years is

$$\frac{1}{k_3} \ln \left[\left(\frac{k_3 A}{k_1 K} \right) + 1 \right].$$

(f) When the actual measurements were made at the University of California at Berkeley, the age of the volcanic material (and thus the age of the bones) was estimated to be 1.75 million years. What was the value of the measured ratio A/K?

2.7. Storage time for radioactive chemicals.. (Adapted from Borelli and Coleman (1996).) In a biochemical laboratory radioactive phosphorus (^{32}P) was used as a tracer. (A tracer, through its radioactive emission, allows the course followed by a substance through a system to be tracked, which otherwise would not be visible.) ^{32}P decays exponentially with a half-life of 14.5 days and its quantity is measured in curies (Ci). (Although it is not necessary for the calculations, one curie is the quantity of a radioactive isotope undergoing 3.7×10^{-5} disintegrations per second.) After the experiment the biochemists needed to dispose of the contents, but they had to store them until the radioactivity had decreased to the acceptably safe level of 1×10^{-5}Ci. The experiment required 8Ci of ^{32}P. Using a simple model of exponential decay, establish how long they had to store the contents of the experiment before it could be disposed of safely.

2.8. Lake Burley Griffin. Read the case study on Lake Burley Griffin. The average summer flow rate for the water into and out of the lake is 4×10^6 m^3/month.

(a) Using this summer flow, how long will it take to reduce the pollution level to 5% of its current level? How long would it take for the lake with pollution concentration of 10^7parts$/$m^3, to drop below the safety threshold? (Assume in both cases that only fresh water enters the lake.)

(b) Use `Maple` or `MATLAB` to replicate the results in the case study, for both constant and seasonal flow and constant and seasonal pollution concentrations entering the lake. Comment on the solutions.

2.9. Pollution with chemical activity. Consider the concentration, $C(t)$, of some pollutant chemical in a lake. Suppose that polluted water with concentration c_i flows into the lake with a flow rate of F and the well-stirred mixture leaves the lake at the same rate F.

In addition, suppose some chemical agent is present in the lake that breaks down the pollution at a rate r kg/day per kg of pollutant. Assuming that the volume of mixture in the lake remains constant and the chemical agent is not used up, formulate (but do not solve) a mathematical model as a single differential equation for the pollution concentration $C(t)$.

2.10. North American lake system. Consider the American system of two lakes: Lake Erie feeding into Lake Ontario. What is of interest is how the pollution concentrations change in the lakes over time. You may assume the volume in each lake to remain constant and that Lake Erie is the only source of pollution for Lake Ontario.

(a) Write a differential equation describing the concentration of pollution in each of the two lakes, using the variables V for volume, F for flow, $c(t)$ for concentration at time t and subscripts 1 for Lake Erie and 2 for Lake Ontario.

(b) Suppose that only unpolluted water flows into Lake Erie. How does this change the model proposed?

(c) Solve the system of equations to get expressions for the pollution concentrations $c_1(t)$ and $c_2(t)$.

(d) Set $T_1 = V_1/F_1$ and $T_2 = V_2/F_2$, and then $T_1 = kT_2$ for some constant k as V and F are constants in the model. Substitute this into the equation describing pollution levels in Lake Ontario to eliminate T_1. Then show that, with the initial conditions $c_{1,0}$ and $c_{2,0}$, the solution to the differential equation for Lake Ontario is

$$c_2(t) = \frac{k}{k-1}c_{1,0}\left(e^{-t/(kT_2)} - e^{-t/T_2}\right) + c_{2,0}e^{t/T_2}.$$

(One way of finding the solution would be to use an integrating factor. See Appendix A.4.)

(e) Compare the effects of $c_1(0)$ and $c_2(0)$ on the solution $c_2(t)$ over time.

2.11. Smoke in the bar. (Adapted from Fulford et al. (1997).) A public bar opens at 6 p.m. and is rapidly filled with clients of whom the majority are smokers. The bar is equipped with ventilators that exchange the smoke-air mixture with fresh air.

Cigarette smoke contains 4% carbon monoxide and a prolonged exposure to a concentration of more than 0.012% can be fatal. The bar has a floor area of 20 m by 15 m, and a height of 4 m. It is estimated that smoke enters the room at a constant rate of $0.006\,\mathrm{m}^3/\mathrm{min}$, and that the ventilators remove the mixture of smoke and air at 10 times the rate at which smoke is produced.

The problem is to establish a good time to leave the bar, that is, sometime before the concentration of carbon monoxide reaches the lethal limit.

(a) Starting from a word equation or a compartmental diagram, formulate the differential equation for the changing concentration of carbon monoxide in the bar over time.

(b) By solving the equation above, establish at what time the lethal limit will be reached.

2.12. Detecting art forgeries. Based on methods used in the case study describing the detection of art forgeries (Section 2.3), comment on whether each of the paintings below is a possible forgery, based on the time it was painted:

(a) 'Washing of Feet', where the disintegration rate for ^{210}Po is 8.2 per minute per gram of white lead, and for ^{226}Ra is 0.26 per minute per gram of white lead.

(b) 'Laughing Girl', where the disintegration rate for ^{210}Po is 5.2 per minute per gram of white lead and for ^{226}Ra is 4 per minute per gram of white lead.

2.13. Cold pills. In Section 2.7, we developed the model

$$\frac{dx}{dt} = -k_1 x, \qquad x(0) = x_0,$$

$$\frac{dy}{dt} = k_1 x - k_2 y, \qquad y(0) = 0,$$

where $k_1, k_2 > 0$ determine the rate at which a drug, antihistamine or decongestant moves between two compartments in the body, the GI-tract and the bloodstream, when a patient takes a single pill. Here $x(t)$ is the level of the drug in the GI-tract and $y(t)$ is the level in the bloodstream at time t.

(a) Find solution expressions for $x(t)$ and $y(t)$ that satisfy this pair of differential equations, when $k_1 \neq k_2$. Show that this solution is equivalent to that provided in the text.

(b) The solution above is invalid at $k_1 = k_2$. Why is this, and what is the solution in this case?

(c) For old and sick people, the clearance coefficient (that is, the rate at which the drug is removed from the bloodstream) is often much lower than that for young, healthy individuals. How does an increase or decrease in k_2 change the results of the model? Using **Maple** or **MATLAB** to generate the time-dependent plots, check your results.

2.14. Cold pills. In Section 2.7, we also developed a model to describe the levels of antihistamine and decongestant in a patient taking a course of cold pills:

$$\frac{dx}{dt} = I - k_1 x, \qquad x(0) = 0,$$
$$\frac{dy}{dt} = k_1 x - k_2 y, \qquad y(0) = 0.$$

Here k_1 and k_2 describe rates at which the drugs move between the two sequential compartments (the GI-tract and the bloodstream) and I denotes the amount of drug released into the GI-tract in each time step. The levels of the drug in the GI-tract and bloodstream are x and y, respectively. By solving the equations sequentially show that the solution is

$$x(t) = \frac{I}{k_1} \left(1 - e^{-k_1 t}\right), \qquad y(t) = \frac{I}{k_2} \left[1 - \frac{1}{k_2 - k_1} \left(k_2 e^{-k_1 t} - k_1 e^{-k_2 t}\right)\right].$$

2.15. Antibiotics. (Adapted from Borelli and Coleman (1996).) *Tetracycline* is an antibiotic prescribed for a range of problems, from acne to acute infections. A course is taken orally and the drug moves from the GI-tract through the bloodstream, from which it is removed by the kidneys and excreted in the urine.

(a) Write word equations to describe the movement of a drug through the body, using three compartments: the GI-tract, the bloodstream and the urinary tract. Note that the urinary tract can be considered as an absorbing compartment, that is, the drug enters but is not removed from the urinary tract.

(b) From the word equations develop the differential equation system that describes this process, defining all variables and parameters as required.

(c) The constants of proportionality associated with the rates at which *tetracycline* (measured in milligrams) diffuses from the GI-tract into the bloodstream, and then is removed, are $0.72 \, \text{hour}^{-1}$ and $0.15 \, \text{hour}^{-1}$, respectively (Borelli and Coleman (1996)). Suppose, initially, the amount of *tetracycline* in the GI-tract is 0.0001 milligrams, while there is none in the bloodstream or urinary tract.

Use **Maple** or **MATLAB**(with symbolic toolbox) to solve this system analytically, and thus establish how the levels of *tetracycline* change with time in each of the compartments. In the case of a single dose, establish the maximum level reached by the drug in the bloodstream and how long it takes to reach this level with the initial conditions as given above.

(d) Suppose that, initially, the body is free from the drug and then the patient takes a course of antibiotics: 1 unit per hour. Use **Maple** or **MATLAB** to examine the level of *tetracycline* (expressed as units) in each of the compartments over a 24-hour period. Use the constants as given above.

2.16. Alcohol consumption. Use the model from the case study on alcohol consumption (Dull, dizzy or dead, Section 2.8), to establish, for the case of drinking on an empty stomach, the following:

(a) Use `Maple` or `MATLAB` to generate graphs to investigate the effects of alcohol on a woman of 55 kg, over a period of time.

(b) Compare these results with those for a man of the same weight.

(c) Assuming the legal limit to be 0.05 BAL (the Australian limit), establish roughly how much alcohol the man and woman above can consume each hour and remain below this limit.

(d) Repeat (a)–(c) for the case of drinking together with a meal.

2.17. Alcohol consumption. *Alcohol is unusual in that it is removed (that is, metabolised through the liver) from the bloodstream by a constant amount each time period, independent of the amount in the bloodstream. This removal can be modelled by a Michaelis–Menten type function $y' = -k_3 y/(y + M)$, where $y(t)$ is the 'amount' (BAL) of alcohol in the bloodstream at time t, k_3 is a positive constant and M a small positive constant.*

(a) If y is large compared with M, then show that $y' \simeq -k_3$. Solve for y in this case.

(b) Alternatively, as y decreases and becomes small compared with M, show that then $y' \simeq -k_3 y/M$. Solve for y in this case.

(c) Now sketch the solution function for $y' = -k_3 y/(y + M)$ assuming that, initially, y is much greater than M. Indicate clearly how the graph changes in character when y is small compared with M, compared with when y is large compared with M. Show how the solution behaves as $t \to \infty$.

(d) When and why would this function be more suitable than simply using $y' = -k_3$ to model the removal rate?

2.18. Solving differential equations. *Consider the differential equations*

$$t \frac{dx}{dt} = x, \qquad x(t_0) = x_0,$$

and

$$y^2 \frac{dx}{dy} + xy = 2y^2 + 1, \qquad x(y_0) = x_0.$$

Put each equation into normal form and then use the integrating factor technique to find the solutions. Establish whether these solutions are unique, and which part of each solution is a response to the initial data and which part a response to the input or forcing.

2.19. Formulating DEs for alcohol case study. *Read over the case study in Section 2.8. Consider two compartments, one for the GI-tract and one for the blood. Let $C_1(t)$ be the concentration of alcohol in the GI tract and $C_2(t)$ be the concentration in the blood, with both concentrations measured in BAL (g per 100 ml). Also let F_1 be the flow rate of fluid from the GI-tract and let F_2 be the flow rate of fluid from the blood to the tissues. Finally, we let i_0 be the rate of ingestion of alcohol (in g/hr). Use conservation of mass of alcohol to deduce the equations in the form*

$$\frac{dC_1}{dt} = I - k_1 C_1,$$
$$\frac{dC_2}{dt} = k_2 C_2 - k_4 C_2$$

and determine I, k_1, k_2 and k_3 all in terms of i_0, F_1, F_2 and V_g, the volume of the fluid in the GI-tract, V_b the volume of fluid in the blood, and α, where α is the proportion of the alcohol leaving the GI-tract goes into the bloodstream.

Note: In the case study we let the rate constant k_4 depend on the blood alcohol concentration

$$k_4 = \frac{k_3}{M + C_2},$$

where k_3 and M are positive constants, k_3 with the same units as k_1 and k_2, namely hours^{-1} and M with the same units as C_2, namely BAL.

2.20. Economic growth. *Read over the case study on a model of economic growth in Section 2.12. In this model the Cobb–Douglas function was used to model production. An alternative model is the Harrod–Domar model of fixed proportions, $Y = \min\{K/a, L/b\}$ is the minimum of the two values, with a units of capital and b units of labour required to produce a unit of output. The expression for Y describes the 'bottlenecks' for the system, that is, whether it is limitations in capital or labour that determine the outcome for production.*

(a) *For the case $r/a < 1/b$, show that*

$$\frac{dr}{dt} = \left(\frac{s}{a} - n\right) r$$

   and solve this to obtain

$$r(t) = r_0 e^{(s/a - n)t}.$$

   where $r(0) = r_0$.

(b) *Consider the case when $n > s/a$ and $r_0 > a/b$. Provide an interpretation of what this scenario means in terms of capital and the demand for labour.*

2.21. Return to scale property. *Show that the Cobb–Douglas function, from Section 2.12,*

$$Y = F(K, L) = K^a L^{1-a}$$

has the return to scale property.

2.22. Stability of equilibrium solution. *Consider Figure 2.16(b), in Section 2.12. Establish the stability of each of the equilibrium points, $r_e^{(1)}$, $r_e^{(2)}$ and $r_e^{(3)}$, from the underlying equation.*

2.23. Equilibria. *Each of the following differential equations has only one equilibrium solution. Find that equilibrium solution and determine if it is stable or unstable?*

(a) $\dfrac{dy}{dt} = y - 1.$

(b) $\dfrac{dC}{dt} = \dfrac{F}{V} c_i - \dfrac{F}{V} C$, *where F, V, c_i are positive constants.*

Chapter 3

Models of single populations

In this chapter, we develop models describing the growth and decline of single populations with continuous breeding programs. Initially we model exponential growth. A more realistic model would include the effects on population growth of limited resources, and thus we extend our model to take account of density-dependent growth (logistic growth), which describes the population size as stabilising after an initial exponential growth spurt. We examine the effects of harvesting the population and see that there is a critical harvesting rate, above which extinction ensues. Furthermore, we consider single populations with discrete breeding seasons and examine how the concept of 'chaos' can arise in population dynamics. Finally, we briefly introduce the notion of time-delayed models through a case study.

3.1 Exponential growth

We begin by developing a very simple mathematical model that describes the growth of a population. This leads to exponential growth.

Background

In many cases, mathematical modelling is applied to understand population growth dynamics for animal and human populations. For example, modelling the way fish populations grow, and accounting for the effect of fishing is essential to the fishing industry, as we cannot afford to deplete this resource. Another use of modelling is to understand the manner in which human populations grow: in the world, in individual countries, in towns and in organisations.

The population of the world was estimated in 1990 to be approximately 5.3 billion individuals. Estimates from census data show that this has grown from approximately 1.6 billion at the turn of the century. What will the population be in 5 years' time or even 100 years' time? By developing simple mathematical models, we attempt to predict this, based upon certain assumptions about birth rates and death rates.

One important factor in modelling populations is whether the population grows continuously with time or in discrete jumps. Many animal populations grow in discrete times, due to having well-defined breeding seasons, whereas human populations grow continuously in time. Some insect populations also have non-overlapping generations, where the adults all die directly after giving birth. Even if a population grows in discrete time jumps, it may still be reasonable to use a continuous time model provided the time interval between jumps is small compared with the overall time in which we are interested.

Another important factor in deciding on how to model population growth is the size of the population. Small populations are subject to random fluctuations in that we cannot predict with certainty when a parent will give birth, or what the size of an animal litter will be. For small populations it makes sense to talk about the probability of giving birth in a certain time interval, and the mean population at a given time. When the population is large, however, random fluctuations between individuals are small compared with the whole population size and so we can model this without the need for probability functions.

For an entire population it would only be appropriate for $X(t)$, a measure of the population size, to be an integer; but for large populations, it is usual to work with a value of $X(t)$ that is continuous in t and round off to the integer value at the end. Of course for population densities, that is, number per unit of area, using fractional numbers poses no problem.

General compartmental model

We can consider this problem as a compartmental model, with the compartment being the 'world', 'town', 'organisation', 'ocean', etc.

Figure 3.1: Input-output compartmental diagram for a population.

This compartmental sketch leads to a word equation describing a changing population,

$$\left\{ \begin{array}{c} \textit{rate of} \\ \textit{change of} \\ \textit{population size} \end{array} \right\} = \left\{ \begin{array}{c} \textit{rate} \\ \textit{of} \\ \textit{births} \end{array} \right\} - \left\{ \begin{array}{c} \textit{rate} \\ \textit{of} \\ \textit{deaths} \end{array} \right\}. \tag{3.1}$$

We now develop equations by making some assumptions and then, under these conditions, describe the birth and death processes in symbols.

Model assumptions

When dealing with large populations, we can ignore random fluctuations between individuals and treat each individual as being identical. Thus we assume that each individual in the population has an equal chance of giving birth and we also assume that each individual has an equal chance of dying within a given time interval. It thus makes sense to talk about a *per-capita birth rate* β per unit time, per member of the population, and a *per-capita death rate* α.

The per-capita birth rate for the world has been estimated at $\beta = 0.027$ per year per individual and the per-capita death rate is $\alpha = 0.010$ per year per individual (1990 estimate, Microsoft (1995)). This is not homogeneous over all countries in the world. For Australia SBS (1998) between 1990 and 1995, the per-capita birth rate was $\beta = 0.014$ per year per individual and the per-capita death rate was $\alpha = 0.007$ per year per individual. The population of Australia in 1993 was approximately 17.8 million people, with a population density of 2.1 persons per km^2. Note that some other countries have much higher per-capita birth and death rates, which is consistent with the higher world per-capita birth and death rates compared with those for Australia.

We make the following assumptions and then build the model on them.

- We assume that the populations are sufficiently large so that we can ignore random differences between individuals.
- We assume that births and deaths are continuous in time.
- We assume that per-capita birth and death rates are constant in time.
- In the model development, we ignore immigration and emigration, which can be included later.

Formulating the differential equation

Let us consider a population whose initial value is x_0, with constant per-capita birth rate given as β, and constant per-capita death rate given as α. Our aim is to find the population size at any time t. The first step is to determine an equation for the population. We assume that the population can only change due to births or deaths, neglecting here any immigration or emigration. Also, we assume that this change in population at any time is proportional to the size of the population at that time. The appropriate word equation was given in (3.1).

Since the per-capita birth rate β is assumed constant, the overall birth rate at any time is the per-capita birth rate multiplied by the current population size. Similarly, the overall death rate is the per-capita death rate multiplied by the population size. Thus we write

$$\left\{ \begin{array}{c} rate \\ of \\ births \end{array} \right\} = \beta X(t), \qquad \left\{ \begin{array}{c} rate \\ of \\ deaths \end{array} \right\} = \alpha X(t). \tag{3.2}$$

Substituting (3.2) into (3.1) we obtain

$$\frac{dX}{dt} = \beta X - \alpha X. \tag{3.3}$$

(Note that $X(t)$ can be written as just X, since it is clear that X is evaluated at t.)

We have obtained a differential equation for the population size $X(t)$. We need one initial condition to ensure a unique solution since this is a first-order, linear differential equation. For the population of the world, if we define the initial time to be 1990 then we have $X(0) = x_0 = 5.3$ billion people.

Solving the differential equation

We now solve the differential equation (3.3) for continuous population growth. Let $r = \beta - \alpha$ and then

$$\frac{dX}{dt} = rX.$$

We call r the *growth rate* or the *reproduction rate* for the population. When $r > 0$ this is a model describing exponential growth, and when $r < 0$ the process is exponential decay. (Note the similarities here with Section 2.2 on exponential decay.)

This differential equation can be solved using the separation of variables technique covered in Appendix A.3. (The Appendices provide a summary of techniques for solving differential equations.) The general solution is

$$X = Ae^{rt}$$

and applying the initial condition $X(0) = x_0$ to obtain the value of the arbitrary constant A, the solution to the differential equation is

$$X(t) = x_0 e^{rt}. \tag{3.4}$$

Clearly this describes exponential growth or decay, depending on the sign of r.

Interpretation of parameters

Returning to the formulation of the differential equation, we can provide an approximate interpretation of the per-capita death rate α. From the rate of deaths we can approximate the number of deaths by multiplying the rate of deaths by the length of the time interval. This approximation would be better if the time interval were short. Thus we can write

$$\left\{ \begin{array}{c} number \\ of\ deaths \\ in\ time\ interval\ \Delta t \end{array} \right\} \simeq \alpha X(t)\Delta t.$$

Let us now suppose that x_0 people will die in time t_1, that is, t_1 is the *average life expectancy*. Then, let $X(t) = x_0$ and $\Delta t = t_1$ so that (from above) we have

$$x_0 \simeq \alpha x_0 t_1.$$

Hence we obtain

$$\alpha \simeq \frac{1}{t_1}$$

giving an *estimate* for α as the reciprocal of the average life expectancy.

For human populations, in developed countries, typically the per-capita death rate is quoted as $\alpha \simeq 0.007$ (year^{-1}, or 7 deaths per 1,000 persons per year (see, for example, SBS (1998)). This approximates an average life expectancy of $1/\alpha = 1/0.007 = 140$ years. This value is too high for humans, but nevertheless is of the correct order of magnitude compared with the measured average life expectancy of humans, which is 70–80 years. The reason for the discrepancy is due to the fact that the real age distribution does not approximately follow an exponential distribution; instead, the population tends to fall off rapidly at older ages. An exponential distribution is the natural distribution for compartment models (see exercises). For many animal populations, with shorter lifespans and greater probability of dying at younger ages, the reciprocal of the per-capita death rate is usually a more accurate approximation of the average life expectancy.

Using the model we can predict the time taken for the population size to double. Note the similarity between this concept of doubling time and the half-life of radioactive substances discussed in Chapter 2.

Example 3.1:   *Find an expression for the time for the population to double in size.*

Solution: *We solve*
$$X(t + T) = 2X(t),$$
where T is the time taken to double the current size. So, using the solution (3.4),

$$\frac{X(t+T)}{X(t)} = 2 = \frac{x_0 e^{r(t+T)}}{x_0 e^{rt}},$$

$$whence \quad T = \frac{\ell n\, 2}{r}.$$

So the time taken for a population to double in size is $T = \ell n\, 2/r$.

Model validation

Let us now see what this model predicts for known values of the parameters. Taking the 1990 world population values $r = 0.017$ and $x_0 = 5.3$ billion, we apply equation (3.4) to predict the population in 1995 as $X(5) = 5.77$ billion. In 10 years we have $X(10) = 6.28$ billion and in 100 years, $X(100) = 29.01$ billion. In just 100 years this predicts that the world population would increase by around 500%, which is hard to believe!

We should also compare our predictions with population sizes at previous times for which data are available. Looking back to 1960, the world's population was 3.005 billion and in 1900 it was 1.608 billion. Our model predicts $X(-30) = 3.18$ billion in 1960, which is not too far out and in 1900, $X(-90) = 1.15$ billion. The predictions appear to get worse the further back we go. One possible contributing factor towards this discrepancy is that the per-capita birth and death rates have changed gradually due to improvements in technology and changing attitudes. (In 1970, the reproduction rate was somewhat higher, at $r = 2.0$.)

Another way to test the model is to measure the population of some very simple organism that breeds fast and is sufficiently small, for example, a yeast culture. A comparison with data from an experiment growing a yeast culture is given in Renshaw (1991), and the results illustrated in the diagram of Figure 3.2. Note that the model appears to be in good agreement in the earlier stages of the growth, but for later times the population levels out rather than continuing to grow exponentially.

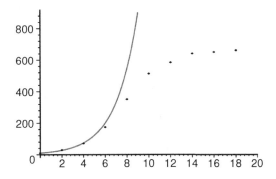

Figure 3.2: Illustration of the difference between the trend of experimental data for a yeast culture, and the continuous growth model predictions, over a period of hours (t). $X(t)$ is a measure of the yeast population size. (Adapted from Renshaw (1991))

The cause of this is that the per-capita death rate increases due to crowding and competition for limited resources, and thus the overall population size growth rate decreases. In certain animal populations, the per-capita birth rate (number of births per unit of time) decreases when the death rate increases. For example, rabbits reabsorb their embryos when their population density is high. However, in the model we develop to describe this observed growth pattern in Section 3.2, we consider only a change in the per-capita death rate dependent on the population density.

There are several other ways in which this basic population model could be extended. It is easy to incorporate immigration and emigration, which is appropriate for models of populations in specific countries, rather than the world. This is done by adding terms to the word equation for the change in population, and making assumptions about the rates of immigration and emigration. A similar concept is the inclusion of harvesting in a population, which we cover in Section 3.3. Some other important extensions (which are not addressed in this book) include discrete growth for animals with distinct breeding seasons, stochastic (random) growth and age-dependent growth.

Summary of skills developed here:

- *Formulate differential equations for single populations, including immigration, emigration or harvesting.*
- *Obtain exact solutions by solving the differential equations.*
- *For simple solutions, draw general sketches.*

3.2   Density-dependent growth

Realistically, populations cannot continue growing exponentially over time due to limited resources and/or competition for these with other species. If populations are observed over long periods they often appear to reach a limit, or to *stabilise*. We modify the exponential growth model of the previous section to account for competition or limited resources and to include the stabilising effect observed in populations.

Background

As a population grows, individuals eventually will compete for the limited resources available. In principle, this competition means that a given environment can support only a limited number of individuals. This number is called the *carrying capacity* for the population and is usually denoted by the symbol K in biological literature. Technically, we define it as the population size (or density) for which the per-capita birth rate is equal to the per-capita death rate, excluding external factors such as harvesting or interaction with another population. We need to extend the model to include an additional death rate due to the resource limitations, and thus curb the exponential growth and allow the population to stabilise.

Formulating the differential equation

The population is described by the same word equation as before,

$$\left\{\begin{array}{c} rate\ of \\ change\ in \\ population \end{array}\right\} = \left\{\begin{array}{c} rate \\ of \\ births \end{array}\right\} - \left\{\begin{array}{c} rate \\ of \\ deaths \end{array}\right\}, \tag{3.5}$$

and, as in the previous section, we assume a constant per-capita birth rate of β. Thus,

$$\left\{\begin{array}{c} rate \\ of \\ births \end{array}\right\} = \beta X(t).$$

Instead of assuming a constant per-capita death rate, we allow the per-capita death rate to increase as the population increases, as can be observed in some populations. We can model this behaviour by assuming a linear dependence of the per-capita death rate on the population size,

$$\left\{\begin{array}{c} per\text{-}capita \\ death \\ rate \end{array}\right\} = \alpha + \gamma X(t),$$

where α (positive) is the per-capita death rate due to natural attrition, and γ (positive) is the per-capita dependence of deaths on the population size. Note that as $X \to 0$, the per-capita death rate tends to α, while for increasing population size the per-capita death rate increases. This linear form is the simplest for a population dependent per-capita death rate that increases with increasing population size. The overall death rate is thus given by multiplying the per-capita death rate by the population size, so that

$$\left\{\begin{array}{c} rate \\ of \\ deaths \end{array}\right\} = \alpha X(t) + \gamma X^2(t). \tag{3.6}$$

The word equation (3.5) then translates to

$$\frac{dX}{dt} = \beta X - \alpha X - \gamma X^2.$$

Writing $r = \beta - \alpha$, which is the reproduction rate, we obtain the model for density-dependent growth,

$$\frac{dX}{dt} = rX - \gamma X^2. \tag{3.7}$$

An alternative formulation of this equation comes from splitting the death rate into a normal death rate and an extra death rate due to members of the population competing with each other for limited resources. Thus we can write

$$\left\{ \begin{array}{c} rate\ of \\ change\ in \\ population \end{array} \right\} = \left\{ \begin{array}{c} rate \\ of \\ births \end{array} \right\} - \left\{ \begin{array}{c} normal \\ rate\ of \\ deaths \end{array} \right\} - \left\{ \begin{array}{c} rate\ of \\ deaths\ by \\ crowding \end{array} \right\}.$$

For the birth rate and normal death rate, we assume constant per-capita rates β and α, respectively.

For the extra deaths by crowding, we assume that the per-capita death rate increases with population size. If we assume that it is proportional to the population size and thus given by γX (where γ is a positive constant), then the extra overall death rate is the extra per-capita death rate multiplied by the current population size. Thus we have

$$\left\{ \begin{array}{c} extra\ rate \\ of\ deaths \\ by\ crowding \end{array} \right\} = \gamma X^2$$

with γ a positive constant. The result is again (3.7).

Without actually solving the differential equation, we can infer how the solution would behave. The population has an initial value x_0. The derivative dX/dt is always positive for $X < K$, and thus the population is increasing. The rate of growth increases initially and then slows down as the population approaches the carrying capacity K. The sketch is illustrated in Figure 3.3.

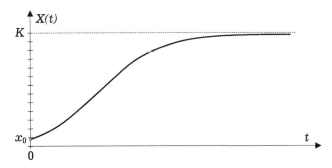

Figure 3.3: Sketch of the general solution to the continuous logistic equation, provided $x_0 < K$.

The logistic equation

With $K = r/\gamma$, the differential equation (3.7) becomes

$$\frac{dX}{dt} = rX - \frac{r}{K}X^2,$$

which can be written as

$$\frac{dX}{dt} = rX\left(1 - \frac{X}{K}\right). \tag{3.8}$$

This model leads to a nonlinear differential equation. It is the *logistic equation* and is also referred to as the limited growth model or the density-dependent model. We consider only $r > 0$ and $K > 0$ to ensure positive population values.

Interpretation of the parameters

Recall the differential equation for unrestricted population growth (see Section 3.1)

$$\frac{dX}{dt} = rX.$$

We can write a general differential equation for population growth as

$$\frac{dX}{dt} = R(X)X,$$

where $R(X)$ represents a population dependent per-capita growth rate. It is interesting to interpret the logistic equation in terms of this population dependent per-capita growth rate, and thus from equation (3.8) we identify $R(X)$ as

$$R(X) = r\left(1 - \frac{X}{K}\right).$$

Note that $R(X)$, a linear function of X, tends to zero as the population approaches its carrying capacity K, while as the population size tends to zero, $R(X)$ approaches r. This form corresponds to a straight line, which passes through the points $R = 0$ when $X = K$ and $R = r$ when $X = 0$, as illustrated in Figure 3.4. If $R < 0$, then we have that $X > K$ and the population (X) is decreasing as it approaches the carrying capacity.

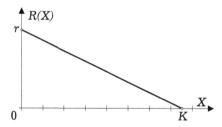

Figure 3.4: The simplest assumption for a population dependent per-capita growth rate is a straight line.

Equilibrium solutions and stability

If we observe the levelling of a population over time, this implies that the rate of change of the population approaches 0, that is, $X' \to 0$. Any value of X that gives a zero rate of change is called an *equilibrium point* or *equilibrium solution*.

Equilibrium solutions are constant solutions where, here, the rate of increase (births) exactly balances the rate of decrease (deaths). Equilibrium solutions satisfy

$$\frac{dX}{dt} = 0, \quad \Rightarrow \quad rX\left(1 - \frac{X}{K}\right) = 0. \tag{3.9}$$

There are two possible equilibrium solutions, $X_e = 0$ and $X_e = K$, that satisfy equation (3.9). We are interested in which of these are stable. For stable solutions, this means that if we start near the equilibrium solution then we are attracted towards it. The condition for local stability is $f'(X_e) < 0$, where f is the RHS of the differential equation (see Section 2.11). Here $f(X) = rX(1 - X/K)$ and so

$$f'(X) = r - \frac{2rX}{K},$$

so $f'(0) = r > 0$ and $f'(K) = -r < 0$, for all positive values of r. So the equilibrium solution $X = 0$ is always unstable and the equilibrium solution $X = K$ is always stable. We can further show that, even if we are not close to the equilibrium solutions, these stability conclusions still hold here.

Returning to the differential equation in factored form,

$$\frac{dX}{dt} = rX(1 - X/K),$$

we see that if $X < K$ then the rate of change of X is always positive. This implies the population is increasing with time and will approach $X = K$. Similarly, if $X > K$, then $dX/dt < 0$ which implies that the population is always decreasing towards $X = K$. When the population *always* approaches the equilibrium population, we say the equilibrium is *globally stable*. Conversely, if it is repelled from the equilibrium, we say the equilibrium is *unstable*. (Note that for $0 < X < K$, $dX/dt > 0$ and thus the equilibrium point at $X = 0$ is globally unstable.)

Thus the model predicts that all populations approach the equilibrium value K, which we have already defined as the carrying capacity for the population. The stable equilibrium point and carrying capacity coincide in this case.

Solving the logistic equation

We can solve the logistic equation in two ways, first, using `Maple` to obtain a numerical solution, and second, analytically using the separation of variables technique.

`Maple` is a symbolic language, which means that it can use analytic methods (where possible) to solve such equations. Often analytic solutions cannot be found and numerical schemes are required. (We take a closer look at some of these numerical methods in Chapter 4.) We can use `Maple` to solve the equation and draw the family of graphs in Figure 3.5 with the following commands (see Listing 3.1)

Listing 3.1: Maple code: c_cp_logistic.mpl

```
restart:with(plots):
r:=1;K:=1000;
de1:=diff(x(t),t)=r*x(t)*(1-x(t)/K);
soln:=x0->dsolve({de1,x(0)=x0},x(t),numeric):
plot1:=x0->(odeplot(soln(x0),[t,x(t)],0..8)):
list1:=seq(plot1(i*50),i=1..24):
line1:=plot([[0,K],[8,K]],colour=gray):
plot11:=display(list1,line1,view=[0..8,0..1200]):
display(plot11);
```

Analytic solution

The next example illustrates how to obtain the analytic solution. (For details on partial fractions and separable techniques, see the Appendices.)

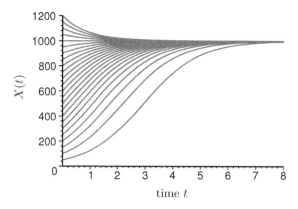

Figure 3.5: Logistic growth model with $r = 1$ and $K = 1,000$, and for a number of different initial conditions.

Example 3.2: *Solve the logistic differential equation (3.8) given the initial condition $X(0) = x_0$.*

Solution: *The logistic equation is not linear, but it is separable,*

$$\frac{dX}{dt} = rX\left(1 - \frac{X}{K}\right) = \frac{rX(K - X)}{K}, \tag{3.10}$$

which is in the form $X' = F(X)G(t)$ where $G(t) = 1$. Separating the variables,

$$\frac{K}{X(K - X)}\frac{dX}{dt} = r,$$

assuming that $X \neq 0$ and $X \neq K$. Integrating gives

$$\int \frac{K}{X(K - X)}\, dX = \int r\, dt.$$

For the integral on the left-hand side (LHS), we need to use partial fractions (see Appendix B.5) and thus

$$\frac{K}{X(K - X)} = \frac{a}{X} + \frac{b}{K - X} = \frac{a(K - X) + bX}{X(K - X)}.$$

Solving for the constants a and b gives

$$aK = K,$$
$$(-a + b)X = 0,$$

which implies that $a = b = 1$, and then

$$\frac{K}{X(K - X)} = \frac{1}{X} + \frac{1}{K - X}.$$

Now

$$\int \frac{K}{X(K - X)}\, dX = \int \frac{1}{X}\, dX + \int \frac{1}{K - X}\, dX = \int r\, dt$$

and we can integrate to get

$$\ell n\,|X| - \ell n\,|K - X| = rt + c \quad \text{(c an arbitrary constant)},$$
$$\left|\frac{X}{K - X}\right| = c_1 e^{rt},$$

where $c_1 = e^c$. Assuming $0 < X < K$, then

$$X = c_1 e^{rt}(K - X). \tag{3.11}$$

Using the initial condition $X(0) = x_0$, we deduce that $c_1 = x_0/(K - x_0)$ and then solving (3.11) for X we obtain

$$X = \frac{K}{1 + me^{-rt}} \qquad \text{where } m = \frac{K}{x_0} - 1. \qquad (3.12)$$

Alternatively, in the case where $0 < K < X$,

$$X = c_1 e^{rt}(X - K).$$

With the initial condition $X(0) = x_0$, we have that $x_0 = c_1(x_0 - K)$ and hence $c_1 = x_0/(x_0 - K)$. The solution is as in (3.12).

It is also possible to use `Maple` to find the analytic solution. The commands are given in Listing 3.2.

Listing 3.2: Maple code: c_cp_logistic_analytic.mpl

```
restart;
de := diff(X(t),t) = r*X(t)*(1-X(t)/K);
dsolve( {de, X(0)=x0}, X(t));
```

Solutions over time

As usual, we are interested in the long-term behaviour of the population as predicted by our model. The following example shows how to sketch the graph and draw conclusions about the long-term behaviour forecast by this model.

Example 3.3: *Using the differential equation and considering the second derivative, sketch the graph of the solution.*

Solution: *Since $r > 0$, as t increases we have $\lim_{t \to \infty} X(t) = K$, which is the carrying capacity of the population. If $0 < x_0 < K$, then $X(t)$ is strictly increasing (a monotonically increasing function). If $x_0 > K$, then $X(t)$ is strictly decreasing (a monotonically decreasing function). What happens if $x_0 = K$?*

Considering the second derivative,

$$X'' = rX' - \frac{2XrX'}{K} = X'r\left(1 - \frac{2X}{K}\right) = rX\left(1 - \frac{X}{K}\right)r\left(1 - \frac{2X}{K}\right),$$

which changes sign at $X = 0$, $K/2$, and K. These are inflection points of the curve, that is, where the curve changes from concave to convex or vice versa. This, together with the limit above and nature of the slope, provides the information required to sketch the graph that was illustrated earlier in Figure 3.3 and generated with `Maple` in Figure 3.5.

With the logistic model developed above we have incorporated an instantaneous reaction to the environment. That is, increased pressure on the resources produces an immediate response from the system in terms of, for example, more deaths. This is often not realistic in that the response usually takes effect after some time delay, or time lag. Vegetation needs time to recover and changed environmental conditions, which may lead to increased birth rates, will take time to appear in the numbers of an adult population. This leads us to a model that includes a time delay resulting from a multitude of sources, such as maturation times, food supply, resources, or crowding — each a measurable quantity. If the time lag is small compared with the natural response time ($1/r$), then there is a tendency to overcompensate, which may produce oscillatory behaviour. This is exemplified in Section 3.7 and Section 3.8.

> **Summary of skills developed here:**
> - *Understand how to modify the differential equation to account for a change in reproduction rate due to crowding.*
> - *Be able to find the equilibrium populations for a given differential equation.*
> - *Determine if the population approaches or is repelled from the equilibrium.*

3.3   Limited growth with harvesting

The effect of harvesting a population on a regular or constant basis is extremely important to many industries. One example is the fishing industry. Will a high harvesting rate destroy the population? Will a low harvesting rate destroy the viability of the industry?

Formulating the equation

Including a constant harvesting rate in our logistic model gives

$$
\begin{Bmatrix} \text{rate of} \\ \text{change in} \\ \text{population} \end{Bmatrix} = \begin{Bmatrix} \text{rate} \\ \text{of} \\ \text{births} \end{Bmatrix} - \begin{Bmatrix} \text{normal} \\ \text{rate of} \\ \text{deaths} \end{Bmatrix}
$$
$$
- \begin{Bmatrix} \text{rate of} \\ \text{deaths by} \\ \text{crowding} \end{Bmatrix} - \begin{Bmatrix} \text{rate of} \\ \text{deaths by} \\ \text{harvesting} \end{Bmatrix}.
$$

(3.13)

Assuming the harvesting rate to be constant, equation (3.13) translates to the differential equation

$$
\frac{dX}{dt} = rX\left(1 - \frac{X}{K}\right) - h.
$$

(3.14)

Here h is included as the constant rate of harvesting (total number caught per unit time, or deaths due to harvesting per unit time) and it is independent of the population size and thus could be interpreted as a quota.

Solving the differential equation

We can infer much useful information about the solution from the differential equation, as we see below and in the exercises, or we can obtain an explicit solution for specific parameter values, which we do first, using `Maple`.

The code associated with Figure 3.5 was adapted to produce the figures below (Figure 3.6 and Figure 3.7), taking a reproduction rate of $r = 1$, a carrying capacity of $K = 1,000$ and harvesting rates of $h = 100$ and $h = 500$.

The following example illustrates how to gain information about the dynamics of the model from the differential equation for certain numerical values of the parameters. (The purely theoretical case, which covers all possible parameter values, is examined in Exercise 3.7.)

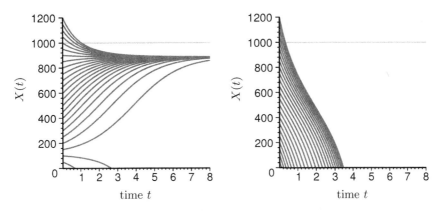

Figure 3.6: Logistic growth including harvesting, with $h = 100$ in the first diagram and with $h = 500$ in the second. The grey line is the carrying capacity K. (The parameter values are $K = 1,000$ and $r = 1$.)

First, we can write (3.14) in factored form

$$\frac{dX}{dt} = -\frac{r}{K}\left(X^2 - KX + \frac{Kh}{r}\right)$$

with r, K and h positive constants.

Example 3.4: Let $r = 1$, $K = 10$, $h = 9/10$ and $X(0) = x_0$. *Use the differential equation to investigate the behaviour of the solution and compare this with the* Maple *solution.*

Solution: *The differential equation becomes*

$$\frac{dX}{dt} = -\frac{1}{10}(X^2 - 10X + 9) = -\frac{1}{10}(X - 1)(X - 9).$$

We have the following cases to consider when sketching a graph. If $x_0 < 1$, then $X' < 0$ and the population declines. If $1 < x_0 < 9$, then $X' > 0$ and the population increases. If $x_0 > 9$, then $X' < 0$ and the population declines. If $x_0 = 1$ or $x_0 = 9$, then the population does not change ($X' = 0$). The Maple solution is illustrated in Figure 3.7 and clearly satisfies these predictions.

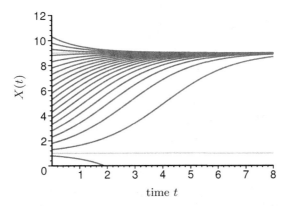

Figure 3.7: The logistic equation with harvesting (3.14), with $r = 1$, $K = 10$ and $h = 9/10$. The grey lines are at $x = 9$ and $x = 1$, across which there is a change in the sign of X'.

Using partial fractions we can solve the differential equation for X explicitly, as shown in the following example.

Example 3.5: *Solve the differential equation.*

Solution: *Separating the variables gives*

$$\int \frac{1}{(X-1)(X-9)}\, dX = -\int \frac{1}{10}\, dt,$$

and using partial fractions,

$$\frac{1}{8} \int \left(\frac{1}{X-9} - \frac{1}{X-1} \right) dX = -\frac{1}{10} \int dt.$$

Carrying out the integration gives

$$\ell n \left| \frac{X-9}{X-1} \right|^{1/8} = -\frac{1}{10}t + c,$$

where c is some arbitrary constant. Then rearranging with constant $b = e^{8c}$,

$$\left| \frac{X-9}{X-1} \right| = b e^{-4t/5}.$$

Now substituting for the initial condition,

$$b = \left| \frac{x_0 - 9}{x_0 - 1} \right|$$

and the explicit solution is

$$X(t) = \frac{9 - b e^{-4t/5}}{1 - b e^{-4t/5}}.$$

From this solution (illustrated in Figure 3.7) we can conclude that the harvesting rate causes the population to stabilise to a value less than the carrying capacity. In the case of harvesting, or some other disaster, causing the population to decrease below a critical level, the threshold level (which in this example is 1), our model predicts that the population will become extinct.

This concept of a threshold level is critical to many industries, and thus harvesting rates should be set with extreme caution as the following scenarios and case study indicate.

Summary of skills developed here:

- *Be able to model a population following exponential growth or logistic growth and include harvesting.*
- *Understand the concept of a threshold harvesting level.*
- *Sketch graphs of the changes in a population over time and ratify them with* Maple *generated solutions.*
- *Modify the model developed to include immigration and/or emigration, or include the impact of a toxin that kills a percentage of the population.*

3.4 Scenario: Anchovy wipe-out

In the following scenario, we cite a case in which a population was fished annually. Although measures were taken to ensure that overfishing did not occur, so that there would always be sufficient stock for recovery, a natural event combined with the fishing resulted in the total destruction of the industry associated with that species. The scenario is adapted from correspondence with Hearn (1998), and from reference May (1981). A discrete model of the system can be found in Caulkins et al. (1985).

Some decades ago, anchovies were a very cheap fish dominating the supermarket shelves: little guys in little tins and very salty. They were a major source of high protein food for people as well as animals. It was noted that the world price of anchovies showed a correlation with the *El Niño* cycle, that is, *El Niño* events increased prices due to shortages in the resource. Ironically it was to be a combination of *El Niño* and fishing practices that destroyed the industry.

In the decades after World War II and the introduction of the nylon fishing nets, the anchovy fishery expanded exponentially to become, by the 1960s, the largest fishery in the world and a major part of the Peruvian economy. The anchovies inhabited a narrow strip on the west coast of Peru, where nutrients are available in abundance from strong upwelling in the Pacific Ocean. The upwelling brings nutrients to the surface (into the photic zone) and starts a food chain in that it allows plants to grow on which fish may then feed. Off the coast of Peru, the upwelling is caused by the trade winds, and during the *El Niño* events these weaken, hence reducing the fishery. Most of the world's major fisheries are in regions of coastal upwelling and this region is particularly remarkable in that it accounts for 22% of the worlds' production of fish. Furthermore, the anchovies themselves were within a food chain and in turn provided 80 to 95% of the diet of guano birds. (Each bird consuming approximately 35,500 fish a year, or close to 100 per day!) Needless to say, the population of these birds was extremely sensitive to that of the anchovies.

Over the years the anchovy catch had increased steadily from an annual 2 million tons in 1959 to 12.3 million tons in 1970. A small *El Niño* event in 1965, together with heavy fishing, was sufficient to wipe out the guano birds, and thus the associated guano industry collapsed. Fishery 'experts' advised the Peruvian government, now concerned about overfishing, that a maximum sustainable yield was about 10 million tons annually. This was immediately adopted as the quota. No regulation was placed on the number of boats involved or the number of fish taken per boat: it was a free-for-all and the anchovies declined alarmingly. Still no action was taken to conserve the resource. In 1971, with hugely expanded fishing and processing, the quota was reached in just 3 months and the fishery had to be closed for the remainder of the year. The catch fell dramatically to 4.5 million tons and then along came a particularly severe *El Niño*. The anchovy population slumped still further and has never recovered. The little anchovies have gone.

To be fair, this was before we understood much about *El Niño*, but the case emphasises the concept of population thresholds for the sustainability of both industry and environment. With discrete breeding seasons, the population lends itself to discrete modelling, and one such model that includes the effect of *El Niño* as well as predators such as the guano birds and the efficiency of fishermen with a variety of net types can be found in Caulkins et al. (1985). Unfortunately, it is with hindsight that we see that the adoption of the recommendations from these improved models may have changed anchovy history. Still, one is left wondering as fishing in the region continues, with authorities too afraid of retribution to improve restrictions.

3.5   Scenario: How can 2×10^6 birds mean rare?

Perhaps it is not always easy to judge just what level of harvesting is safe; however, the following scenario emphasises the importance of biological research into understanding the species' behaviour before such levels can be set legitimately. The following is adapted from Quammen (1997).

When it comes to the extinction of species, there is no magic number below which a population ends with extinction. In fact, rarity may not mean extinction, just as abundance may not imply survival.

For example, the passenger pigeon, *Ectopistes migratorius*, was probably once the least rare bird on Earth. Inhabiting the eastern half of North America, the rookeries covered as much as 500 square kilometres and the birds literally darkened the sky when they took off. Around 1810, a flock was seen in Kentucky taking three days to pass overhead! Biologically this species was a huge success. Then, within the very short space of 36 years, its population went from roughly 3 billion to zero. It was people who reduced its number to a few million, enough to tip the balance, and extinction ensued. The decline happened suddenly during the 1880s and the unbelievably numerous pigeons simply ceased to exist.

The massacre/harvest peaked after the Civil War. However, during the last years of the pigeons' existence, their population decline was too steep to attribute to hunting alone. Habitat destruction would have been a further cause for population decline, but even so this was not enough to cause the crash. The explanation proposed by biologist T.R. Halliday in 1980 was in terms of a critical colony size. He argued the benefits to the population of crowding, as protection from predators, cooperative food source location, the maintenance of mating and nesting rhythms, and hence also its breeding rate. He suggested that a remnant population of a couple of million occurring in small flocks would be inadequate to sustain the population. That is, abundance was essential to the species and human interference had pushed the numbers below its particular threshold. Although two or so million sounds like abundance, for *Ectopistes migratorius* this was rare.

At the opposite extreme we have the Kestrels of Mauritius, *Falco punctatus*. In that case the population declined slowly. Originally the population would have been roughly 700 to 800 birds but by the 1950s it had decreased to well below 50 and by 1971 only 4 surviving birds were known to exist. With reduced and fragmented habitat, DDT, introduced exotic predators and plants, human hunters and then a cyclone, the population appeared doomed. It was the most rare bird in the world. Carl Jones, a Welshman and a 'bird person', arrived in 1979 having been sent by the International Council for Bird Preservation (which later withheld funding). He incubated eggs he collected from precarious ledges (replacing them with dummy eggs), used artificial insemination to improve the breeding rates and successfully reared birds in captivity. Furthermore, he supplemented the feed of those in the wild. Single handedly, and learning on the run, he eventually turned the population decline around.

By 1988 the population had climbed to 80 and although, due to the fragmentation of habitat and exotic predators and plants, the numbers will never reach their original estimates, the population today is relatively healthy.

3.6   Case Study: It's a dog's life: The control of stray dogs

Stray dogs are abandoned dogs and their offspring living wild in city environments. We now consider how the theory for single populations developed so far could contribute to the design of practical programs for the control of stray dogs, with a case study adapted from Amaku et al. (2010). That paper models a single population of stray dogs, which are known to cause health problems in cities and damage to livestock in rural areas, with a focus on a comparison between possible control strategies. Here we provide an insight into how to incorporate one control strategy (euthanasia, which involves killing stray dogs) into a differential equation model for the population. Later (in Chapter 5), we will extend these results to include a second management strategy (sterilisation).

Stray dogs can cause a number of problems. In many countries around the world they have become a major public health hazard, with dogs abandoned for numerous reasons, including financial and social (Amaku et al., 2010). After abandonment, dogs typically group in packs and breed, with shelters often unable to cope with the numbers roaming the streets. Another problem identified is that of stray dogs in rural areas where packs attack livestock and significantly impact on productivity.

Typically, euthanasia is the preferred strategy targeted at stray dogs. It impacts on population growth rates but total eradication is unlikely with the ongoing abandonment of animals. Here we apply a mathematical model to establish how euthanasia may be effective in the control of these animals.

We start by making some general assumptions about a population of stray dogs. Let $N(t)$ be the density of the population, that is, the number of animals per km^2. We assume logistic growth so that

$$\frac{dN}{dt} = rN\left(1 - \frac{N}{K}\right),$$

where $N = N(t)$ is a function of time, r is the intrinsic growth rate and K the carrying capacity. Parameter r can be expressed, in a crude sense, as the difference between birth rate a and death rate b so that $r = a - b$.

To incorporate euthanasia, we assume that, in each time-step, a constant proportion ϵ of stray dogs is euthanised in each km^2 — that is, at rate $\epsilon N(t)$ per km^2. This results in a differential equation for the strategy of euthanasia:

$$\frac{dN}{dt} = rN\left(1 - \frac{N}{K}\right) - \epsilon N,$$

where it is assumed that the removal rate is not dependent on density dependence, and thus not multiplied by $(1 - N/K)$. Re-writing the above, the euthanasia model becomes

$$\frac{dN}{dt} = -\frac{Nr}{K}\left[N - K\left(1 - \frac{\epsilon}{r}\right)\right]. \tag{3.15}$$

To generate results we require parameter estimates from the literature that are relevant to stray dogs. The carrying capacity (dogs per km^2) was taken as 250, close to the average estimate in Baltimore (United States) in 1970-71 (Beck, 2002). Further, following Amaku et al. (2010), appropriate instantaneous birth and death rates per year (based on the expected number of litters and puppy survival rates) are $a = 0.34$ and $b = 0.12$.

With these values, Figure 3.8 illustrates dog density (number per km^2) for a range of values for the control parameter ϵ, and starting with a number of different initial dog populations (5 solid curves). Note that, with $\epsilon > a$, the population is forced to extinction within a relatively short time (approximately 20 years) (first plot of Figure 3.8), while with $r < \epsilon < a$, extinction takes much longer (approximately 60 years) (second plot of Figure 3.8). Also, for $\epsilon > r$, the dog population decreases steadily to extinction (first and second plots of Figure 3.8), while for $\epsilon < r$ the population approaches a stable and positive density (third plot of Figure 3.8).

The same results can be established theoretically by setting $dN/dt = 0$ and finding the equilibrium points: $N_1 = 0$ and $N_2 = K(1 - \epsilon/r)$. The latter point defines the steady state

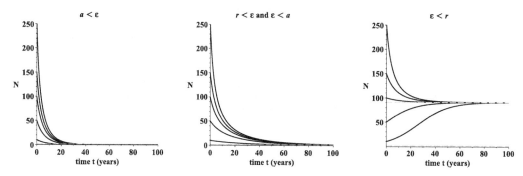

Figure 3.8: Euthanasia model (3.15). The threshold dynamics of stray dog numbers per km^2 over time ($N(t)$) are illustrated as the euthanasia parameter ϵ varies, for a number of initial conditions (solid lines). The grey dashed lines are the equilibrium values. Other parameter values are: $\epsilon = 0.36$, 0.25 and 0.14, $K = 250$, $a = 0.34$, $b = 0.12$, (Amaku et al., 2010).

illustrated in the third plot of Figure 3.8, with the population increasing over time when $0 < N < K(1 - \epsilon/r)$ and decreasing otherwise.

These results (Figure 3.8) infer that intrinsic growth rate r is a threshold for the proportion euthanised — a program has the best chance of success if the proportion of euthanised animals is above the intrinsic growth rate, although the population can also be controlled with lower proportions. While this is intuitive, the model exposes the nonlinear nature of how the proportion euthanised impacts on population size over time, expected times until extinction, and the likely size of dog populations in the event effective euthanasia is below the intrinsic growth rate. Since uncertainty exists around intrinsic growth rates and proportions euthanised, this 'big picture' view of the dynamics that modelling provides is likely to contribute to improved strategies for the control of stray dogs.

3.7   Discrete population growth and chaos

Many animal populations have distinct breeding seasons. It is often more appropriate to model such population growth using *difference equations* rather than differential equations. In this section we see how discrete growth models with crowding can predict oscillatory or chaotic growth in populations.

Background

Some observations of the growth of two beetle populations are shown in Figure 3.9. There appears to be quite a large variation in the type of growth. In the first example the effect of overcrowding appears to be levelling out the population. However the population fluctuations in the second example are more complex. The challenge is to find a single model that is capable of predicting both types of population growth in Figure 3.9.

Formulating a difference equation

In discrete growth we assume that the population does not change except at discrete intervals, corresponding to breeding seasons. Instead of thinking about the rate of change of the population at an arbitrary time, we consider the time interval as the time from one breeding season to the next. Let us choose to measure time in units of this period. Furthermore, we assume here that the population changes occur from births or deaths alone.

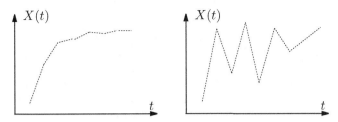

Figure 3.9: Schematic diagrams of some experimental data for growth of two different populations of beetle, where $X(t)$ is the number of beetles over a number of generations t. (From May (1981).)

We assume that the deaths are due to normal causes and also to crowding effects. Over a single breeding interval, we thus write

$$\left\{\begin{array}{c} change\ in \\ population \\ size \end{array}\right\} = \left\{\begin{array}{c} no.\ of \\ births \end{array}\right\} - \left\{\begin{array}{c} no.\ of \\ deaths \end{array}\right\}. \qquad (3.16)$$

Let us define X_n to be the population size in the n-th breeding interval. Here n is an integer, representing time, but taking only discrete values. Consequently, X_{n+1} is the population in the next breeding interval.

Assuming a constant per-capita birth rate β as before, the overall birth rate is βX_n. Since the birth rate is constant over the current breeding interval, the number of births for that interval is obtained by multiplying the birth rate by the length of the time interval (which is 1). Thus, for a single breeding interval,

$$\left\{\begin{array}{c} no.\ of \\ births \end{array}\right\} = \beta X_n. \qquad (3.17)$$

For the per-capita death rate we use $\alpha + \gamma X_n$, where the extra term represents an increased per-capita death rate due to competition for limited resources. The number of deaths is obtained by multiplying the current population size by the time interval. Thus for a single breeding interval,

$$\left\{\begin{array}{c} no.\ of \\ deaths \end{array}\right\} = \alpha X_n + \gamma X_n^2. \qquad (3.18)$$

From one breeding interval to the next the change in population is $X_{n+1} - X_n$. Substituting equations (3.17) and (3.18) into the word equation (3.16), we obtain

$$X_{n+1} - X_n = \beta X_n - \alpha X_n - \gamma X_n^2, \qquad n = 0, 1, 2, 3, \dots. \qquad (3.19)$$

This type of equation is known as a difference equation. It gives the population at one time in terms of the population at the previous time.

Suppose we write $r = \beta - \alpha$ and $\gamma = r/K$ as before, then equation (3.19) becomes

$$X_{n+1} = X_n + r X_n \left(1 - \frac{X_n}{K}\right), \qquad n = 0, 1, 2, 3, \dots. \qquad (3.20)$$

This equation is sometimes called the *discrete logistic equation*[1]. Here K is the carrying capacity and r is the per-capita reproduction rate.

---

[1]Another version of this equation $X_{n+1} = r X_n (1 - X_n/K)$ is also called the discrete logistic equation. This form arises for populations with non-overlapping generations — where all the adults die after they have given birth.

Solving the equation

We solve equation (3.20) iteratively, using a computer or calculator, but we first need to choose values for r, K and x_0. One way of gaining an understanding of the population growth is to slowly vary one of the parameters r, K or x_0, observing any changes in the behaviour, or dynamics of the population. In this case we examine the effect on the population growth of varying only r, the intrinsic reproduction rate. We fix the carrying capacity K at 1,000 and the initial population x_0 at 100. Using Maple, we can generate the time-dependent values and plot them. A variety of growth patterns emerge as we increase the per-capita growth rate r: from logistic growth through oscillatory growth to chaotic growth. These are described below together with the diagrams generated using the code in Listing 3.3. Similar MATLAB code is given in Listing 3.4.

Listing 3.3: Maple code: c_cp_logistic_discrete.mpl

```
restart:with(plots):
X[0]:=100;r:=0.2;K:=1000;
for n from 0 to 50 do
  X[n+1]:=X[n]+r*X[n]*(1-X[n]/K):
end:
points:=[seq([n,X[n]],n=0..50)]:
plot1:=plot(points,style=point,symbol=circle):
plot2:=plot(points,style=line,colour=red):
display(plot1,plot2);
```

Listing 3.4: MATLAB code: c_cp_logistic_discrete.m

```
N = 50; %number of iterations
r = 0.2; K = 1000; x0 = 100;
X = zeros(N+1,1); t = zeros(N+1,1);
X(1) = x0; t(1) = 0; %initial values
for n=1:N %loop over interations
    t(n+1) = n; %set time values
    X(n+1) = X(n) + r*X(n)*(1-X(n)/K);
end
plot(t, X, '.');
axis([0, N, 0, K*1.4]);
```

Logistic growth

Figure 3.10 illustrates the population growth for the small value of r, $r = 0.2$. Note that the growth is very similar to that for the continuous case, with the population growing exponentially at first and then levelling off as the reproduction rate declines due to crowding effects. We see that the population appears to approach its carrying capacity $K = 1,000$.

As we increase the reproduction rate from $r = 0.2$ to $r = 0.8$, we observe that the population curve is initially much steeper and then levels off more quickly. This is illustrated in Figure 3.10.

Oscillatory growth

When the intrinsic reproduction rate r is increased further so that $r > 1$, it becomes possible for the population to increase above the carrying capacity, but in the next iteration it then falls below K. A *damped* oscillation results, where the oscillations become smaller with time, as seen in Figure 3.11. As r increases, the amplitude of these damped oscillations increases.

Once the population is above the carrying capacity K the reproduction rate $r(X_n)$ now is negative (corresponding to the death rate being higher than the birth rate). Thus, in the next breeding season, the population drops below the carrying capacity. In the continuous model, the reproduction rate changed instantaneously; with the discrete model however,

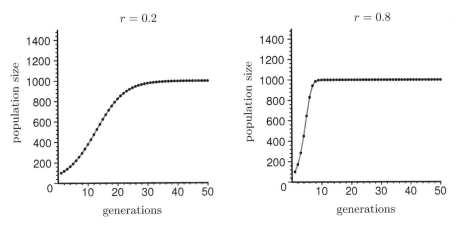

Figure 3.10: Computer-generated solutions for the discrete logistic equation with $K = 1,000$, $x_0 = 100$. Values of r used were $r = 0.2$ and $r = 0.8$. The growth follows logistic curves.

there is a delay of one breeding season before the reproduction rate can adjust to the change in population. This is why it is possible for the population to jump above the carrying capacity in the discrete model, which is not the case in the continuous model.

As we increase r further (see Figure 3.11), for $r = 2.2$ the population is again oscillating about the carrying capacity, but the amplitude of the oscillations appears to be constant. We call this oscillation a 2-cycle since the population size is repeated every second breeding cycle once the initial transients have died out. As r increases further, the amplitude of the oscillation increases. It is possible to prove, using the difference equation (3.20), that stable 2-cycles persist if $2 \leq r < 2.4$. This illustrates an increasingly common approach in mathematics, where computer experiments suggest results that are then proved using analysis. Somewhere between $r = 2.4$ and $r = 2.5$, the 2-cycles become unstable.

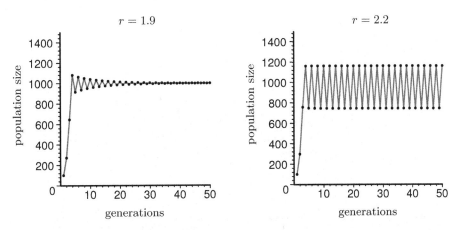

Figure 3.11: Computer-generated solutions of the discrete logistic equation with $K = 1,000$, $x_0 = 100$ and with $r = 1.9$ and $r = 2.2$. The graph for $r = 1.9$ shows a damped oscillation. The graph for $r = 2.2$ exhibits a 2-cycle.

Period doubling and chaotic growth

Again, the trend here is for the amplitude of the 2-cycles to increase but they do not increase indefinitely. We can see from Figure 3.12 that the 2-cycle has become a 4-cycle by $r = 2.5$, where the values repeat themselves every 4 breeding cycles. As r increases further, this becomes an 8-cycle and then a 16-cycle, and so on.

When we try $r \geq 2.6$, some entirely new behaviour is observed. The population does not grow in ordered cycles but seems to change randomly in each breeding period. We call this type of growth chaotic, and it is illustrated in Figure 3.12 for $r = 2.7$.

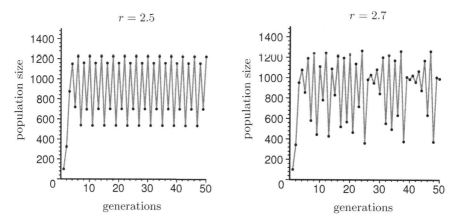

Figure 3.12: Computer-generated solutions of the discrete logistic equation with $K = 1,000$, $x_0 = 100$ and $r = 2.5$ and $r = 2.7$. For $r = 2.5$, we get a 4-cycle. For $r = 2.7$, we get chaotic growth where there is seemingly a random pattern to the growth.

Another feature of chaotic growth is that it is very sensitive to a change in the initial population. This feature of chaotic growth is illustrated in Figure 3.13 with two initial population sizes of 100 and 101.

Although the difference between the populations is only 1 initially, after some time they become very different. Both populations are increasing and decreasing in a random-like pattern but there are significant differences in the graphs.

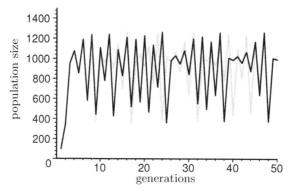

Figure 3.13: Sensitivity to initial conditions. Computer-generated solutions of the discrete logistic equation for $r = 2.7$ and $K = 1,000$, and for two different initial populations: $x_0 = 100$ (black line) and $x_0 = 101$ (grey line). As time increases, the solutions move further apart.

If we try to plot a similar graph using initial conditions $x_0 = 100$ and $x_0 = 101$ for any value of r that gave non-chaotic growth ($r < 2.6$), the graphs of the two populations become indistinguishable after only a few breeding seasons; see Exercise 3.15.

The discovery that simple population models like this can give rise to chaotic growth had an extremely profound effect. Previously, when presented with records of population growth that seemed to have the population growing in a random manner, it was thought that this was due to some external factor such as climate, environment, etc. Now, however, it is known that the random-like behaviour may be a natural feature of the way that the population grows, that is, a nonlinear response to a time delay in compensating the reproduction rate to account for crowding. (It should be noted here that $r = \beta - \alpha$ and thus that large values of r are only valid for certain populations.)

Bifurcation diagram

To summarise the different types of growth as the parameter r varies, it is useful to plot the possible equilibrium solutions against r. One way to do this is to run the simulation until large numbers of generations n, for each different value of r. This is done in Figure 3.14 below. The `Maple` code used to produce the bifurcation diagram is given in Listing 3.5. Similar `MATLAB` code is given in Listing 3.6.

Listing 3.5: Maple code: c_cp_logistic_birf.mpl

```
restart:with(plots):
n:=125: nrep:=32: K:=1000:
r := array(1..n):
X := array(1..n):
y:=array(1..nrep):
ini := 0.1*K:
inir := 1.5: endr := 3:
for i from 1 to n do
    r[i] := evalf(inir + i*(endr-inir)/n):
    for j from 1 to n-1 do
        X[1] := ini:
        X[j+1] := evalf(X[j]+r[i]*X[j]*(1-X[j]/K));
    end:
    for k from 1 to nrep do
        y[k] := X[n-nrep+k]:
    end:
    pp := [[r[i],y[jj]] $jj=1..nrep];
    bif[i] := plot(pp, x=inir..endr, y=0..1.5*K, style=point):
    col[i] := display([seq(bif[j],j=1..i)]);
end:
plot1:=display(col[n]):display(plot1);
```

Listing 3.6: MATLAB code: c_cp_logistic_birf.m

```
Nt = 1000; Nr = 125; Nlast = 20;
K = 1000; x0 = 100;
rv = linspace(1.5,3,Nr); % vector of r values
Xb = zeros(Nr,Nlast); rb = zeros(Nr, Nlast)

for k = 1:Nr; %loop over r values
    r = rv(k);
    X = zeros(Nt,1); t = zeros(Nt,1);
    X(1) = x0; t(1) = 0;
    for n=1:Nt % loop over iterations
        t(n+1) = n;
        X(n+1) = X(n) + r*X(n)*(1-X(n)/K);
    end
    Xb(k,:) = X(end-Nlast+1: end);
    rb(k,:) = r*ones(1,Nlast);
```

```
end

plot(rb, Xb, 'r.');
axis([rb(1), rb(end), 0, K*1.4]);
```

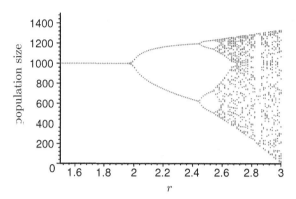

Figure 3.14: Bifurcation diagram for the discrete logistic model. This is a plot of the solutions, after a period of time, as the parameter r increases. They are the equilibrium solutions.

As r increases from 0, there is only a single equilibrium population $N = 1,000$ until $r = 2$. At $r = 2$, the solution curve splits into two and we have two equilibrium solutions. This is called a *bifurcation* and hence the diagram is called a bifurcation diagram. As r continues to increase, the solution undergoes further bifurcations, corresponding to 4-cycles, 8-cycles, and so on. Each bifurcation occurs for relatively smaller increases in r.

The continual bifurcation behaviour eventually results in chaos. It might be useful to think of chaos as an infinite number of bifurcations. As the parameter r increases further some unexpected things happen. We appear to get small regions, or windows, of regular behaviour within the chaos. This includes the appearance of 3-cycles.

Even a simple equation such as the logistic equation exhibits a rich variety of behaviour. For the interested reader an excellent series of articles in *New Scientist* provides a good introduction to the subject, in particular Vivaldi (1989) and Stewart (1989). For an example of chaos in real populations the articles Costantino et al. (1995) and Costantino et al. (1997) describe experiments on flour beetle populations, where manipulating the adult mortality rate can cause the population to exhibit cycles or become chaotic. They used three difference equations to describe the larvae, pupae and adult populations.

Limitations of the model

When $r > 3$, it is possible for the population to become negative. We would interpret this as the population becoming extinct. Of course this may not happen in practice as the stronger members of the population might survive into the next breeding season. Thus this model has a serious weakness if $r > 3$; however, it is useful in demonstrating how extremely complex patterns of growth can evolve from simple assumptions.

A more robust model, which does not have negative populations for any values of r, is the model

$$X_{n+1} = X_n e^{a(1-X_n/K)}, \qquad \text{where} \qquad a = \ln(r+1).$$

This model is derived from simple probabilistic arguments[2] and exhibits similar behaviour to the logistic model, with damped oscillations, 2-cycles, 4-cycles and chaotic growth behaviour (although for different values of r). It is superior to the logistic model for discrete growth since it incorporates a constant probability of death for each individual. This model is discussed in May (1981).

> ### Summary of skills developed here:
>
> - *Explore a difference equation model on a computer and describe the range of phenomena.*
> - *Understand the different modelling approaches of discrete and continuous growth.*
> - *Be able to discuss the properties of chaotic growth.*

3.8   Time-delayed regulation

We discuss, briefly, how to introduce a time lag response into the logistic model of population growth.

Background

With the logistic model developed in Section 3.2, we incorporated an instantaneous reaction to the environment. That is, increased pressure on the resources produces an *immediate response* from the system (in terms of, for example, more deaths).

Often this is not realistic in that the response usually takes effect after some time delay, or time lag. Vegetation needs time to recover and changed environmental conditions that may lead to increased birth rates will take time to show up in the numbers of an adult population. This leads us to a model that includes a time delay which could result from a multitude of measurable sources (maturation times, food supply, resources, crowding, etc.). If the time lag is small compared with the natural response time ($1/r$), then there is a tendency to overcompensate, which may produce oscillatory behaviour. This is what we observe in the following case study (Section 3.9).

Logistic equation with time lag

From Section 3.2, the logistic equation for population growth was

$$\frac{dX}{dt} = rX(t)\left(1 - \frac{X(t)}{K}\right),$$

where $X(t)$ is the population size, r the intrinsic reproduction rate and K the carrying capacity for the environment. The term $r(1 - X(t)/K)$ represents the density-dependent reproduction rate. This states that the reproduction rate decreases with population size owing to increased deaths from overcrowding. If we now suppose that this is applied at

---

[2]The model assumes interactions between two different individuals are randomly distributed (as a Poisson distribution) with a rate of contact proportional to $(1 - X_n/K)$.

the earlier time $t - \tau$, where τ represents the time delay between increased deaths and the resulting decrease in population reproduction, then we obtain

$$\frac{dX}{dt} = rX(t)\left(1 - \frac{X(t - \tau)}{K}\right). \tag{3.21}$$

This type of equation is known as a *differential-difference* equation and also known as a differential-delay equation. While some linear equations of this type can yield analytic solutions, equation (3.21) is nonlinear and we must use numerical techniques. These techniques are similar to the techniques for solving ordinary differential equations (see Chapter 4) but also take account of the fact that the term $X(t - \tau)$ is evaluated at an earlier time.

The MATLAB software has a built-in routine for dealing with differential-delay equations, dde23, similar to ode45 for solving ordinary differential equations. In addition to specifying the RHS of the differential equation, we also need to specify the lag time τ and separately any terms that are lagged (that is, evaluated at an earlier time). Some MATLAB code for solving (3.21) is given in Listing 3.7. The graph of the solution is given in Figure 3.15. The effect of the time delay has been to induce some oscillations in the population. Again, as with the discrete logistic equation, it occurs because the regulating effect is delayed, meaning that the population can grow larger than the carrying capacity, but when the regulating effect can respond to this larger population, it causes a larger death rate, consequently allowing oscillations. For smaller values of the intrinsic growth rate r and the time delay τ, the population growth is more like the typical logistic growth, as in Figure 3.5.

Listing 3.7: MATLAB code: c_cp_logistic_dde.m

```
function c_cp_logistic_dde
global r K tau;
r=2.0; K=100;
tau=1.0; % delay amount
tend = 30; %end time
X0 = 50; %initial value

sol = dde23(@rhs, tau, X0, [0 tend]);
plot(sol.x, sol.y);


function Xdot = rhs(t, X, Xlag)
global r K;
Xdot = r*X*(1-Xlag/K);
```

We cannot use Maple directly to solve equation (3.21); however, we can approximate it by writing

$$\frac{dX}{dt} \simeq \frac{X(t + \Delta t) - X(t)}{\Delta t}.$$

Defining our time scale so that $\Delta t = 1$ day and write $X(t) = X_n$ and $X(t + \Delta t) = X_{n+1}$. Choose the delay τ to be an integer number of time periods. The differential-difference equation (3.21) becomes the difference equation

$$X_{n+1} = X_n + rX_n\left(1 - \frac{X_{n-\tau}}{K}\right). \tag{3.22}$$

This can be iterated from an initial condition using Maple by adapting the code provided in Section 3.7.

We do not examine the model further here, but the interested reader will find details in May (1981). What follows is a case study adopting one such model.

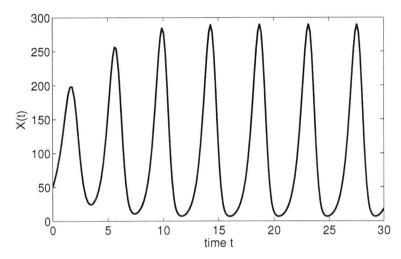

Figure 3.15: Solution of the delayed logistic equation with initial population $X(0) = 50$, and with parameter values $r = 2.0$, $K = 100$.

3.9 Case Study: Australian blowflies

We now present an example of a time-delayed model, and, although we do not examine it in any depth, it serves to illustrate how the inclusion of a time lag produces oscillatory behaviour. These model results closely replicate Nicholson's data, which recorded the size of Australian blowfly populations. The following is adapted from May (1981).

Blowflies reproduce by laying their eggs in carcasses and meat or in the open wounds or sores of live animals. In the latter case, the maggots hatch, and after eating the rotten flesh burrow into the animal, literally eating it alive. The open wound becomes worse, attracting more flies and so the process can snowball within a single day. Over the years, blowflies have caused a particular problem in the Australian sheep industry where farms may comprise many thousands of hectares, and the fly population is huge.

In the early part of this century, A.J. Nicholson conducted single-species experiments on this Australian sheep-blowfly population (Lucilia cuprina). His results are well approximated by the model described by a differential-difference equation, which itself can be approximated by the difference equation

$$X_{n+1} = X_n + rX_n \left(1 - \frac{X_{n-\tau}}{K}\right).$$

Here X is the fly population size, r its reproduction rate and K its carrying capacity. This formulation assumes the time for a newly laid egg to mature into a blowfly is τ days and further assumes that the density dependence occurs when the eggs are laid rather than the current time.

Using `Maple` *or* `MATLAB` *this can be solved iteratively with the following relevant parameter values $r = 0.212 \, \text{days}^{-1}$, $K = 2.8 \times 10^3$ flies and $\tau = 9$ days. Different from the delay equation in Section 2.9, here we do not assume that the population is constant for $t < 0$ but instead simulate the population from $t = \tau$ and use historical data to calculate $X(t - \tau)$ for $t \leq \tau$.*

The match between the data collected and the time-delayed model predictions can be seen in Figure 3.16 where the model output is imposed on Nicholson's data (obtained from May (1981)). The model performs well in spite of being somewhat simplistic in not including accurate details of reproduction, age classes or other details associated with the flies.

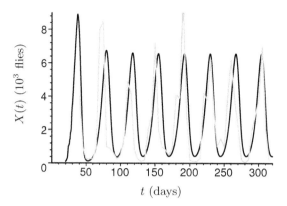

Figure 3.16: A plot of Nicholson's measurements (grey) with the model output (black). The model parameter values used are $r = 0.212\,\text{days}^{-1}$, $K = 2.8 \times 10^3$ flies and $\tau = 9$ days. Time steps of $1/2$ day have been used.

3.10 Exercises for Chapter 3

3.1. Fish farm. *In a fish farm, fish are harvested at a constant rate of $2,100$ fish per week. The per-capita death rate for the fish is 0.2 fish per day per fish, and the per-capita birth rate is 0.7 fish per day per fish.*

(a) *Write a word equation describing the rate of change of the fish population. Hence obtain a differential equation for the number of fish. (Define any symbols you introduce.)*

(b) *If the fish population at a given time is $240,000$, give an estimate of the number of fish born in one week.*

(c) *Determine if there are any values for which the fish population is in equilibrium. (That is, look for values of the fish population for which there is no change over time.)*

3.2. Modelling the spread of technology. *Models for the spread of technology are very similar to the logistic model for population growth. Let $N(t)$ be the number of ranchers who have adopted an improved pasture technology in Uruguay. Then $N(t)$ satisfies the differential equation*

$$\frac{dN}{dt} = aN \left(1 - \frac{N}{N_T} \right),$$

where N_T is the total population of ranchers. It is assumed that the rate of adoption is proportional to both the number who have adopted the technology and the fraction of the population of ranchers who have not adopted the technology.

(a) Which terms correspond to the fraction of the population who have not yet adopted the improved pasture technology?

(b) According to Banks (1994), $N_T = 17,015$, $a = 0.490$ and $N_0 = 141$. Determine how long it takes for the improved pasture technology to spread to 80% of the population.

Note: This same model can be used to describe the spread of a rumour within an organisation or population.

3.3. Quadratic population model. Consider the population model

$$\frac{dN}{dt} = aN - bN^2,$$

where a and b are positive constants. Here bN^2 represents a death term due to overcrowding (i.e., proportional to N^2 due to interactions of the population with itself).

(a) Find all the equilibrium points. Are there any conditions on the parameters a and b for the equilibrium population to remain positive?

(b) Determine the stability of each of the equilibrium points.

(c) It is claimed that this model is exactly the same as the logistic growth model. If this claim is true, then express the constants a and b in terms of the intrinsic growth rate r and carrying capacity K. If it is not true, explain why.

3.4. Allee effect model. Consider the population model

$$\frac{dN}{dt} = rN(N - b)\left(1 - \frac{N}{K}\right),$$

where r is the intrinsic growth rate, K is the carrying capacity and b is a positive constant where $b < K$.

(a) It is claimed that this model has the property that for small populations the net reproduction becomes negative but for larger populations behaves like the logistic model in that the reproduction rate is zero at the carrying capacity. Sketch the reproduction function (i.e. the per-capita growth rate) as a function of N and hence verify if the claim is true or false.

(b) Find all the equilibrium populations.

(c) Determine the stability of each of the three equilibrium points.

(d) What happens if the initial population is below b?

3.5. Density-dependent births. Many animal populations have decreasing per-capita birth rates when the population density increases, as well as increasing per-capita death rates. Suppose the density-dependent per-capita birth rate $B(X)$ and density-dependent death rate $A(X)$ are given by

$$B(X) = \beta - (\beta - \alpha)\delta\frac{X}{K}, \qquad A(X) = \alpha + (\beta - \alpha)(1 - \delta)\frac{X}{K},$$

where K is the population carrying capacity, β is the intrinsic per-capita birth rate, α is the intrinsic per-capita death rate and δ, where $0 \leq \delta \leq 1$, is a parameter describing the extent that density dependence is expressed in births or deaths.

Show that this still gives rise to the standard logistic differential equation

$$\frac{dX}{dt} = rX\left(1 - \frac{X}{K}\right).$$

3.6. Mouse population model. *A population, initially consisting of 1,000 mice, has a per-capita birth rate of 8 mice per month (per mouse) and a per-capita death rate of 2 mice per month (per mouse). Also, 20 mouse traps are set each week and they are always filled.*

First, write a word equation describing the rate of change in the number of mice and then formulate a differential equation for the population with an initial condition.

3.7. Harvesting model. *Consider the harvesting model from Section 3.3,*

$$\frac{dX}{dt} = rX\left(1 - \frac{X}{K}\right) - h.$$

(a) *Show there can be two non-zero equilibrium populations, with the larger value given by*

$$X_e = \frac{K}{2}\left(1 + \sqrt{1 - \frac{4h}{rK}}\right),$$

For the parameter values used in Figure 3.7, calculate the value of this equilibrium population.

(b) *If the harvesting rate h is greater than some critical value h_c, the non-zero equilibrium values do not exist and the population tends to extinction. What is this critical value h_c?*

(c) *If the harvesting rate is $h < h_c$, the population may still become extinct if the initial population x_0 is below some critical level, perhaps due to an ecological disaster. Show that this critical initial value is*

$$x_c = \frac{K}{2}\left(1 - \sqrt{1 - \frac{4h}{rK}}\right).$$

(Hint: Consider where $X' < 0$.)

3.8. Fishing with quotas. *In view of the potentially disastrous effects of overfishing causing a population to become extinct, some governments impose quotas that vary depending on estimates of the population at the current time. One harvesting model that takes this into account is*

$$\frac{dX}{dt} = rX\left(1 - \frac{X}{K}\right) - hX.$$

(a) *Show that the only non-zero equilibrium population is*

$$X_e = K\left(1 - \frac{h}{r}\right).$$

(b) *At what critical harvesting rate can extinction occur?*

Although extinction can occur with this model, as the harvesting parameter h increases towards the critical value the equilibrium population tends to zero. This contrasts with the constant harvesting model in Section 3.3 and Exercise 3.7, where a sudden population crash (from a large population to extinction) can occur as the harvesting rate increases beyond a critical value.

3.9. Predicting population size. *In a population, the initial population is $x_0 = 100$. Suppose a population can be modelled using the differential equation*

$$\frac{dX}{dt} = 0.2X - 0.001X^2,$$

with an initial population size of $x_0 = 100$ and a time step of 1 month. Find the predicted population after 2 months. (Use either an analytical solution or a numerical solution from `Maple` *or* `MATLAB`*.)*

3.10. Plant biomass. *Let the dry weight of some plant (that is, its biomass) at time t be denoted by $x(t)$. And suppose this plant feeds off a fixed amount of some single substrate, or a nutrient medium, for which the amount remaining at time t is denoted by $S(t)$. Assume that the growth rate of the plant is proportional to its dry weight as well as to the amount of nutrient available, and that no material is lost in the conversion of S into x.*

(a) *Starting with a word equation, model the rate of plant growth dx/dt with an initial plant biomass of x_0 and with x_f the amount of plant material associated with $S = 0$. (Use the fact that S can be written as a function of x and x_f: $S(t) = x_f - x(t)$.)*

(b) *Solve the equation using analytical techniques or numerically with Maple or MATLAB. (Separable techniques together with partial fractions are one way to solve the equation.)*

(c) *Using this model, why can the plant biomass of x_f never be attained? (In spite of this, and the fact that plants do reach a maximum biomass with finite time, this model does give a reasonable prediction of annual plant growth.)*

3.11. Modelling the population of a country. *Consider the population of a country. Assume constant per-capita birth and death rates, and that the population follows an exponential growth (or decay) process. Assume there to be significant immigration and emigration of people into and out of the country.*

(a) *Assuming the overall immigration and emigration rates are constant, formulate a single differential equation to describe the population size over time.*

(b) *Suppose instead that all immigration and emigration occurs with a neighbouring country, such that the net movement from one country to the other is proportional to the population difference between the two countries and such that people move to the country with the larger population. Formulate a coupled system of equations as a model for this situation.*

In both (a) and (b), start with appropriate word equations and ensure all variables are defined. Give clear explanations of how the differential equations are obtained from the word equations.

3.12. Newly abandoned dogs. *Read the case study (Section 3.6) for a model that incorporates euthanasia as a control for stray dogs.*

(a) *Let A denote the number of newly abandoned dogs per km^2 each year. Assuming this is a constant annual rate, and is not density dependent or dependent on stray dog numbers, include this in the euthanasia model (3.15).*

(b) *The first plot in Figure 3.8 illustrates that for $\epsilon = 0.36$ ($K = 250$, $a = 0.34$, $b = 0.12$) euthanasia results in the eradication of stray dogs. Using the model in (a), establish that with the regular abandonment of dogs ($h > 0$) extinction is no longer a possible outcome, but with control the population approaches a non-zero stable population.*

3.13. Scaling. *Scaling is a technique commonly applied in mathematical modelling to reduce the number of parameters while retaining the dynamical properties of the original system.*

(a) *Consider the standard logistic equation,*

$$\frac{dX}{dt} = rX\left(1 - \frac{X}{K}\right).$$

Using the substitution $x = X/K$, show that this equation can be written as

$$\frac{dx}{dt} = rx\left(1 - x\right).$$

This latter equation has been scaled. The resulting equation is independent of parameter K but retains the same dynamical properties, and its dynamics can now easily be compared with other logistic models, with different carrying capacities.

(b) Show that model ((3.15)) from the case study on stray dog control (Section 3.6) can be scaled using the transformation $n = N/K$ to give,

$$\frac{dn}{dt} = rn\,(1-n) - \epsilon n.$$

3.14. Population density. It is often convenient to measure population abundance (size) as a population density (number of animals per unit area). What difference does it make to the population equations? To find out, let $n(t) = N(t)/A$, where A is the fixed area where the population resides. Given the population logistic equation

$$\frac{dN}{dt} = rN\left(1 - \frac{N}{K}\right),$$

what is the differential equation for the density $n(t)$?

3.15. Sensitivity to initial conditions. Referring to the results generated in Figure 3.13 for two separate initial conditions, $x_0 = 100$ and $x_0 = 101$, generate the results with these initial conditions for $r < 2.7$, $r = 2.5$ and $r = 1.9$. What do you notice about the distance between the two graphs at each time step? (The relevant code is given in Section 3.7.)

3.16. Investigating parameter change. Using Maple or MATLAB, examine the effect of increasing the parameter r on the solution to the equation

$$X_{n+1} = X_n e^{a(1-X_n/K)}, \qquad \text{where} \qquad a = \ell n(r+1).$$

Establish (roughly) for what values of r the system undergoes its first two bifurcations. (Code can be adapted from that in Section 3.7.)

3.17. Finding equilibrium solutions. For the following discrete population models find all the equilibrium solutions by setting $N_{k+1} = N_k = N$. Determine the stability of each of the equilibrium populations.

(a) $N_{k+1} = 5N_k$ (b) $N_{k+1} = 0.8N_k - 0.1N_k^2$

3.18. Adults die after laying eggs. A difference equation describing female insects with a periodic breeding time is

$$X_{k+1} = rX_k(1 - X_k/K),$$

where all the parent insects die after laying their eggs.

(a) Where is this model different from the discrete logistic equation in lectures?

(b) Find all the equilibrium solutions, for $r > 1$, and determine their stability.

3.19. Modelling insect populations. Many insect populations breed only at specific times of the year and all the adults die after breeding. These may be modelled by a difference equation, such as

$$X_{n+1} = r(X_n - 0.001X_n^2).$$

Using Maple or MATLAB, investigate what happens as the parameter r (the growth rate) is varied from $r = 0$ to $r = 3$. Sketch all the different types of growth patterns observed, labelled with the corresponding value of r.

3.20. Discrete growth with harvesting. Consider the discrete model for linear population growth with a constant positive number h harvested each time period. In this model all adults die after giving birth. The difference equation is

$$X_{k+1} = rX_k - h,$$

where r is the per-capita net growth rate (per time step). Find all the equilibrium solutions and determine their stability.

3.21. Ricker model. *The Ricker model is sometimes used for fish populations as an alternative to the discrete logistic equation. The difference equation is*

$$N_{k+1} = N_k e^{r(1-N_k/K)}, \qquad r > 0,$$

where N_k is the population size in generation k, r is the intrinsic growth rate and K is the carrying capacity, which are positive constants. Find all the equilibrium solutions and determine their stability.

3.22. Stability of 2-cycles. *Consider the discrete logistic equation (with $K = 1$)*

$$X_{n+1} = X_n + rX_n(1 - X_n).$$

(a) *Show that every second term in the sequence $X_0, X_1, X_2, \ldots$ satisfies the difference equation*

$$X_{n+2} = (1 + 2r + r^2)X_n - (2r + 3r^2 + r^3)X_n^2$$
$$+ (2r^2 + 2r^3)X_n^3 - r^3 X_n^4.$$

(b) *For equilibrium solutions, let $S = X_{n+2} = X_n$ and obtain a quartic equation (that is, an equation with the unknown raised to the fourth power, at most).*

Explain why $S = 0$ and $S = 1$ must be solutions of this quartic equation.

Hence show that the other two solutions are

$$S = \frac{(2 + r) \pm \sqrt{r^2 - 4}}{2r}.$$

[Note: Comparing with Figure 3.11 ($r = 2.2$) we see that these two values of the two non-zero equilibrium solutions are the values between which the population oscillates in a 2-cycle. Furthermore, when r increases to where these two equilibrium solutions become unstable, this corresponds to where the 2-cycle changes to a 4-cycle.]

3.23. Linear differential-delay equation. *Consider the linear differential-delay equation*

$$\frac{dX}{dt} = X(t - 1), \qquad X(0) = 1.$$

Look for an exponential solution of the form $X(t) = Ce^{mt}$, where m is a constant to be determined and C is an arbitrary constant.

3.24. Chemostat. *A chemostat is used by microbiologists and ecologists to model aquatic environments, or waste treatment plants. It consists of a tank filled with a mixture of some medium and nutrients, which microorganisms require to grow and multiply. A fresh nutrient-medium mixture is pumped into the tank at a constant rate F and the tank mixture is pumped from the tank at the same rate. In this way the volume of liquid in the tank remains constant. Let $S(t)$ denote the concentration of the nutrient in the tank at time t, and assume the mixture in the tank is well stirred. Let $x(t)$ denote the concentration of the microorganism in the tank at time t.*

(a) *Draw a compartmental diagram for the amount of nutrient.*

(b) *In the absence of the organism, suggest a model for the rate of change of $S(t)$.*

(c) *If the microorganisms' per-capita uptake of the nutrient is dependent on the amount of nutrient present and is given by $p(S)$, and the per-capita reproduction rate of the microorganism is directly proportional to $p(S)$, extend the model equation above to include the effect of the organism. (The per-capita uptake function measures the rate at which the organism is able to absorb the nutrient when the nutrient's concentration level is S.)*

(d) *Now develop an equation describing the rate of change of the concentration of the live organism (x') in the tank to derive the second equation for the system.*

(e) *The nutrient uptake function $p(S)$ can be shown experimentally to be a monotonically increasing function bounded above. Show that a Michaelis–Menten type function*

$$p(S) = \frac{mS}{a + S},$$

with m and a positive, non-zero constants, satisfies these requirements. What is the maximum absorption rate? And why is a called the half-saturation constant? (Hint: The maximum absorption rate is the maximum reached by $p(S)$. For the second part, consider $p(a)$.)

This system of equations is known as the Monod Model for single species growth and was developed by Jaques Monod in 1950.

Chapter 4

Numerical solution of differential equations

This chapter provides a brief overview of numerical procedures on which we rely when employing software as a tool in the solution and analysis of mathematical models. While developments have ensured that in most cases the methods are robust, it is important to understand the trade-offs we must accept when using them, and the possible errors that could accumulate, making interpretation inaccurate. Perhaps the most important part of this chapter is the final discussion.

4.1 Introduction

In finding solutions to differential equations, plotting trajectories and displaying time-dependent graphs, we have come to rely on the performance of computers and software packages. Some packages use analytical solutions where possible, a symbolic approach such as in `Maple`, but many problems cannot be solved analytically and there is a need to employ numerical schemes to find a solution. In this case the numerical solution, while possibly extremely accurate, is an approximation to the exact analytic solution.

There are some major drawbacks in accepting whatever is produced by a computer, particularly when numerical solvers are used, as errors may accrue for the following reasons:

- There are *round-off errors* and these increase with an increase in the number of calculations performed.

- There are *discretisation errors* resulting from the estimation of a solution for a discrete set of points; these may decrease with an increase in the number of points used.

- There are errors due to the *estimating procedure*, or method of approximation, used by the numerical solver.

Many different numerical procedures are available for finding derivatives, integrals, sums, etc. First we provide a very brief outline of a few used in finding solutions to differential equations (the main use of solvers in this book), and we then consider some of the drawbacks in using these numerical approximations.

4.2 Basic numerical schemes

To examine some numerical schemes used in solving differential equations, we consider the general problem $y' = f(t, y)$ with $y(t_0) = y_0$, and apply the schemes that follow to calculate (estimate) $y(t)$ with $t_0 < t < T$. We begin with a very simple method, Euler's method, to gain a basic understanding of the process any method needs to adopt, and then look briefly at a well-known and widely used method, the Runge–Kutta method.

Euler's Method

Clearly a computer cannot calculate every point on a curve as there are infinitely many, so an approximation is made for a discrete set of points. Euler's method provides a very simple way of approximating the solution of a differential equation at a discrete set of points.

The steps in Euler's method are as follows:

- Divide the interval into N equal sections, then $t_n = t_0 + nh$ for $n = 0, 1, \ldots, N - 1$ and $h = (T - t_0)/N$ is the step size.

- We know that (t_0, y_0) is on the curve and we approximate $y_n = y(t_n)$. To approximate y_1 we follow the tangent of the curve through (t_0, y_0) (which is known), extending it to t_1. Then

$$y_1 \approx y_0 + hf(t_0, y_0)$$

or

$$y' = f(t_0, y_0) \approx \frac{y_1 - y_0}{h} = \frac{y_1 - y_0}{t_1 - t_0}.$$

- We follow this procedure repeatedly: $y_2 \approx y_1 + hf(t_1, y_1)$ and so on for $y_3, \ldots, y_n$.

Figure 4.1 illustrates this procedure.

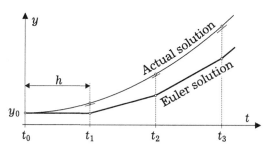

Figure 4.1: Diagram of Euler's method for solving differential equations.

In general, Euler's method is the recursive scheme

$$y_{n+1} = y_n + hf(t_n, y_n)$$

with

$$t_{n+1} = t_n + h \quad \text{and} \quad 0 \le n \le N - 1.$$

The following example shows how to apply Euler's method.

Example 4.1: Use Euler's method, with step size $h = 0.1$, to find $y(0.2)$ for the differential equation

$$\frac{dy}{dt} = y^2 + t, \qquad y(0) = y_0 = 1.$$

Solution: For this equation, $f(t, y) = y^2 + t$. Using Euler's method, with $h = 0.1$ and letting $n = 0$, we obtain

$$y_1 = y(0.1) = y_0 + hf(t_0, y_0) = 1 + 0.1 \times (1^2 + 0) = 1.1.$$

Now, with $n = 1$, we obtain

$$y_2 = y(0.2) = y_1 + hf(t_1, y_1) = 1.1 + 0.1 \times (1.1^2 + 0.1) = 1.231.$$

Taylor's Theorem states that

$$y_{n+1} = y_n + h\frac{dy_n}{dt} + \frac{h^2}{2!}\frac{d^2 y_n}{dt^2} + \dots.$$

(Taylor's Theorem provides an accurate polynomial approximation to a large group of functions. For details, see Appendix B.2.)

Comparing this approximation with Euler's method, it is clear that the latter consists of the first two terms in this expansion. We say that Euler's method is a *first-order approximation*. (Including further terms in this series would produce second-order, third-order, etc., approximations.)

With any approximation, and thus with each numerical scheme, comes an associated error term. Since the approximation in Euler's method consists of only the first two terms of the Taylor series, it would be useful to be able to calculate the error term and then perhaps control the size of the error by choice of the step size. From Euler's method we have the estimated value

$$y_{n+1} = y_n + hf(t_n, y_n) \quad n = 0, \dots, N-1.$$

From Taylor's Theorem we have the actual value

$$y(t_{n+1}) = y(t_n) + hf(t_n, y(t_n)) + \frac{h^2}{2}f'(\xi_n, y(\xi_n)),$$

where ξ_n lies between t_n and t_{n+1}.

Subtracting the estimate from the true value, we get

$$y(t_{n+1}) - y_{n+1} = y(t_n) - y_n$$
$$+ h[f(t_n, y(t_n)) - f(t_n, y_n)] + \frac{h^2}{2}f'(\xi_n, y(\xi_n))$$

which is the error term E_{n+1}. So, for each step of the method, an error E_{n+1} is incurred. We can find an upper bound for this error term and it can be shown that

$$E_{n+1} \leq \frac{Dh}{2L}\left[e^{T-t_0} - 1\right],$$

where L is an upper bound for f and D is an upper bound for f' on the interval $[t_0, T]$, where $f(t, y)$ is continuously differentiable and $T = t_0 + (N-1)h$.

Clearly $\lim_{h\to 0} E_{n+1} = 0$, indicating increasing accuracy of the estimation with decreasing step size. (For a full analysis see, for example, Kincaid and Cheney (1991).)

However, it should be remembered that every computational operation (for example, addition, division, etc.) comes at a cost, as each may incur an error. This is a result of the fact that a computer can only store a fixed number of digits and so a number may be rounded off at each step. Thus, while decreasing the step size h (or equivalently increasing the number of points in the interval N), we are simultaneously increasing the number of operations or calculations and are thus increasing the impact of round-off errors. For best performance then, we should aim at finding some optimum h such that the combined effect of these two errors is minimised.

Runge–Kutta Methods

One-step algorithms that use averages of the slope function $f(t, y)$ at two or more points over the interval $[t_{n-1}, t_n]$ in order to calculate y_n are called *Runge–Kutta methods*. They are also examples of *predictor-corrector methods* as they make predictions for 'next' values, and then with a series of weights, correct them.

The fourth-order Runge–Kutta method (RK4) is one of the most widely used methods of any step algorithms. It involves weighted averages of slopes at the midpoint and end points of the subinterval. For the IVP $y' = f(t, y)$, $y(t_0) = y_0$, the fourth-order RK method is a one-step method (that is, y_n depends only on y_{n-1}) with the constant step size h and is given by

$$y_{n+1} = y_n + \frac{h}{6}(k_1 + 2k_2 + 2k_3 + k_4),$$

where

$$k_1 = f(t_n, y_n),$$
$$k_2 = f(t_n + \frac{h}{2}, y_n + \frac{h}{2}k_1),$$
$$k_3 = f(t_n + \frac{h}{2}, y_n + \frac{h}{2}k_2),$$
$$k_4 = f(t_n + h, y_n + hk_3).$$

Another Runge–Kutta method, using a simpler scheme of averaging than the RK4 above, is Heun's method. This is illustrated in Figure 4.2.

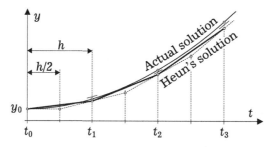

Figure 4.2: Diagram of Heun's method for solving differential equations.

Like Euler's scheme, this method is a one-step method but it is more accurate. With a constant step size h, the general form of Heun's method is given by

$$y_{n+1} = y_n + \frac{h}{2}\left(k_1 + k_2\right),$$

where

$$k_1 = f(t_n, y_n),$$
$$k_2 = f(t_n + h, y_n + hk_1).$$

4.3 Computer implementation using Maple and MATLAB

Most good ODE solver packages use sophisticated methods of controlling errors, and being efficient most of the time they have widespread acceptance. Many schemes choose their own step sizes in order to optimise the results, and some vary this step size on an interval in order to provide an optimal result.

Note, once again, that there is a trade-off between decreasing the step size and increasing the round-off error, due to an increased number of computations. Another problem is that of instability over long time periods: sometimes, while providing an accurate approximation over an initial period of time, the solution may 'become' very inaccurate at later times.

Using `Maple` or `MATLAB`, the method of integration can be stipulated and the results compared. `Maple` or `MATLAB` also allow you to change and choose the *initial* step size of the default numerical scheme that, as we see below, is sometimes required for adequate resolution. Note, however, that with the standard DE solvers `ode45` for `MATLAB` and `DEplot` for `Maple`, adaptive stepping is used. This means the step size is varied, becoming smaller when the function is changing fast, so as to maintain a specified error, and where the step size increases when the function is changing slowly, so generally, specifying the step size only really is useful if the *initial* default step size happens to be too large. In practice, the default values for `MATLAB` appear to be adequate for most problems, but for `Maple`, setting the `stepsize` parameter to 0.1, for example, is sometimes required, as discussed below.

In Chapter 2, in the discussion on finding a computer solution to a radioactive element decay cascade, we mentioned that it would be difficult to find a suitable numerical scheme because of the vastly different half-lives. (For Uranium-238, the half-life is billions of years, whereas for the next element in its decay series, Thorium-234, the half-life is in days.) This problem requires large time steps to resolve the Uranium decay process and (comparatively) very small time steps to resolve the Thorium decay: such a problem is known as a *stiff problem*. This discrepancy between rates can be seen clearly in the following example. Suppose a solution to some differential equation $X' = f(x)$ is

$$X(t) = e^{-t} + e^{-1000t}, \qquad t \geq 0.$$

Clearly, the second component decays at a much faster rate than the first. When t is small, the value of $X(t)$ is dominated by e^{-1000t}, and small step sizes are required for resolution of this behaviour. Alternatively, for t away from 0, the solution is dominated by e^{-t} and large step sizes (in comparison) can be used for high accuracy. Attempting to solve a stiff problem with a standard adaptive time-stepping method, such as the default in `DEplot` of `Maple` or `ode45` in `MATLAB`, will result in the method taking smaller and smaller time steps, and usually being unable to complete the calculation. For some software this may result in a prompt of an error message, triggered when a specified maximum number of function evaluations has been reached.

Alternative numerical methods, involving implicit solution methods, have been developed to deal specifically with stiff problems, but typically they increase the computation time incurred when the problem is not stiff. Thus, they are usually employed only once a problem is known to be stiff. `Maple` provides a stiff solver that can be used for the solution of such problems by stipulating the numerical method (always after `numeric`) in `dsolve`. Applying this to the example of lake pollution (Section 2.5), which is not a stiff problem, it is easy to establish a substantial increase in computation time. The code below gives an example of one way to specify a stiff solver in `Maple`.

```
> restart:with(plots):
> cin:=3;V:=28;F:=4*12;threshold:=4;init_c:=10;
> de1:=diff(c(t),t)=(F/V)*(cin-c(t));
> soln:=c0->dsolve({de1,c(0)=c0},c(t),numeric,method=lsode):
> plot1:=c0->odeplot(soln(c0),[t,c(t)],0..8):
> list1:=seq(plot1(i/2),i=1..12):
> line1:=plot([[0,threshold],[8,threshold]]):
> display({list1,line1});
```

With `MATLAB`, to use a stiff solver, simply replace the call to `ode45` with one to `ode15s`. `MATLAB` provides a few different stiff solvers, of different accuracy, but the `ode15s` is a good first method of choice.

A problem that frequently arises when using `Maple` to sketch solutions in a plane is a very odd-shaped or angular 'scribble', rather than a smoothly varying solution. By now, if you have experimented with `Maple`, you will no doubt have encountered this problem in some implementation. Consider a pair of differential equations in time, `de1`=dX/dt and `de2`=dY/dt. Then the following code fragment below, *without* the step size parameter stipulated or set too large, could produce the 'mess' in the first diagram of Figure 4.3, while, with the appropriate step size, we obtain the second diagram. When the step size is specified, the problem 'rights' itself as displayed in the second diagram of Figure 4.3.

```
plot1:=DEplot([de1,de2],[X,Y],t=0..80,{inits},X=0..Xlimit,Y=0..Ylimit,
   scene=[P,H],linecolor=black,stepsize=0.1,arrows=none):
display(plot1);
```

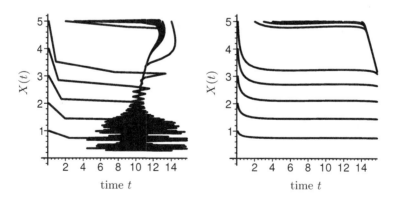

Figure 4.3: Solution of a differential equation for different initial conditions. In the diagram on the left a large step size was used, and instability of the numerical scheme prevents convergence. In the diagram on the right the step size of the solver has been reduced.

This problem is that of *instability* of the numerical scheme. Typically this can be improved with a decrease in the step size, but sometimes the values of a step size for which convergence to the solution ensues can be extremely small and depend on the differential equation itself. (We see an example of this in Section 4.4.) Furthermore, as the step size decreases, there is an increasing error due to a larger number of operations that once again becomes an issue.

It is possible to control the absolute and relative error using `Maple` or `MATLAB` as well as the initial step size, and several other parameters. The details of how to do this can be found in the help systems for either software package.

4.4 Instability

Recall that Euler's method is a recursive application of

$$y_{n+1} = y_n + hf(t_n, y_n),$$

where $t_{n+1} = t_n + h$ and $0 \le n \le N - 1$.

Note also that the logistic equation is given by

$$y' = ry\left(1 - \frac{y}{K}\right) = ry(1 - y) \quad \text{when} \quad K = 1,$$

with $y(0) = y_0$ and $r > 0$, $K > 0$. We know that the resulting solution is a smooth monotonically increasing, or decreasing, function approaching $K = 1$ (see Chapter 3).

What we now consider is how well Euler's method predicts or approximates the logistic equation solution as h (the step size) increases. Applying Euler's method to the logistic equation (with $K = 1$) gives successive approximations for $y_1, y_2, \ldots$ with

$$y_{n+1} = y_n + hry_n(1 - y_n) \quad n = 0, 1, 2, \ldots. \tag{4.1}$$

Note here that Euler's equation is a difference equation estimating a differential equation. Furthermore, this equation is identical to that considered in Section 3.7 (to describe discrete population growth) if rh of (4.1) is set equal to parameter r of that section, and the carrying capacity is taken as $K = 1$.

Applying the results of that section to this case, for small rh the estimation approaches the analytic solution smoothly; however, for a small increase in rh, the estimation will oscillate about the analytic solution. And for some further increase, the numerical solution predicts chaos.

This means that when solving the logistic equation with parameter r using Euler's method, if the parameter rh is sufficiently small then the method will converge to the analytic solution. If not, then the method will never converge and the sequence of generated values may oscillate or behave chaotically. For any fixed r, there is not one single value for h (the step size of the method) that will ensure convergence to the analytic solution in all cases. Choice of the step size would need to depend on the parameter r, as otherwise the solution would not necessarily be stable, regardless of how small the step size is taken.

4.5 Discussion

Thus numerical methods need to be chosen carefully for accurate results. Euler's method, for example, while simple to use, is not a good choice for situations in which high accuracy is desired. Likewise, it brings us to the realisation that it would be very valuable to have some prior knowledge of expected behaviour/results in order to interpret and evaluate numerical solutions correctly.

To this end we have already developed some tools, and will develop others, both analytical and graphical, to establish aspects concerning the behaviour of the solution. These can then be compared with the numerical results. When using computational methods, it is commonplace to run experiments determining values that are already known exactly (through some theoretical or experimental means) in order to establish consistency with expected results.

Another essential aspect of generating numerical solutions to be borne in mind is that a single solution to a single set of parameter values, and for a single set of initial values, may impart very little information about the general behaviour of a system. Analytical tools are essential in providing a framework for choosing which numerical solutions to generate in order to gain an understanding of the system dynamics. In Chapter 6, in particular, we see how well analytical and numerical methods can combine to provide this information.

Summary of skills developed here:

- *Understand the general principles of applying numerical methods and how the process contributes towards generating errors in computed solutions.*
- *Understand the notion of a stable or unstable numerical scheme.*
- *Understand the difference between an exact and an estimated solution.*
- *Know how to choose a numerical scheme offered by* `Maple`.

4.6   Exercises for Chapter 4

4.1. Round-off errors. *A simple example can illustrate how the round-off errors accumulate to provide inaccuracy. Using* `Maple`, *set*

$$z = (x+1)^{12} - (4x+3)^6.$$

(a) *Set* $x = 1 + \sqrt{3}$ *and evaluate* z.

(b) *Simplify* z *and compare this solution with that in (a).*

Their difference arises through round-off errors that accumulate from the calculations required for the evaluation in (a), compared with the analytic approach of (b) in which no numerical calculations occur.

Some useful `Maple` *commands are:*

    x:=(1+sqrt(3));   evalf(z);   and simplify(z).

4.2. Numerical schemes. *A variety of numerical methods can be used to solve the same equation and thus give an indication of their differences, in terms of estimation, compared with the analytical solution. Consider the simple IVP*

$$\frac{dy}{dt} = y, \qquad y_0 = y(0) = 1.$$

(a) *Using* `Maple` *(which uses a symbolic approach), find and plot the analytical solution.*

(b) *Now create further plots choosing particular numerical methods as follows:*

    (i) *Using the code*

```
plot2:=DEplot(eqn1, {[y(0)=1]}, [y(t)],
    t=0..4,y=0..maxy,iterations=5,
    stepsize=0.1, arrows=NONE,linecolour=blue,
    method=classical[foreuler]):
```

    (ii) *and, as above, with*

```
> plot3:=DEplot(...
    linecolor=red, method=classical[heunform]):
```

```
> plot4:=DEplot(...
    linecolor=green, method=classical[rk4]):
```

Display these on the same axes to see how they differ from the analytical solution.

```
> display(plot1, plot2, plot3, plot4, view = [0 .. 4, 0 .. 30]);
```

(c) *Try changing the step size in the methods above and compare the results.*

4.3. Numerical and symbolic solutions. *One can get an idea of how the error between the analytical and numerical solutions grows by plotting that error. To this end, for the IVP in Exercise 4.2, use the following code in Listing 4.3 to examine these errors, comparing a variety of time steps and methods. Notice the rate of growth in the error functions as t increases.*

Listing 4.3: Maple code: c_cn_num_sym.mpl

```
restart:with(plots):
lasttime:=10:
timelist:=seq(i,i=1..lasttime):
deq:=diff(y(t),t)=y(t):
init:=y(0)=1:
Digits:=20:
# three methods of solution

ans1:=dsolve({deq,init},y(t),type=numeric,
    method=classical[foreuler],output=array([timelist]),stepsize=0.1):
ans2:=dsolve({deq,init},y(t),type=numeric,
    method=classical[heunform],output=array([timelist]),stepsize=0.1):
ans3:=dsolve({deq,init},y(t),type=numeric,
    method=classical[rk4],output=array([timelist]),stepsize=0.1):
# compare the solutions using the difference between solutions

plot1:=plot([seq([ans1[2,1][i,1],ans3[2,1][i,2]-ans1[2,1][i,2]],i=1..lasttime)],
    style=line):
plot2:=plot([seq([ans2[2,1][i,1],ans3[2,1][i,2]-ans2[2,1][i,2]],i=1..lasttime)],
    style=line,colour=blue):
display(plot1,plot2,view=[0..lasttime,0..10]);
# compare the numerical solutions with the analytical solution

plot3:=plot([seq([ans1[2,1][i,1],exp(i)-ans1[2,1][i,2]],i=1..lasttime)],
    style=line,colour=red):
plot4:=plot([seq([ans2[2,1][i,1],exp(i)-ans2[2,1][i,2]],i=1..lasttime)],
    style=line,colour=blue):
plot5:=plot([seq([ans3[2,1][i,1],exp(i)-ans3[2,1][i,2]],i=1..lasttime)],
    style=line,colour=green):
display(plot3,plot4,plot5,view=[0..lasttime,0..10]);
```

4.4. Discretisation and round-off errors. *The number of steps chosen for a numerical method will also have an impact on the accuracy of the solution, with an increase in the number of steps reducing the discretisation error. However, when the step size approaches the accuracy of the machine (the smallest number representable by the machine), the error produced by the increased number of round-off errors becomes dominant, and the combined error increases reducing the accuracy of the solution. To illustrate this we can use the same IVP as above (Exercise 4.2) and set the machine accuracy through the variable* `Digits`.

(a) *Using the* `Maple` *code as given below in Listing 4.4, compare the errors in the RK4 method (fourth-order Runge–Kutta method) as the step size decreases. Use step sizes of 0.1, 0.001 and 0.0005 and set the machine accuracy* (`Digits`) *to 10.*

(b) *What happens when the step size is close to the maximum accuracy of the machine? To see this, compare the output with the machine accuracy (through* `Digits`) *set to 10 and set to 20.*

Listing 4.4: Maple code: c_cn_roundoff.mpl

```
restart:with(plots):
lasttime:=5:Digits:=10:
timelist:=seq(i,i=1..lasttime):
deq:=diff(y(x),x)=y(x):
init:=y(0)=1:
# Analytical solution

ans0:=dsolve({deq,init},y(x));
# Numerical solutions with different stepsizes

ans1:=dsolve({deq,init},y(x),type=numeric,
    method=classical[rk4],output=array([timelist]),stepsize=0.1):
ans2:=dsolve({deq,init},y(x),type=numeric,
    method=classical[rk4],output=array([timelist]),stepsize=0.001):
ans3:=dsolve({deq,init},y(x),type=numeric,
    method=classical[rk4],output=array([timelist]),stepsize=0.0005):
# Compare the solutions with the analytical solution

plot1:=plot([seq([ans1[2,1][i,1],exp(i)-ans1[2,1][i,2]],i=1..lasttime)],
    style=line,colour=red):
plot2:=plot([seq([ans2[2,1][i,1],exp(i)-ans2[2,1][i,2]],i=1..lasttime)],
    style=line,colour=blue):
plot3:=plot([seq([ans3[2,1][i,1],exp(i)-ans3[2,1][i,2]],i=1..lasttime)],
    style=line,colour=green):
display(plot1,plot2,plot3);
display(plot2, plot3, view = [0 .. 3, -10^(-10) .. 10^(-10)]);
```

4.5. MATLAB comparison of methods. *Consider the differential equation and initial condition*

$$\frac{dy}{dt} = 3y, \qquad y(0) = 1.$$

(a) *Solve this using the* `MATLAB` *standard* `ode45` *function and plot the results. Also plot the exact solution* $y = e^{3t}$ *on the interval* $0 > t > 2$.

(b) *The code in Listing 4.5 shows how to write a function to solve the differential equation on* $[0, 2]$, *using Euler's method, with* $N = 20$ *equally spaced time-steps* $h = 0.1$. *Using this code, compare, on the same figure, the solutions graphically for* $h = 0.1$, $h = 0.05$, $h = 0.01$ *and the exact solution.*

Listing 4.5: MATLAB code: c_cn_eulersolve.m

```
function c_cn_eulersolve
tend = 1; trange = [0, tend];
```

```
Npts = 10; %number of time-steps
y0 = 1;
[tsol, ysol] = odeEuler(@rhs, trange, y0, Npts);
plot(tsol, ysol,'b'); hold on;
plot(tsol, exp(3*tsol),'g');

function ydot = rhs(t, y)
ydot = 3*y;

function [t, y] = odeEuler(fcn, trange, y0, Npts)
h = trange(end)/Npts; % the step size
t = zeros(1,Npts); y = zeros(1,Npts);
y(1) = y0; t(1)=trange(1);
for k=1:Npts
    y(k+1) = y(k) + h*fcn(t(k),y(k));
    t(k+1) = t(k) + h;
end
```

(c) *Modify the code to solve the equation using Huen's method and compare the solution with Euler's method and the exact solution for $n = 20$ points, $h = 0.1$.*

Chapter 5

Interacting population models

Interacting population models are relevant where two or more populations depend on each other. We study, in detail, four examples of interacting populations: (1) an epidemic model, (2) a predator-prey interaction, (3) a competing species interaction, and (4) a model of a battle between two opposing groups. In this chapter, we concentrate on formulating the differential equation model governed by two simultaneous first-order differential equations. We solve the differential equations numerically using `Maple`*. In Chapter 6, we return to the models and develop analytic techniques that provide general insights into the models developed here.*

5.1 Introduction

Ecological systems may be extraordinarily complex — an inter-related system of plants and animals, predators, prey, flowering plants, insects, parasites, pollinators, seed-dispersing animals, etc. In such systems there is a constant stream of arrivals and departures involving time periods of millions of years. New species evolve or arrive, others decline to extinction or migrate. Human interference has impacted hugely on most of the world's ecosystems, particularly over the past 200 years, with one seemingly insignificant species extinction able to spark a cascade of effects throughout the trophic levels.

Interacting populations

Large numbers of one species may be unaffected by others; however, in some instances, removing, introducing or modifying one resource or species through (for example) harvesting or poisoning may have wide-ranging ramifications for the system. Evidence suggests that, typically, communities with many interacting species have greater stability than those comprising much simpler systems. That is, while rainforests are stable, cultivated land and orchards are relatively unstable and the populations of species in laboratory controlled predator-prey systems undergo large oscillations.

The relationships between the species within a system are often highly nonlinear, such that it is extremely difficult to establish, with certainty, a precise mathematical model describing the processes involved. However, there is clearly a need to understand these systems, or aspects of them. To this end we develop simple mathematical models where the independent variable is time and the dependent variables are the numbers or densities (numbers per area) of the various different interacting populations. We include case studies of interactions where these simple models, or extensions of them, appear to reflect the main processes involved.

The most obvious examples come from nature, where different species interact with each other. Examples of populations that interact include spread of a disease between population groups, predator-prey interactions, battles between opposing groups and interactions between different age or gender groups within a population.

One type of interaction, which occurs within a single species, is the interaction between those who are infected with a *disease* and those who are not. This effectively divides the population into two separate sub-populations, where the growth or decline of one sub-population is affected by the size of the other sub-population. We study, in detail, a simple model for the growth and then decline of the number of individuals in a population infected with a disease, and compare the model predictions with data recorded during a flu epidemic in a boarding school.

There are many different types of interactions between populations in different trophic levels. One such interaction, where some species use other species as their food supply, is the *predator-prey* interaction. We study the simplest example of it in this chapter. Another interaction process between populations on the same trophic level is that of *competing species*, where two or more populations compete for a limited resource, such as food or territory. We model this system in a manner very similar to that developed for the predator-prey system.

The fourth example we study is that of a *battle* between two opposing groups, such as may occur between two insect or human populations. The rate at which soldiers are wounded, or killed, in a battle depends largely on the number of enemy soldiers. We focus on a very simple human war model for aimed fire, where a soldier aims at a target with a given probability of hitting it. The model is compared with an actual battle in World War II for which daily records of the number of casualties were kept.

Modelling assumptions and approach

Different populations are expected to have different birth and death rates. For some populations the size of another population will affect these parameters, and the populations are then said to interact with each other.

In each of the examples that follows, we make some simplifying assumptions and then build the mathematical model on these. We assume that the populations are sufficiently large so that we can neglect random differences between individuals. Furthermore, we also assume the growth to be continuous, rather than discrete, over time. The model we derive for the rate of change of each population over time is based upon the input-output principle of the balance law: the compartmental model technique. This approach leads to two, or more, simultaneous differential equations, or a system of equations.

Models can describe the number of individuals (population size), as will be the case for the epidemic and battle models we develop. Alternatively, the population density, number per unit area, can be modelled as in the predator-prey and competing species models that follow. For both, the approach is the same.

Systems of differential equations

Many processes are described by more than one differential equation. When these equations need to be satisfied simultaneously, the set of equations is known as a *system of differential equations*. Although systems can comprise many differential equations and many unknowns, in this book we consider mainly systems of first-order equations where the number of unknowns is the same as the number of equations in the system. This ensures a unique solution to the system when the initial conditions are specified.

The system of equations is known as a *dynamical system* if it allows prediction of future states given present and/or past states. In most of the models we develop, such as in the growth and decay processes examined in previous sections and the population models we develop here, this is the case.

We also develop systems in which the equations are *nonlinear*: that is, they include products of the dependent variables or their derivatives. While appearing simple, they do not all have analytic solutions and we rely on the approximate solution to these systems obtained by `Maple` or `MATLAB`, which employs numerical solvers to find them. Later, in Chapter 7, we develop some theory about linearisation that allows us to predict the behaviour of the system by considering linear approximations to the nonlinear equations.

The systems of equations may be *coupled*, which implies that their solutions are interdependent. For example, in the case of two equations $x'(t) = F(x, y, t)$ and $y'(t) = G(x, y, t)$, if the solution of x' depends on the value of y and the solution of y' depends on x, then the equations are coupled. Alternatively, as was the case in Section 2.7 in the example of drug assimilation, the first equation could be solved independently of the second, and this system is said to be *uncoupled*.

5.2 Model for an influenza outbreak

We develop a model to describe the spread of a disease in a population and use it to describe the spread of influenza in a boarding school. To do so, the population is divided into three groups: those susceptible to catching the disease, those infected with the disease and capable of spreading it, and those who have recovered and are immune from the disease. Modelling these interacting groups leads to a system of two coupled differential equations.

Background

Over the centuries, there have been dramatic examples of how epidemics of various diseases have had a significant effect on the human population. One of the most well known is the Black Death in Europe in the fourteenth century. Today epidemics are still prevalent, the most notable being possibly AIDS and the Ebola virus. If we can understand the nature of how a disease spreads through a population, then we are better equipped to contain it through vaccination or quarantine. Or in the case of the biological control of pests, we may wish to determine how to increase the spread of the disease (for example, myxomatosis or calicivirus in rabbits) so as to find an efficient way of reducing the population. Unfortunately, humans themselves have been subjected to this means of 'control'. In colonial times, the spread of European diseases, such as measles and smallpox, had a disastrous impact on certain indigenous populations who had no resistance to them.

Many diseases are spread by infected individuals in the population coming into close contact with susceptible individuals. These include influenza, measles, chickenpox, glandular fever and AIDS. On the other hand, malaria is transmitted through a host, a mosquito, which carries the disease from individual to individual. Certain diseases are more contagious than others. Measles and influenza are highly contagious, whereas glandular fever is much less so. Many diseases, such as mumps and measles, confer a lifelong immunity; however, influenza and typhoid have short periods of immunity and can be contracted more than once.

There are some natural definitions that we require in order to proceed with our modelling. The *incubation period* of a disease is the time between infection and the appearance of visible symptoms. This should not be confused with the *latent period*, which is the period of time between infection and the ability to infect someone else with the disease. The latent period is shorter than the incubation period so that an individual can be spreading the disease, yet be unaware of having it. For measles, the incubation period is approximately 2 weeks and the latent period is approximately 1 week. Below we consider a simple mathematical model for a flu (influenza) epidemic at a boarding school over a period of about 15 days. For this period it is reasonable to assume that reinfection does not occur.

Model assumptions

When considering a disease, the population can be divided into distinct groups: susceptibles $S(t)$ and contagious infectives $I(t)$, where t denotes time. The susceptibles are those liable to catch the disease, while the infectives are those infected with the disease who are capable of giving it to susceptibles. There are also those who have recovered from the disease and are no longer susceptible, who form a further separate group.

Initially, we make some assumptions and then build the model based on them:

- We assume that the populations of susceptibles and contagious infectives are large so that random differences between individuals can be neglected.

- We ignore births and deaths in this model and assume the disease is spread by contact.

- We neglect the latent period for the disease, setting it equal to zero.

- We assume all those who recover from the disease are then immune (at least within the time period considered).

- We also assume that, at any time, the population is homogeneously mixed, that is, we assume that the contagious infectives and susceptibles are always randomly distributed over the area in which the population lives.

As mentioned above, we assume perfect immunity where the individual cannot be reinfected for the time of the model simulation. This is not generally true for influenza, where the virus mutates and people can get reinfected year after year. However, the time scale we are interested in here is about 15 days for a single outbreak, so perfect immunity is a reasonable assumption for this time scale.

Formulating the differential equations

We start with an input-output compartment diagram and then describe the rate of change in the number of susceptibles and contagious infectives with word equations. The following example illustrates this process.

Example 5.1: *Construct a compartmental diagram for the model and develop appropriate word equations for the rates of change of susceptibles and contagious infectives.*

Solution: *The only way the number of susceptibles can change is the loss of those who become infected, as there are no births and none of those who become contagious infectives can become susceptibles again. The number of infectives changes due to the susceptibles becoming infected and decreases due to those infectives who die, become immune or are quarantined. The latter cannot become susceptibles again (from the assumptions made). This is illustrated in the compartmental diagram of Figure 5.1.*

The appropriate word equations are

$$\left\{ \begin{array}{c} \text{rate of} \\ \text{change in no.} \\ \text{susceptibles} \end{array} \right\} = - \left\{ \begin{array}{c} \text{rate} \\ \text{susceptibles} \\ \text{infected} \end{array} \right\}$$

$$\left\{ \begin{array}{c} \text{rate of} \\ \text{change in no.} \\ \text{infectives} \end{array} \right\} = \left\{ \begin{array}{c} \text{rate} \\ \text{susceptibles} \\ \text{infected} \end{array} \right\} - \left\{ \begin{array}{c} \text{rate} \\ \text{infectives} \\ \text{have recovered} \end{array} \right\} \qquad (5.1)$$

$$\left\{ \begin{array}{c} \text{rate of} \\ \text{change in no.} \\ \text{recovered} \end{array} \right\} = \left\{ \begin{array}{c} \text{rate} \\ \text{infectives} \\ \text{have recovered} \end{array} \right\}$$

Note that the term on the RHS of the equation for the susceptibles is exactly the same term as the first term on the RHS for the infectives.

Figure 5.1: Input-output diagram for the epidemic model of influenza in a school, where there is no reinfection.

We must account for those removed from the system, in this case, those who have recovered from the disease. More generally, the removed can also consist of fatalities due to the disease, those who become immune to the disease, and those infectives who are quarantined. The number of infectives removed in the time interval should not depend in any way upon the number of susceptibles, but only on the number of infectives. We assume that the rate at which infectives recover is directly proportional to the number of infectives and write

$$\left\{ \begin{array}{c} \text{rate} \\ \text{infectives} \\ \text{recovered} \end{array} \right\} = \gamma I(t), \qquad (5.2)$$

where γ is a positive constant of proportionality, called the *recovery rate*, or more generally, the removal rate. The rate γ is a per-capita rate. Its reciprocal, γ^{-1}, can be identified with the residence time in the infective compartment, that is, the mean time that an individual is infectious. For influenza, the infectious period is typically 1 to 3 days.

To model the total rate of susceptibles infected, first consider the susceptibles infected by a single infective. It is evident that the greater the number of susceptibles, the greater the increase in the number of infectives. Thus the rate of susceptibles infected by a single infective will be an increasing function of the number of susceptibles. For simplicity, let us assume that this rate is directly proportional to the number of susceptibles.

If we denote the number of susceptibles at time t by $S(t)$, then the rate susceptibles are infected is $\lambda(t)S(t)$. However, it is not reasonable to treat λ as a constant since the more infectives there are, the higher the risk that a single susceptible will become infected. The total rate that susceptibles are infected is

$$\left\{ \begin{array}{c} \text{rate} \\ \text{susceptibles} \\ \text{infected} \end{array} \right\} = \lambda(t)S(t). \qquad (5.3)$$

We later find a suitable dependence for λ on the number of infectives. The term $\lambda(t)$ is called the *force of infection*. It is the per-capita rate at which susceptible individuals become newly infected. It has units of time^{-1}. We can also interpret $\lambda(t)$ as the instantaneous probability per unit time of a single susceptible becoming infected, and thus, for very small time intervals, $\lambda(t)\Delta t$ as the probability (risk) of a susceptible becoming infected.

Putting together the assumptions for rate of new infections and rate of recoveries gives the system of differential equations

$$\frac{dS}{dt} = -\lambda(t)S, \qquad \frac{dI}{dt} = \lambda(t)S - \gamma I, \qquad \frac{dR}{dt} = \gamma I. \tag{5.4}$$

The force of infection, $\lambda(t)$, depends on the current number of infectives $I(t)$ and increases as the proportion of infectives in the population increases. It also depends on the rate that individuals make contacts, which we denote by c (number of contacts per time) and the probability, p, that a contact results in an infection, given the contact is between an infective and a susceptible.

One suitable model for the force of infection is

$$\lambda(t) = cp\frac{I(t)}{N(t)},$$

where $N(t)$ is the total population size. In this model note that $N(t) = S(t) + I(t) + R(t)$ so $N'(t) = S'(t) + I'(t) + R'(t)$, and from (5.4), adding the three equations gives $N'(t) = 0$ so the total population size N is constant. Assuming the contact rate c and the probability of infection p are constant, then write $\beta_f = cp$ so

$$\lambda(t) = \beta_f \frac{I(t)}{N}.$$

The constant β_f is called the *transmission coefficient* and it also has the same units as the force of infection, time^{-1}. However, we cannot interpret β_f^{-1} as the time a susceptible remains a susceptible because β_f is not the exit rate from the S compartment. An alternative model for the force of infection is $\lambda(t) = \beta I(t)$, where β is a different constant with different units; this is discussed later in this section.

Our model for the spread of influenza now becomes

$$\frac{dS}{dt} = -\beta_f S\frac{I}{N}, \qquad \frac{dI}{dt} = \beta_f S\frac{I}{N} - \gamma I, \qquad \frac{dR}{dt} = \gamma I, \tag{5.5}$$

subject to the initial conditions $S(0) = s_0$, $I(0) = i_0$ and $R(0) = 0$.

Since the total population size is constant, we can also write the governing equations as

$$\frac{dS}{dt} = -\beta SI, \qquad \frac{dI}{dt} = \beta SI - \gamma I, \qquad \frac{dR}{dt} = \gamma I, \tag{5.6}$$

where we define $\beta = \beta_f/N$, which is also constant. The new parameter β is also called a transmission coefficient but has different units from β_f. This change simplifies the system of equations but has the disadvantage that the units of β are no longer time^{-1}.

Equations (5.6), a coupled system of nonlinear differential equations, were originally derived by Kermack and McKendrick in 1927 (Kermack and McKendrick, 1927) as a special case of a more general model. Since variable R does not appear in the first two differential equations, the first two differential equations can be studied as a system on its own.

Numerical solution

We can solve the differential equations (5.6) using a numerical method (e.g., Euler's method, Runge–Kutta methods, or `Maple` or `MATLAB` built-in solvers). However, we need to specify initial conditions and parameter values.

Using data from an influenza outbreak at a British boarding school of $N = 763$ students, parameters β_f and γ were estimated in Appendix E.1 for equations (5.5). The data is also provided in this appendix. We obtained parameter estimates $\gamma = 0.44$ and $\beta_f = 1.67$. For solving (5.6) we convert β_f to β, $\beta = \beta_f/N = 1.67/763 = 2.18 \times 10^{-3}$. For initial conditions we assume the outbreak was started by a single infective at $t = 0$ so $I(0) = 1$ and $S(0) = 762$.

Graphs of the the numerical solution are given in Figure 5.2. `Maple` code to generate the graph is given in Listing 5.1 and `MATLAB` code is given in Listing 5.2.

Listing 5.1: Maple code: c_pe_epidemic.mpl

```
restart:with(plots):with(DEtools):
unprotect(gamma); gamma:='gamma':
interface(imaginaryunit=i);

beta:=2.18*10^(-3): gamma:=0.44:
de1 := diff(S(t),t)=-beta*S(t)*I(t);
de2 := diff(I(t),t)=beta*S(t)*I(t)-gamma*I(t);
inits:=[S(0)=762,I(0)=1]:
myopts:=stepsize=0.1,arrows=NONE:
plot1:=DEplot([de1,de2],[S,I],t=0..30,[inits],scene=[t,S],linecolour=black,myopts):
plot2:=DEplot([de1,de2],[S,I],t=0..30,[inits],scene=[t,I],linecolour=red,myopts):
display(plot1,plot2);
```

Listing 5.2: MATLAB code: c_pe_epidemic.m

```
function c_pe_epidemic
global beta gamma N;

tend = 15; %set the end time to run the simulation
u0 = [762; 1]; %set initial conditions as a column vector
beta=2.18*10^(-3); gamma=0.44;
[tsol, usol] = ode45(@rhs, [0, tend], u0);
Ssol = usol(:, 1); Isol = usol(:, 2);
plot(tsol, Ssol, 'r'); hold on; plot(tsol, Isol, 'b');

function udash = rhs(t, u)
global betaf gamma N;
S=u(1); I=u(2);
lambda = beta*I; %force of infection
Sdash = -lambda*S;
Idash = lambda*S - gamma*I;
udash = [Sdash; Idash];
```

The data appear to agree well with the model's predictions. The number of infectives starts small, increases substantially over 6 days, and then decreases gradually. What is happening is that the number of susceptibles is being 'used up'. Thus, in the latter stages of the outbreak, there is a much smaller chance of any given infective coming into contact with someone who has not yet been infected.

Limitations of the model

There are a number of different variations of the basic model to consider: for example, the effect of births, which continually introduce more susceptibles into the population, or diseases without immunity, where infectives become susceptible again after removal.

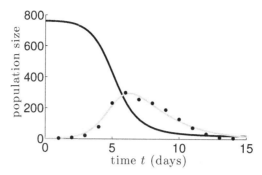

Figure 5.2: Numerical solution of the differential equations for an influenza outbreak in an English boarding school in 1978, system (5.6). The outbreak was started by one infective, thus $I_0 = 1$ and the total number of susceptibles was $s_0 = 762$. The values of the transmission coefficient and removal rate were estimated as $\beta = 2.18 \times 10^{-3}\,\text{day}^{-1}$ and $\gamma = 0.44\,\text{day}^{-1}$, respectively. The black dots correspond to the original data with the black line, the susceptibles and the grey line, the infectives. (Data from Murray, 1990.)

Births will provide a source of additional susceptibles and this can lead to oscillations in the number of infectives. Indeed, before the days of vaccinations for diseases such as measles, regular outbreaks occurred every 2 years in some countries, and every year in others.

While some diseases confer permanent immunity, (e.g., measles, chickenpox) many others do not (e.g., colds, influenza over a longer period than 25 days). This can easily be modelled, by allowing the infectives to become susceptibles again. Thus the compartmental diagram Figure 5.1 changes as the output from the contagious compartment becomes the input for the susceptible compartment. Another implicit assumption made here is that the rate of new infections is directly proportional to the product $S(t)I(t)$. If the number of infectives becomes a large proportion of the population, then it is likely that there would be a maximum limit to the rate of new infections since there is a limited number of contacts that could be made in any given time interval. The most suitable way to take this into account is by using a 'contact function' $\beta = pc(N)/N$, where p is the probability of an infection given a contact and $c(N)$ is the rate of contacts between individuals, which is dependent on the population size (or population density). See Roberts and Heesterbeek (1993) and Diekmann et al. (2012) for an in-depth discussion.

The basic reproduction number

From exploring the numerical solution we observe that the dynamics has a threshold behaviour where, if the initial number of susceptibles, s_0, is below a certain amount, then the number of infectious $I(t)$ decreases. On the other hand, if s_0 is greater than the threshold, then $I(t)$ increases before decreasing again. This appears to be independent of the initial number of infectious individuals, i_0. We can determine this threshold quantity by defining an important quantity called the *basic reproduction number*, denoted by R_0.

The basic reproduction number is defined as

> the number of new secondary infections resulting from a single infectious individual placed in a completely susceptible population, over the time that individual is infectious.

If $R_0 < 1$, we would expect the disease outbreak to die out ($I(t)$ to decrease) and if $R_0 > 1$, then it would increase initially.

Using the above definition of R_0 we can refer back to our model differential equations and determine a simple formula for R_0 for this SIR model. The instantaneous rate of new infections for the population was βSI. Thus the rate of new infections caused by a single infectious individual is $\beta s_0 \times 1$, where we have set $I = 1$ and $S = s_0$, the initial number of susceptibles. To obtain the number of new infections, we multiply this rate by the average time that an individual is infectious for. This time is given by the residence time in the infectious compartment, γ^{-1} (see Section 2.2). Thus we can define R_0 as the quantity

$$R_0 = \frac{\beta s_0}{\gamma}. \tag{5.7}$$

The basic reproduction number is thus a measure of how rapidly an infectious disease spreads through a population when $R_0 > 1$. Some typical estimates of R_0 for some common infectious diseases (in the pre-vaccination era) include $R_0 \simeq 3\text{--}4$ for influenza, $R_0 \simeq 16\text{--}18$ for measles, and whooping cough, $R_0 \simeq 4$ for smallpox and $R_0 \simeq 10\text{--}12$ for chickenpox (see Keeling and Rohani, 2008 and Anderson and May, 1991).

From the differential equation for I,

$$\frac{dI}{dt} = \beta SI - \gamma I = \gamma I \left(\frac{\beta S}{\gamma} - 1 \right),$$

we also see that dI/dt is positive only if $\beta S/\gamma > 1$ and dI/dt is negative only if $\beta_1 S/\gamma < 1$. Recalling the formula $R_0 = \beta S/\gamma$, then this also gives $dI/dt > 0$ if $R_0 > 1$ and $dI/dt < 0$ if $R_0 < 1$. This says the number of infectives increase if the basic reproduction number is greater than 1 and decreases otherwise, which is consistent with the interpretation of R_0.

Vaccination

As an application of the basic reproduction number we consider the vaccination of a population. Assuming we could instantaneously vaccinate a proportion of a population, what proportion would result in the eradication of the infectious disease?

This is easy to answer using the concept of the basic reproduction number R_0. Since $R_0 = \beta_1 s_0/\gamma$, if we vaccinate a proportion P of the population of susceptibles then this means that the basic reproduction number changes to

$$R_v = \frac{(1 - P)\beta_1 s_0}{\gamma} = (1 - P)R_0,$$

since there are now only $(1 - P)s_0$ susceptibles who could potentially catch the disease. Setting $R_v < 1$ for eradication and solving for P, we obtain the simple formula

$$P > 1 - \frac{1}{R_0}. \tag{5.8}$$

For smallpox, for example, where $R_0 \simeq 4$, we calculate $P \simeq 75\%$ of the population, but for measles, in unvaccinated communities, where typically $R_0 \simeq 15$, then $P \simeq 93\%$. It is clear from this calculation why smallpox was chosen to be targeted worldwide for eradication.

Endemic diseases

Endemic diseases persist in the population; they are characterised by outbreaks that occur from time to time. There are many examples, including childhood diseases such as measles and chickenpox. These models are usually studied on a larger scale, for example, a population the size of a city or a country.

To model an endemic disease we need to include in the model some of the underlying population dynamics, that is, we need to include the effects of natural births and deaths. The following model is a relatively simple model for exploring an endemic disease — it is a variant of the SIR model. A compartment diagram is given in Figure 5.3.

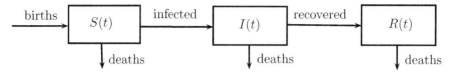

Figure 5.3: Input-output diagram for an endemic model of a disease in a city or country over a long time-scale, where there is no reinfection.

Let a and b denote the natural per-capita death rate and birth rate of population, where deaths are due to natural causes. The differential equations are

$$\frac{dS}{dt} = bN - \beta SI - aS,$$
$$\frac{dI}{dt} = \beta SI - \gamma I - aI, \qquad (5.9)$$
$$\frac{dR}{dt} = \gamma I - aI,$$

where $N(t) = S(t) + I(t) + R(t)$. If we add the three differential equations together, we obtain a differential equation for $N(t)$,

$$\frac{dN}{dt} = (b - a)N.$$

If $b = a$, then the population remains constant. This is sometimes a useful assumption to make provided the time-scale over which the model is applied is sufficiently short so that a constant population is a reasonable assumption (but also long enough that births create a sufficient source of new susceptibles).

A numerical solution of equations (5.9) is given in Figure 5.4. We have assumed a population size of 1 million, with an initial number of infectives, $i_0 = 10$, initial number of susceptibles, $s_0 = 10^5$ and the initial number of people who have previously had the disease and are immune, $r_0 = N - i_0 - s_0$. These are typical values for a disease such as measles in a medium sized city and where many of the population are immune through vaccination or previous exposure, but a significant proportion (10%) are not. For parameter values we have used $\gamma = 52$, years ($\gamma^{-1} = 1\,\text{week} = 1/52\,\text{years}$, and a life expectancy, a^{-1}, of 80 years. For β we have assumed $R_0 = 10$, typical for measles. This leads to a transmission coefficient $\beta = 5.2 \times 10^{-4}$ since here $R_0 = \beta N/(\gamma + a)$.

The number of infectives initially has similar dynamics to that in Figure 5.2, but after some time the number of infectives rebounds instead of dying out. Recurrent outbreaks occur every two to three years. Initially there are sufficient susceptibles for the infectious disease to spread in the population (i.e., $R_0 > 1$). After a time, the number of susceptibles falls so that there are insufficient susceptibles available to sustain the increase (i.e., a single infective cannot infect more than one susceptible while they are still infectious), and the number of infectives begins to fall. However, due to births of new susceptibles, there will come a time when the number of susceptibles again reaches a critical value so that the number of infectives begins to rise again, thus causing a new outbreak. In this model we obtain damped oscillations, and the number of infectives tends to a steady state.

In practice, the transmission coefficient β can be seasonally dependent and this can cause sustained oscillations rather than damped oscillations. Additional oscillatory forcing of this type often produces chaotic behaviour. This is discussed in Keeling and Rohani (2008), and see also Roberts and Tobias (2000) for an example of the same behaviour for recurrent epidemics of measles in New Zealand.

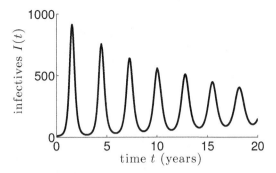

Figure 5.4: An endemic disease such as measles. Number of infectives from a numerical solution of the differential equations (5.9) for an infectious disease in a hypothetical population of $N = 10^6$. Parameter values used are $\beta = 5.2*10^{-4}$ years^{-1} persons, $\gamma = 52$ years^{-1}, $b = a = 1/80$ years^{-1} with initial populations $s_0 = 10^5$, $i_0 = 10$ and $r_0 = N - i_0 - s_0$.

Frequency-dependent and density-dependent transmission

Recall that the force of infection is the product of the contact rate and the probability a contact leads to an infection, given the contact is a susceptible with an infective. The contact rate (number of contacts each individual makes per unit time) can depend on the population size or density. We might assume the rate of contacts is independent of the population density or, alternatively, we might assume it is proportional to the population size or density, that is, more crowded populations have more frequent contacts.

When the total population N is constant, as it was with the influenza model considered here, the contact rate is constant so the two systems of equations (5.5) and (5.6) are exactly the same with $\beta = \beta_f/N$. However, there are many circumstances where the population is not constant. These include endemic models over a long time scale where the per-capita birth rate is greater than the per-capita death rate and diseases that result in a significant number of deaths can cause a decreasing total population. For this situation it becomes important to focus on the type of transmission. There are two main types: frequency dependent transmission and density-dependent transmission.

For frequency-dependent transmission, we assume the contact rate is constant, $c(N) = c_0$, and the force of infection is then

$$\lambda(t) = pc_0 \frac{I(t)}{N(t)} = \beta_f \frac{I(t)}{N(t)}, \qquad \beta_f = pc_0.$$

The terminology 'frequency-dependent' comes from the assumption that the contact rate depends only on the frequency of contacts. For density-dependent transmission, $c(N)$ is assumed proportional to N, $c(N) = \kappa N$, and the force of infection is now

$$\lambda(t) = p\kappa N(t) \frac{I(t)}{N(t)} = \beta I(t), \qquad \beta = p\kappa.$$

It is called density-dependent transmission because the contact rate depends on the population size (or density).

Frequency-dependent and density-dependent transmission represent two extremes. Typically, frequency-dependent transmission is more appropriate for human populations (where humans usually have a fixed circle of social contacts mostly independent of population size). On the other hand, density-dependent transmission can be more appropriate for some animal populations where numbers of contacts per unit time are higher with more crowded populations due to more likely chance encounters.

In practice, contact rates often show linear dependence on population size for smaller population sizes, and constant dependence for large population sizes, and can be modelled by a suitable Michaelis–Menten type function, e.g. $c(N) = \kappa N/(1 + \epsilon N)$. Roberts and Heesterbeek (1993) and Begon et al. (2002) provide a good discussion of these issues.

Discussion

One question we might ask is whether a rapid increase in the number of infectives is always followed by a decrease? Also, by adjusting any of the parameters, could we limit the increase or even prevent it? Changing parameters could correspond to, for example, adopting certain vaccination strategies.

To answer these questions it is useful to gain more qualitative information about this infectious disease model. We would like to be able to say what happens for any values of the parameters. An exact solution of the simultaneous equations, however, is not easily obtained because the differential equations are nonlinear. An alternative approach is to use the chain rule to eliminate time and reduce the pair of differential equations to a single first-order differential equation, from which some insight can be gained. This analysis is covered in the next chapter.

There are a number of extensions of the basic SIR infectious disease model. Some of these are developed in the exercises. One extension is to incorporate a latent period, which is neglected in the basic SIR model. A latent period is the time from contact to when an individual is infectious to others. The simplest way to model the latent period is to include an additional compartment, with population size $E(t)$, consisting of those *exposed* who are infected but not yet infectious. The exposed then become infectious at a constant per-capita rate. This leads to an additional differential equation in the system. Such models are known as SEIR models.

Further extensions include continuous vaccination, where susceptibles move into a vaccinated compartment; sexually transmitted diseases; and disease spread by an animal agent, such as malaria spread by mosquitoes.

A classic reference in the field of mathematical epidemiology is Anderson and May (1991). Murray (1990) discusses the same model developed above. Also considered are the Black Plague, and rabies in foxes, among other examples. Braun (1979) outlines an extension to the model for sexually transmitted diseases. For an introduction to stochastic approaches to modelling human epidemics, see Daley and Gani (1999), Allen (2003) and Keeling and Rohani (2008).

For some further extensions of the basic models, see Grenfell and Dobson (1995). Keeling and Rohani (2008) give a comprehensive treatment of the modelling of infectious diseases in human and animal populations, including a chapter on vaccination and other means of controlling infectious diseases and a discussion of density-dependent transmission coefficients (where the rate of contacts between individuals is proportional to population size) and frequency-dependent transmission coefficients (where the rate of contacts is independent of the size of the population). These are particularly significant when the total population size changes with time. Diekmann et al. (2012) discuss many fundamental ideas in infectious disease modelling and, in particular, how to compute the basic reproduction number for more complicated models that involve populations structured into several groups, such as age groups or social groups.

Summary of skills developed here:

- *Formulate differential equations for variations on the two models presented here, such as diseases with a latent period, continuous vaccination and diseases without immunity.*
- *Obtain a numerical solution for a system of differential equations for the SIR model and its variations.*
- *Calculate the basic reproduction number for the SIR model*
- *Understand the different assumptions underlying the SIR model*

5.3 Case Study: Cholera

Cholera is a particularly dangerous disease. Modelling can provide an understanding of circumstances under which an outbreak can occur. Here we formulate a model that includes interacting susceptible and infectious populations. However, what is different from the usual approach is that it involves transmission from the environment. The case study is based on Codeço (2001) and Grad et al. (2012).

Cholera is a serious water-borne gastrointestinal disease that is contracted through the ingestion of contaminated water or food. In severe cases, and without treatment, it can kill victims through dehydration within hours of infection. Infection occurs from water contaminated with untreated sewerage, where the infectious agent responsible for cholera (bacterium Vibrio cholerae) forms a disease reservoir in the water supply.

Cholera poses a real and serious public health problem in communities with poor sanitation infrastructure, and one of the reasons special attention is paid to clean drinking water in camps setup to house refugees from war-torn areas is to minimise the risk of cholera outbreaks. Outbreaks of cholera can also occur after natural disasters, when infrastructure fails and water supplies become contaminated, such as in Haiti after the 2010 earthquake.

The statistical study of cholera began with the work of physician John Snow in the suburb of Soho in London, UK, in 1854. By mapping sites of infections, Snow traced the cause of cholera back to a certain water pump used by most residents. He managed to have the contaminated pump disabled, but it was a very controversial decision as it occurred well before the discovery of bacteria as a cause of disease. Snow's study is considered the beginning of the science of epidemiology.

Governing equations. *Because cholera has a short latent period, the variables needed to describe the prevalence of cholera in the population are $S(t)$, susceptibles, and $I(t)$, infectives, where t is time. Individuals who recover from cholera have immunity from reinfection lasting approximately two years. As long as the time scale of interest in the model is less than two years, it is reasonable to assume that infected individuals recover without becoming susceptible again.*

Another important variable is the concentration of cholera bacteria in the water supply. This influences how easily cholera is spread to susceptibles as they make contact with the water through food preparation or drinking. We use the variable $B(t)$ to represent the bacterial concentration measured as a cell count per ml in the water supply, also called the bacterial count. This will change with time as more bacteria enter the water supply through ongoing sewerage contamination, which then increases with an increasing number of infectives shedding cholera bacteria.

The differential equations for the model are

$$\frac{dS}{dt} = -\lambda(t)S,$$

$$\frac{dI}{dt} = \lambda(t)S - \gamma I,$$

$$\frac{dB}{dt} = eI + (n_b - m_b)B,$$

where differentiation is with respect to time t, $\lambda(t)$ is the force of infection, discussed below, γ is the recovery rate and $(n_b - m_b)$ is the net per-capita growth rate of bacteria in the water supply. Normally the bacteria population will become extinct if not for the introduction of new bacteria by infected individuals, so $n_b - m_b < 0$. The parameter e represents the rate of excretion of bacteria into the water supply from a single infection, so that $eI(t)$ is the total rate of increase of bacteria (per unit volume of water per unit time). We do not include deaths due to cholera, but this could be easily incorporated.

Following Codeço (2001), we assume that cholera is only contracted through contact with the environment and not through person-to-person contact (generally not important) or by food contact. The force of infection $\lambda(t)$ is the probability per unit time of a susceptible being infected. This is the contact rate (c contacts with the water supply per day) multiplied by the probability of infection, which depends on the bacterial concentration $B(t)$. While we could assume the probability is proportional to $B(t)$, it is more realistic to assume that it is linear for small $B(t)$, tending to one as $B(t)$ becomes large — that is, for large bacteria concentrations a contact with the water supply always results in infection. A suitable functional form for the probability of getting infected, given contact with the water supply, is $p(B) = B/(k_{50} + B)$, where the constant k_{50} represents the bacterial concentration that leads to a 50% chance of becoming infected. We therefore assume a force of infection

$$\lambda(t) - cp(B) = c\frac{B}{k_{50} + B}.$$

Substituting for $\lambda(t)$ we obtain the governing equations for the model

$$\frac{dS}{dt} = -c\frac{B}{k_{50} + B}S,$$

$$\frac{dI}{dt} = c\frac{B}{k_{50} + B}S - \gamma I, \qquad (5.10)$$

$$\frac{dB}{dt} = eI + (n_b - m_b)B.$$

To examine the system graphically, appropriate parameter values are required. These have been taken from Codeço (2001) and are given in Table 5.1. We specify initial conditions of $S(0) = 10{,}000$ (the size of a a small town), $I(0) = 1$ (one infective individual introduced) and $B(0) = 0$ (initially the water supply is not contaminated with cholera bacteria).

The result of running this model is shown in Figure 5.5, where it is evident that with this parameter combination an outbreak occurs. The outbreak lasts for about 200 days with the number of infectives increasing from 1 to about 400 people, and peaking at around 140 days after the introduction of the single infective.

The bacterial count is also plotted, and it also peaks around the 140-day mark. However, careful inspection of the graph shows that the maximum of the bacterial count occurs about 2.7 days after the maximum in the number of infections occurs.

Table 5.1: Parameters used in the model with units. The units for the parameter e are cells ml^{-1} day^{-1} person^{-1}.

Symbol	Parameter	Value	Units
c	Rate of contact with water supply	1	day^{-1}
k_{50}	Bacterial concentration for 50% chance of infection	10^6	cells ml^{-1}
γ	Recovery rate for infected person	0.2	day^{-1}
e	Excretion rate	10	*
$(n_b - m_b)$	Net per-capita growth rate of bacteria in water	0.33	day^{-1}

*The units for e are cells ml^{-1} day^{-1} person^{-1}

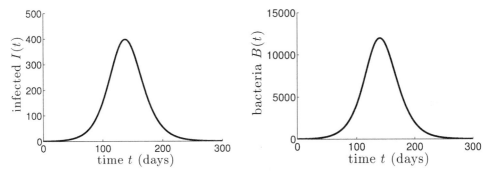

Figure 5.5: Result of running the cholera model (5.10) with parameter values given in Table 5.1, and initial conditions $S(0) = 10^4$, $I(0) = 1$ and $B(0) = 0$.

Basic reproduction number. *For diseases, it is of particular interest to establish interventions that could prevent an outbreak. The measure R_0, the basic reproduction number, predicts that an outbreak could occur when $R_0 > 1$, but will not occur when $R_0 < 1$.*

The definition of R_0 is the number of new infections produced directly from a single infective introduced into a fully susceptible population. For this model, the time the infective is infectious is given by 1/exit-rate from the I compartment (i.e. $1/\gamma$) and the rate of new infections at the start is the term $cBs_0/(K + B)$, where $s_0 = S(0)$ is the initial number of susceptibles.

Suppose we introduce $I = 1$ infective into a susceptible population of size s_0. This one infective will shed a number of bacteria into the water supply determined by substituting $I = 1$ into the equilibrium equation for B:

$$0 = e \times 1 + (n_b - m_b)B,$$

which gives

$$B = \frac{e}{m_b - n_b}$$

bacteria. This will be positive only if the bacteria death rate m_b is greater than the bacteria birth rate n_b.

We now calculate the rate of new infections from an initial number of susceptibles s_0 as $cBs_0/(k_{50} + B)$, but we also linearise this for small B to cBs_0/k_{50}. The number of new infectives produced during the time $1/\gamma$ for which the introduced infective is infectious is given by

$$R_0 = \frac{cs_0}{k_{50}} \times \frac{e}{m_b - n_b} \times \frac{1}{\gamma} = \frac{ces_0}{\gamma k_{50}(m_b - n_b)}.$$

This is the same as the value derived in Codeço (2001).

From the formula for R_0, increasing contact with the water-supply, c, or increasing the rate of excreted bacteria contamination of the water supply, e, or increasing the initial population, s_0, all contribute to an increased R_0. Increasing the recovery rate, γ, (i.e. decreasing the length of time infected) or increasing the net per-capita death rate of bacteria, m_b, each lead to a decreased R_0. These interpretations are as expected. With the parameters given in Table 5.1 the calculated value of R_0 is $R_0 \simeq 1.5$. Since $R_0 > 1$, this means that one introduced infected individual produces more than one new infection. Thus an outbreak will occur.

Since R_0 depends on the initial number of susceptibles, we can set $R_0 = 1$ to find a critical town size, S_c, below which an outbreak will not occur. We obtain

$$S_c = \frac{\gamma k_{50}(m_b - n_b)}{ce}.$$

With the parameter values given in Table 5.1 the critical town size is $S_c = 6{,}600$. The value of $s_0 = 10{,}000$ used in Figure 5.5 is above this critical value and an outbreak does occur.

We can use the formula for S_c to investigate the impact of possible interventions to prevent cholera outbreaks. Figure 5.6 illustrates values for e (the unit rate bacteria enters the water source) for each initial number of susceptibles (s_0), such that $R_0 = 1$ or equivalently $s_0 = S_c$. This diagram illustrates the impact of improved sanitation (modelled by reducing the value of the parameter e for contamination of water with cholera bacteria). For each s_0, if the point (e, s_0) is above the curve an outbreak occurs, while if the point (e, s_0) is below the curve then there is no outbreak.

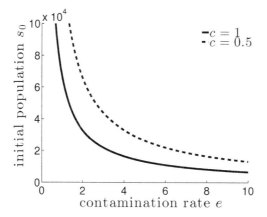

Figure 5.6: Plot of critical town size against excretion rate e for model (5.10) and two values for the contact rate c, $c = 1$ and $c = 0.5$; each curve represents $R_0 = 1$. Points (e, s_0) above the curve predict an outbreak and points below the curve predict no outbreak. Parameter values used are $k_{50} = 10^6$, $\gamma = 0.2$, $m_b - n_b = 0.33$.

This analysis illustrates that as the population increases, only a narrow range of e (very low rates of bacteria entering the water source) can prevent an outbreak. And only for relatively small populations is this not the case. While the result is intuitive, Figure 5.6 quantifies the nonlinear nature of the relationship, and can inform an understanding of risks and the design of control programs.

The above model is quite simple, and an obvious drawback is that births and deaths in the population are not included. Codeço (2001) considers births and deaths for susceptibles only and omits deaths of infected and recovered, while all are included by Fung (2014).

Grad et al. (2012) review further developments of the Codeço (2001) model. Some of these include allowing for waning immunity, asymptomatic cases, vaccines and spatial aspects. The latter models allow more detailed investigations of interventions and control, and show that cholera can remain endemic within populations with outbreaks possibly triggered by weather events. Nevertheless, the methodology developed for the simple model above is also relevant for these more complex models and the provision of information for disease control.

5.4 Predators and prey

In this section, we develop a simple predator-prey model for carnivores using the growth of a population of small insect pests that interact with another population of beetle predators. An example of a model for herbivores is examined in a case-study in Section 8.6, while models for parasitic interactions or cannibalism will be simple to derive from these given examples.

Background

There are several types of predator-prey interactions: that of herbivores, which eat plant species, that of carnivores, which eat animal species, that of parasites, which live on or in another species (the host), and that of cannibals, which eat their own species and which is often an interaction between the adults and young.

One interesting example of a predator-prey interaction occurred in the late nineteenth century when the American citrus industry was almost destroyed by the accidental introduction from Australia of the cottony cushion scale insect. To combat this pest, its natural predator, the Australian ladybird beetle, was also imported, but this did not solve the problem and finally DDT was used to kill both predator and prey in a bid to eradicate the pest. Surprisingly, application of DDT to the orchards led to an increase in the scale insects, the original pest, suggesting that the use of pesticide is advantageous to the pest!

Model assumptions

We make a few preliminary assumptions on which to build the model:

- Initially we assume the populations are large, sufficiently large to neglect random differences between individuals.

- We ignore the effect of DDT initially, but modify the model later to incorporate its impact on the system.

- We also assume there are only two populations, the predator and the prey, that affect the ecosystem.

- We assume that the prey population grows exponentially in the absence of a predator.

Compartmental model

There are two separate quantities that vary with time: the number of prey and the number of predators. For populations of animals, it is common to consider the population density, or number per unit area, rather than population size as we do here. We need to develop two word equations, one for the rate of change of prey density and one for the rate of change of predator density.

Example 5.2: *Determine a compartment diagram and appropriate word equation for each of the two populations, the predator and the prey.*

Solution: *The only inputs for each population are births and the only outputs are deaths. However, the prey deaths occur due to the predators capturing and eating them. This is illustrated in the input-output compartmental diagram of Figure 5.7.*

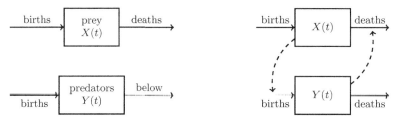

Figure 5.7: Input-output diagram for the 2-species predator-prey model. The additional diagram on the right includes dashed arrows illustrating how the different compartments influence the input and output rates.

Here we further distinguish between natural prey deaths and prey deaths due to predators. We also distinguish between natural predator births, occurring in the absence of prey, and the additional births that would occur due to the predators having more food from eating prey. The appropriate word equations are

$$
\left\{\begin{array}{c} \text{rate of} \\ \text{change of} \\ \text{prey} \end{array}\right\} = \left\{\begin{array}{c} \text{rate of} \\ \text{natural prey} \\ \text{births} \end{array}\right\} - \left\{\begin{array}{c} \text{rate of} \\ \text{natural prey} \\ \text{deaths} \end{array}\right\} - \left\{\begin{array}{c} \text{rate of} \\ \text{prey killed} \\ \text{by predators} \end{array}\right\},
$$

$$
\left\{\begin{array}{c} \text{rate of} \\ \text{change of} \\ \text{predators} \end{array}\right\} = \left\{\begin{array}{c} \text{rate of} \\ \text{predator} \\ \text{births} \end{array}\right\} - \left\{\begin{array}{c} \text{rate of} \\ \text{natural predator} \\ \text{deaths} \end{array}\right\}.
$$

(5.11)

Let us assume that the per-capita birth rate for the prey (the scale insect) is a constant b_1. Remember that the per-capita birth rates give the rate of births from an individual prey. The rate of births for each *individual* scale insect does not depend on the predator density. Similarly, the natural per-capita death rate of the scale insect is a constant a_1. Alternatively, the per-capita death rate of prey due to being killed by the predators will depend on the predator density; the simplest assumption is to assume this per-capita rate is proportional to the predator density, a per-capita rate c_1Y. The greater the density of predators, the more likely it is that an individual prey will be eaten. We assume a constant per-capita death rate for the predators (the ladybird beetles) independent of the prey density. The predator per-capita birth rate is more complicated. We assume the prey are an essential requirement for births of the predator, so the per-capita birth-rate for the predators will be the sum of a natural rate, which is constant, plus an additional rate that is proportional to the rate of prey killed. At any time the per-capita birth rate will increase with more food available and depend heavily on the amount of prey available.

Example 5.3: *Let $X(t)$ denote the number of prey per unit area and $Y(t)$ the number of predators per unit area. Using the above assumptions and the word equations (5.11), formulate differential equations for the prey and predator densities.*

Solution: *First look at the constant per-capita terms, the prey births and predator deaths. (Subscripts 1 and 2 will be used for the parameters associated with X the prey, and Y the predator, respectively.)*

Since the overall rates are the per-capita rates multiplied by the respective population densities, we can write,

$$\left\{\begin{array}{c}\text{rate of}\\\text{prey}\\\text{births}\end{array}\right\} = b_1 X(t), \qquad \left\{\begin{array}{c}\text{rate of}\\\text{prey}\\\text{natural deaths}\end{array}\right\} = a_1 X(t), \qquad \left\{\begin{array}{c}\text{rate of}\\\text{predator}\\\text{deaths}\end{array}\right\} = a_2 Y(t). \qquad (5.12)$$

For the prey deaths we denote the per-capita death rate as $c_1 Y(t)$, since it is proportional to the predator density, with c_1 as the positive constant of proportionality. Thus the rate at which prey are eaten is given by $c_1 Y(t) X(t)$. The predator birth rate has a component that is proportional to this rate of prey eaten, so we write

$$\left\{\begin{array}{c}\text{rate of}\\\text{prey killed}\\\text{by predators}\end{array}\right\} = c_1 Y(t) X(t), \qquad \left\{\begin{array}{c}\text{rate of}\\\text{predator}\\\text{births}\end{array}\right\} = b_2 Y + f c_1 Y(t) X(t), \qquad (5.13)$$

where f is also a positive constant of proportionality.

Now substitute equations (5.12) and (5.13) into the word equations (5.11). We obtain the pair of differential equations

$$\frac{dX}{dt} = b_1 X - a_1 X - c_1 XY, \qquad \frac{dY}{dt} = b_2 Y + f c_1 XY - a_2 Y.$$

We can combine some parameters. Let $\beta_1 = b_1 - a_1$, $-\alpha_2 = b_2 - a_2$ and $c_2 = fc_1$. Then

$$\frac{dX}{dt} = \beta_1 X - c_1 XY, \qquad \frac{dY}{dt} = c_2 XY - \alpha_2 Y, \qquad (5.14)$$

where we assume that β_1, α_2, c_1 and c_2 are all positive constants.

This system of equations is called the *Lotka–Volterra predator-prey system* after the two mathematicians who first worked with them. The parameters c_1 and c_2 are known as *interaction parameters* as they describe the manner in which the populations interact. Since there are positive and negative terms on the RHS of each differential equation, we might anticipate that the populations could either increase or decrease. These differential equations are coupled since each differential equation depends on the solution of the other. The differential equations are also nonlinear since they involve the product XY. One interpretation of the product XY is that it is proportional to the rate of encounters (contacts) between the two species.

We now have a model to which we can apply some simple checks to ascertain whether the equations behave as we might expect. For a two-species model, we would expect that, in the absence of any predators, the prey would grow without bound (since we have not included any growth limiting effects other than the predators). Also, in the absence of prey, we would expect the predators to die out. The following example indicates that the model incorporates this behaviour.

Example 5.4: *Check the Lotka–Volterra model in the limiting cases of prey with no predators, or predators with no prey.*

Solution: *Suppose there are no predators so that $Y = 0$. The equations then reduce to*

$$\frac{dX}{dt} = \beta_1 X,$$

which is the equation for exponential growth. The prey grows exponentially.

If there are no prey, then $X = 0$ and the equations reduce to

$$\frac{dY}{dt} = -\alpha_2 Y,$$

that is, exponential decay, which means that the predator population decreases exponentially and dies out.

Numerical solution

Although the equations may appear simple, they are not simple to solve. In fact we cannot find an analytic solution. However, to solve (5.14) numerically we can use `Maple` or `MATLAB`. We have chosen the values of the parameters β_1, α_2, c_1 and c_2 arbitrarily in order to get a feel for how the model behaves. A sample numerical solution is illustrated in Figure 5.8 with the `Maple` code provided in Listing 5.3, and the `MATLAB` code provided in Listing 5.4.

Listing 5.3: Maple code: c_pe_predprey.mpl

```
restart; with(plots): with(DEtools):
beta[1]:=1.0: alpha[2]:=0.5: c[1]:=0.01: c[2]:=0.005:
de1 := diff(X(t),t) = beta[1]*X(t)-c[1]*X(t)*Y(t);
de2 := diff(Y(t),t) = -alpha[2]*Y(t)+c[2]*X(t)*Y(t);
inits := [X(0)=200, Y(0)=80];
myopts := stepsize=0.1, arrows=none:
plot1 := DEplot([de1,de2], [X,Y], t=0..20,
    [inits],scene=[t,X],linecolor=red,myopts):
plot2 := DEplot([de1,de2], [X,Y], t=0..20,
    [inits],scene=[t,Y],linecolor=blue,myopts):
display(plot1,plot2);
```

Listing 5.4: MATLAB code: c_pe_predprey.m

```
function c_cp_predprey
global beta1 alpha2 c1 c2;

beta1=1.0; alpha2=0.5; c1=0.01; c2=0.005;
tend = 20;    %set the end time to run the simulation
u0 = [200; 80]; %set initial conditions as column vector
[tsol, usol] = ode45(@rhs, [0, tend], u0);
Xsol = usol(:, 1); Ysol = usol(:, 2);
plot(tsol, Xsol, 'b'); hold on; plot(tsol, Ysol, 'r');

function udot = rhs(t, u)
global beta1 alpha2 c1 c2;
X = u(1); Y=u(2);
Xdot = beta1*X - c1*X*Y;
Ydot = -alpha2*Y + c2*X*Y;
udot = [Xdot; Ydot];
```

The prey population oscillates out of phase with the predator population. Further exploration with other values of the parameters appears to indicate that this model always leads to oscillating populations. In Chapter 6 we use some mathematical analysis to prove this.

We also note that the predator population oscillation lags behind the prey cycles. We show (Chapter 6) that the Lotka–Volterra equations (5.14) always predict oscillations and that the predator oscillations always lag behind those of the prey.

If DDT is sprayed on the crops, we need to modify the equations since this represents a different mechanism for prey and predator deaths. We assume the per-capita death rates due to DDT are constant, with different per-capita rates p_1 and p_2 for the prey and predator, respectively. This means the DDT has an equal effect on each individual of a given species. The modified differential equations (5.14), including the additional terms $p_1 X(t)$ for the prey and $p_2 Y(t)$ for the predator, are

$$\frac{dX}{dt} = \beta_1 X - c_1 XY - p_1 X, \qquad \frac{dY}{dt} = c_2 XY - \alpha_2 Y - p_2 Y. \tag{5.15}$$

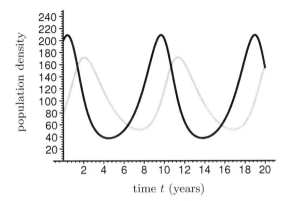

Figure 5.8: Sample numerical solution of the predator-prey equations using `Maple`. The black curve represents the prey and the grey curve the predator. Parameters here are chosen to illustrate the dynamics, and are not based on data from real populations: $\beta_1 = 1$, $\alpha_2 = 0.5$, $c_1 = 0.01$, $c_2 = 0.005$, with initial populations $x_0 = 200$ and $y_0 = 80$.

Note that the differential equations (5.15) are of the same form as the original system (5.14), with the positive constant β_1 replaced by $(\beta_1 - p_1)$ and the positive constant α_2 replaced by $(\alpha_2 + p_2)$. (There is an extra provision here that $\beta_1 - p_1 > 0$ so the prey population still increases in the absence of predators as DDT is not sufficient to kill all the pests.)

To investigate the effect of DDT on the two populations, we run the numerical solution again with $p_1 = p_2 = 0.1$. The results are presented in Figure 5.9. We see that, compared with Figure 5.8, the effect of the pesticide on the predator has been to decrease its overall numbers. It appears, however, that the mean prey population has increased. This is quite the opposite from the desired effect of the DDT and comes as a result of dealing with a nonlinear system. The DDT also reduces the average predator population so that there will be a reduced number of prey deaths due to the prey being eaten by the predators.

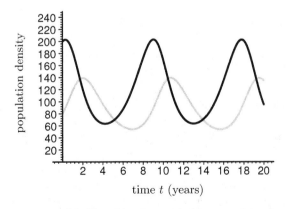

Figure 5.9: The scale insect or prey (black) and beetle population or predator (grey) where DDT has been used. Here we have $\beta_1 = 1$, $\alpha_2 = 0.5$, $c_1 = 0.01$, $c_2 = 0.005$, $p_1 = 0.1$ and $p_2 = 0.1$. The initial conditions are $x_0 = 200$, $y_0 = 80$. (These parameter values are chosen to illustrate the dynamics, and are not based on data from real populations.)

Interpretation of parameters

The differential equations (5.14) have four parameters: the prey per-capita birth rate β_1, the predator per-capita death rate α_2 and the two constants of proportionality c_1 and c_2. Making some further assumptions allows us to express c_1 and c_2 in terms of biologically relevant parameters.

Choose an arbitrary time interval Δt. We assume that Δt is sufficiently small, so that the populations $X(t)$ and $Y(t)$ do not change significantly over the time interval. We assume that each individual predator can search an area $A_s \Delta t$, where A_s is the area searched per unit time. We assume the areas covered by each individual predator do not overlap, and we neglect the time it takes for the predator to eat the prey. The positive constant A_s is called the *searching rate*. For a single predator the number of prey encountered is the prey density (number of prey per unit area) multiplied by the area covered by the predator, $A_s \Delta t$, giving the number of prey encounter as $X(t)A_s\Delta t$. If we assume each encounter results in a kill for the predator, then the rate of prey deaths by a single predator is given by dividing by Δt. So

$$\left\{ \begin{array}{c} \text{rate of} \\ \text{prey killed by} \\ \text{one predator} \end{array} \right\} = A_s X(t),$$

and therefore the total rate of prey deaths (per unit area) is obtained by multiplying by $Y(t)$,

$$\left\{ \begin{array}{c} \text{rate of} \\ \text{prey killed} \\ \text{by predators} \end{array} \right\} = A_s X(t)Y(t).$$

Comparing with equation (5.13) this gives $c_1 = A_s$, for constant A_s. This provides a biological interpretation for the parameter c_1 as the rate of area covered by a predator per unit time. In Chapter 8 we improve the model to also incorporate the time a predator spends eating the prey and resting.

Recall that $c_1 X(t)Y(t)$ represents the rate of prey deaths and $c_2 X(t)Y(t) = fc_1 X(t)Y(t)$ the rate of predator births, so the constant ratio $f = c_2/c_1$ represents the amount of prey required for a unit predator birth. This parameter f is interpreted as the *predator efficiency parameter*.

Limitations and extensions

A number of attempts have been made to validate the Lotka–Volterra equations. These include experiments by the Russian microbiologist Gause with two different types of protozoa, *Paramecium caudatum* and *Dididium nasutum* (see Kormondy (1976)). Gause found that the prey protozoa died out within only a few oscillations. In Renshaw (1991), a modified version of the Lotka–Volterra equations, which incorporate random births and deaths, is investigated. This model exhibits behaviour similar to the Gause experiments. In general, however, the Lotka–Volterra equations are not widely used in practice, but rather form a sound basis for more complicated models, which we investigate in Chapter 8.

Some mathematics texts point to a well-known data set regarding the number of pelts of lynx and snow-shoe hare as a validation of the Lotka–Volterra equations. This data set has 10-year oscillations. However, Murray (1990) observes that these data are not a validation of the Lotka–Volterra equations at all, since the hare oscillations lag one quarter of a period behind those of the lynx, contradicting what we might expect. It would mean that the hare was eating the lynx, which would be a ridiculous conclusion. In Renshaw (1991), it is pointed out that the hares and the grass they eat show the correct predator-prey lag according to the Lotka–Volterra equations. This simple example demonstrates how easily the oversimplification of an ecosystem, into only two interacting species in this case, can lead to misinterpretation.

An obvious limitation of the model is the inclusion of exponential growth, and in the following section we improve on this by including a density-dependent growth rate. Then in Chapter 8 (Section 8.3) we extend this predator-prey model further to improve its performance in predicting the observed behaviour of populations. We consider more realistic functions describing the prey death rate as well as the overall predator growth rate.

Density-dependent growth

The oscillatory nature of the predator-prey model above is what one might expect intuitively, and together with the phase lag of the predator oscillation behind that of the prey it is qualitatively what has been observed in some natural systems. However, our model is based on the assumption that the prey grows exponentially in the absence of the predator and this is clearly unrealistic. No matter how abundant the food supply, disease and/or food shortage will eventually curb the growth.

Thus, returning to the case without DDT, we replace the growth term for the prey population in (5.14) with density-dependent growth and a carrying capacity of K. (For details on density-dependent growth, refer to Section 3.2.) The system becomes

$$\frac{dX}{dt} = \beta_1 X \left(1 - \frac{X}{K}\right) - c_1 XY, \qquad \frac{dY}{dt} = c_2 XY - \alpha_2 Y. \qquad (5.16)$$

Modifying the `Maple` code used to obtain Figure 5.8, and assuming a carrying capacity of $K = 1,000$, we get the time-dependent plot of Figure 5.10. Note that the amplitude of the oscillations now decreases with an increase in time, for both populations. In fact, as time increases further, the figure suggests that each population might settle to a fixed population density, that is, stabilise. We look more closely at these results in the next chapter where we develop some analytical tools leading to a better understanding of the processes.

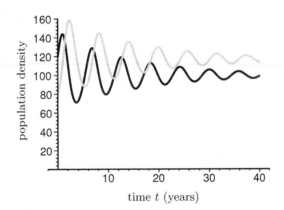

Figure 5.10: The scale insect (black line) and beetle population (grey line) where logistic growth for the scale insect has been included. Here we have $\beta_1 = 1$, $\alpha_2 = 0.5$, $c_1 = 0.01$, $c_2 = 0.005$ and $K = 1,000$. (These parameter values are chosen to illustrate the dynamics, and are not based on data from real populations.)

Discussion

The control of pests (plant or animal) through the introduction of a natural enemy has had some remarkable successes and also some disastrous failures. The reasons why some programs succeed and others fail are not always well understood, and thus with current knowledge the outcome is often difficult to predict. One research paper that discusses this issue is Beddington et al. (1978), in which the authors attempt to identify features of the interaction process that have a significant impact on the outcome.

The scale insect/ladybird beetle system discussed in this chapter is just one example of failure. One example of an extremely successful case occurred in Australia in 1925. Prior to this date, the prickly-pear cactus had been introduced into Australia with disastrous consequences, as it spread rapidly, thriving in the conditions and rendering thousands of hectares of farmland useless. In 1925, a natural predator moth was introduced to halt the spread of the cacti, and was spectacularly successful, nearly wiping out the weed in just two years. The case study is presented in Chapter 8 and provides a mathematical model that forecasts this success.

Some discussion of more realistic predator-prey models is given in Chapter 8 and also in Murray (1990) and Edelstein–Keshet (1988). For a biological perspective of interactions between a variety of species see (Begon et al., 1990, Chapters 7–12) and also May (1981). Further details of the effect of DDT on scale insects and ladybird beetles is given in MacArthur and Connell (1966). A brief discussion of further examples concerning the use of chemical pesticides on predator-prey systems is given in Keeton (1972).

> ### *Summary of skills developed here:*
>
> - *Formulate differential equations for two predators and one prey species, or two prey and one predator.*
> - *Distinguish between models describing different types of interactions, such as models for symbiosis, parasite-host interactions and competition for the same resource.*
> - *Generate numerical solutions for systems of equations.*

5.5   Scenario: Nile Perch catastrophe

We cite here a case where the ramifications of a disturbance within an interacting population system were wide reaching and caused an ecological disaster for the communities living in and around Lake Victoria. It emphasises the responsibility that should accompany any unsuitable modification of environmental processes, and the need for good modelling and forecasting, incorporating a multidisciplinary approach. Adapted from Murray (1990) and Quammen (1997).

Lake Victoria, which feeds into the Nile River, is the largest and most northern of a series of lakes punctuating the Rift Valley in Africa. Round as a pond and nearly as big as Ireland, it is bordered by Kenya, Tanzania and Uganda, and until 1960 it supported many communities along its shores providing them with a large and diverse population of fish as well as fresh water. In particular, it supported an abundance of cichlids (tropical freshwater fish of the family *Cichlidae*) in both quantity and species — the sort of fish known for their garish appearance and collected for aquaria.

Because of its scalloped shoreline, its irregular patterns of depth and shallows, and its subjection to periods of intense drought and low water levels during the centuries long history of the lake, pockets of cichlids were cut off from others for long periods. In this way they diverged and speciated. Later reconnection allowed the species to mix spatially, but they had become incompatible reproductively and were now in competition . In order to survive they radiated; that is, they settled in different niches within the environment. Thus the lake represented a kind of underwater Galapagos archipelago with species for every niche: rock-scrapers, sand-digging-insect eaters, plant-scrapers, scale-eaters, nibbler-of-other-fishes'-fins, fish-eaters and the famed *Haplochromis compressiceps* known for biting out eyeballs! Lake Victoria supported about 200 such cichlid species, all of which are thought to have descended from a single ancestral species.

In 1960, supported by the United Nations Food and Agriculture Organisation, the Nile Perch was introduced into the lake to provide an additional source of protein. (The fish can grow to 100 kg or more.) Objections were voiced; however, they were ignored, the introduction went ahead and the perch thrived. Being large and carnivorous, the perch all but wiped out the smaller fish species, hundreds of varieties unknown elsewhere. Many of these fish also provided the staple foods and economies for local fishing communities. The markets became flooded with perch and the overall productivity of the lake was reduced by 80% (from the 1960 levels) in 25 years.

The ramifications spread far wider than the local ecological and economic disaster. The Nile Perch are oily and cannot be sun dried, the traditional means by which fish had been preserved. Instead they had to be smoked, which resulted in major felling operations in the local environs to provide the fuel.

Furthermore, the cichlids had controlled the level of a particular snail that lives in and around the lake. The snails are an essential link in the spread of the disease bilharzia (liver fluke disease), which can be fatal to humans. Essentially, bilharzia is a disease produced by the larvae of a flatworm. The larvae first infect certain freshwater snails in streams and rivers, but particularly in lakes. They develop in the digestive gland of the snail and are discharged into the lake as larvae. From there they penetrate the skin of animals, such as humans, with a muscular boring action. Once inside the body, they mature to an adult fluke and attach to the intestine or bladder and begin producing eggs. The effect on humans is a range of unpleasant problems of the kidney, liver, lung, intestine or central nervous system and can be seriously debilitating or fatal. Without the cichlids there was a marked increase in the number of snails and hence also the disease.

All in all, the perch introduction was catastrophic. Many of the ramifications could have, and should have, been foreseen and avoided. However, in spite of this, further introductions were being planned in the late 1980s for Nile Perch into other African lakes such as Lake Malawi.

5.6   Case Study: It's a dog's life: More on the control of stray dogs

In Section 3.6 we presented a case study based on part of the research article Amaku et al. (2010), and developed a model for the control of stray dogs through euthanasia . Here we introduce a second control strategy — that of of sterilisation — and show how the use of a coupled system of differential equations can facilitate a comparison between the impact of the two strategies on control. In terms of the modelling developed in this chapter, humans act as 'predators' on stray dogs in two distinct ways. Our purpose is to introduce a means of assessing management strategies and establishing the circumstances under which one strategy may be better than another.

Stray dogs can cause a number of problems in both urban and rural areas across the world, causing major health problems and posing a threat to livestock and property. Dogs are abandoned for numerous reasons, with a natural tendency to group in packs and breed. There is a clear need to control populations in order to minimise damage, and the question becomes: Which management strategy is most effective for the control of stray dogs? Here we apply a mathematical model to compare the strategies of euthanasia and sterilisation, and establish how to determine conditions under which one may be more effective than the other.

Sterilisation is typically targeted at owned dogs, and euthanasia at stray dogs. However, some owners are against sterilisation, and sterilisation programs have also been designed to target stray dogs. One such program in Bangkok (Thailand) failed because the sterilisation rate was too low, while another in Jaipur (India) led to a stable 70% of female dogs sterilised (Amaku et al., 2010). This suggests that there are critical sterilisation rates, above which a program is more likely to succeed. Likewise, euthanasia will impact on population growth rates, although total eradication of stray dogs is unlikely with the ongoing abandonment of animals. A further consideration is cost, not explicitly dealt with here. However, an understanding of how the two strategies (sterilisation and euthanasia) impact on populations of stray dogs will inform any feasibility assessment.

Euthanasia. First we incorporate euthanasia. Following the case study in Section 3.6, this results in a differential equation with ϵ the proportion of dogs euthanised per km^2 (see euthanasia model (3.15)):

$$\frac{dN}{dt} = -\frac{Nr}{K}\left[N - K\left(1 - \frac{\epsilon}{r}\right)\right],$$ (5.17)

where $N = N(t)$ is a function of time, r is the intrinsic growth rate and K the carrying capacity. Parameter r can be expressed as the difference between birth rate a and death rate b so that $r = a - b$ and ϵ represents the additional per-capita death rate for the euthanasia control program.

Sterilisation. Second we consider an alternative control strategy. We adjust the model to incorporate sterilisation that varies with time. We assume only female dogs are sterilised (as is usual for such programs but could be easily changed) and let $Q(t)$ be the proportion of female dogs sterilised. Thus, if $S(t)$ is the number of dogs sterilised per km^2, and ν the proportion of female dogs in the population, then

$$Q(t) = \frac{S(t)}{\nu N(t)}.$$ (5.18)

The differential equation model becomes

$$\begin{aligned}\frac{dN}{dt} &= N\left(a\left(1 - Q\right) - b - r\frac{N}{K}\right) \\ &= rN\left(1 - \frac{N}{K}\right) - aNQ,\end{aligned}$$ (5.19)

since $r = b - a$.

Now let σ be the rate of sterilisation. Then,

$$\frac{dS}{dt} = \left(-b - r\frac{N}{K}\right)S + \sigma(\nu N - S),$$ (5.20)

where the first term in brackets incorporates no births from sterilised animals but density dependence determined by the full population, and the second term, $\sigma(\nu N - S)$, is the rate of change in number of sterilised dogs with only unsterilised female dogs available for sterilisation. (that is, $\nu N - S$).

These equations can be combined to provide an equation in $Q(t)$ alone. Differentiating equation (5.18) and substituting in equations (5.19) and (5.20) gives

$$\frac{dQ}{dt} = (1 - Q)(\sigma - aQ).$$ (5.21)

The two equilibrium solutions are $Q_e = 1$ and $Q_e = \sigma/a$.

In order to examine the impact of sterilisation graphically, we require appropriate parameter values. Estimates that are relevant to stray dogs have been taken from the literature (see case study Section 3.6) with the carrying capacity $K = 250$ (dogs per km^2), and instantaneous birth and death rates per year given by $a = 0.34$ and $b = 0.12$. The proportion of male to female for stray dogs has been typically found to be 64:36 leading to $\nu = 0.36$ (Beck, 2002; Amaku et al., 2010). Parameter σ (like ϵ for euthanasia) is the control parameter for sterilisation, and we consider a number of different values.

Results with the above parameter values are illustrated in Figure 5.11. For the sterilisation of female dogs the intrinsic birth parameter a acts as a threshold, with sterilisation proportions above a leading to a fully sterilised population with almost no potential to reproduce after, approximately, 50 years. For sterilisation proportions below a, the system stabilises to ongoing control with a sterilised population proportion $Q(t) = \sigma/a$. It is evident that the dynamics of the effect of the euthanasia and sterilisation control strategies are similar in certain respects and distinct in others (compare Figure 3.8 and Figure 5.11).

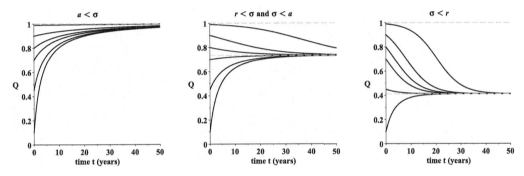

Figure 5.11: Sterilisation model (5.21). The proportion sterilised over time ($Q(t)$) is illustrated as the sterilisation parameter σ varies, for a number of initial conditions (solid lines). The grey dashed lines are the equilibrium values. Parameters are: $\sigma = 0.36$, 0.25 and 0.14, $K = 250$, $a = 0.34$, $b = 0.12$, $\nu = 0.36$, with time intervals of 1 year, and units of area km^2 (Amaku et al., 2010).

Comparison. The purpose of this analysis is to compare the effectiveness of the two control strategies, but the quantities $Q(t)$ (the proportion of all females sterilised in Figure 5.11) and $N(t)$ (the total number per km^2 in Figure 3.8) are very different and cannot be compared directly. Thus we make use of a coupled system of equations that relate $Q(t)$ to $S(t)$ (number of dogs sterilised per km^2) and $N(t)$ (the total number of dogs per km^2). Replacing $Q = S/(\nu N)$ in equation (5.19), and coupling it with equation (5.20), leads to system,

$$\frac{dN}{dt} = rN\left(1 - \frac{N}{K}\right) - \frac{aS}{\nu}$$
$$\frac{dS}{dt} = \left(-b - r\frac{N}{K}\right)S + \sigma(\nu N - S). \tag{5.22}$$

Figure 5.12(a) illustrates the results for the total population density over time, with solid lines illustrating results for sterilisation (system (5.22)), and dashed curves for euthanasia (equation (5.17)), and 3 distinct values for the euthanasia and sterilisation constants (black, fine black and grey curves). Note that, for each comparison (black, fine black and grey dashed and solid pairs in Figure 5.12(a), each with $\sigma = \epsilon$), the dog numbers are consistently lower with euthanasia than with sterilisation. If the effort required for the same rates of *euthanasia and sterilisation are comparable and money no issue, euthanasia would be the preferred option. However, a more helpful comparison for decision-making incorporates the number of animals euthanised ($E(t) = \epsilon N(t)$ per km^2) and the number sterilised ($S(t)$ per km^2), as illustrated in Figure 5.12(b).*

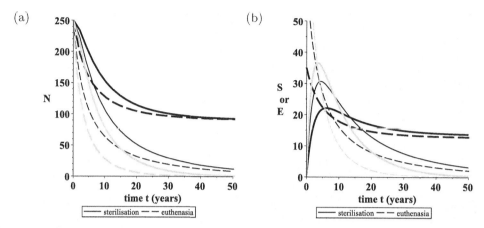

Figure 5.12: Comparison of sterilisation and euthanasia models: (a) The dynamics of stray dog numbers ($N(t)$ per km^2) for the sterilisation and euthanasia models are compared for $\sigma = 0.36 = \epsilon$ (grey lines), $\sigma = 0.25 = \epsilon$ (fine lines) $\sigma = 0.14 = \epsilon$ (thick lines). (b) The number of dogs sterilised ($S(t)$) is compared with the number of dogs euthanised ($E(t) = \epsilon N(t)$ per km^2) with the same parameter values as for plot (a). Other parameters are $K = 250$, $a = 0.34$, $b = 0.12$, $\nu = 0.36$, with time intervals of 1 year, units of area km^2, and initial values $N(0) = K$ and $S(0) = 0 = E(0)$ (Amaku et al., 2010).

Given the costs of each treatment per animal, this allows a direct cost comparison and thus a means of assessing the most affordable and effective strategy. With the same conditions (with $\sigma = \epsilon$ for each pair of curves), the number of animals euthanised early in the process are consistently and considerably larger than those sterilised, although after a few years the number to be sterilised exceeds the number to be euthanised. Without costs this result is not conclusive, but suggests that a combination of the strategies may provide the optimal strategy.

We conclude that these simple models provide a means to compare control strategies in a meaningful way, even with uncertainty around growth rates and population estimates. The impact of other system drivers can easily be included, and as mentioned earlier an important aspect of the dynamics not included here is that dogs are continuously abandoned and join the stray dog population. This modelling approach to comparing management strategies is relevant to many other applications (for example, pests, diseases, weeds, fishing quotas), particularly when data are few, and it can be a very valuable tool in the design of programs for control.

5.7   Competing species

Another simple ecosystem to model is that of competing species, where two (or more) populations compete for limited resources such as food or territory. There are two aspects of competition: *exploitation*, when the competitor uses the resource itself and *interference*, where the population behaves in such a manner as to prevent the competitor from utilising the resource. This system is very similar to the predator-prey model of the previous section; however, the terms describing the interaction between the species differ.

Model assumptions

We start with a list of assumptions on which to build our first model:

- We assume the populations to be sufficiently large so that random fluctuations can be ignored without consequence.
- We assume that the two-species model reflects the ecosystem sufficiently accurately.
- We assume each population grows exponentially in the absence of the other competitor(s), although we later incorporate density-dependent growth for each.

General compartmental model

Let $X(t)$ and $Y(t)$ be the two population densities (number per unit area) where t is again time. As before, we have birth and death rates associated with each population; however, in this case an increase in the number of deaths in one population causes a decrease in the number of deaths in the other population. The compartmental diagram of Figure 5.13 illustrates the process.

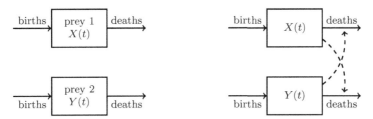

Figure 5.13: Input-output diagram for the two-species competition model.

In words, we have for each population in the system

$$\left\{\begin{array}{c} \text{rate of} \\ \text{change of} \\ \text{population} \end{array}\right\} = \left\{\begin{array}{c} \text{rate of} \\ \text{population} \\ \text{births} \end{array}\right\} - \left\{\begin{array}{c} \text{rate of} \\ \text{population} \\ \text{deaths} \end{array}\right\}. \tag{5.23}$$

Formulating the differential equations

Since neither population is dependent on the other as far as growth rates are concerned (unlike in the predator-prey example of Section 5.4), we let the positive constants β_1 and β_2 describe the per-capita birth rates for species X and Y, respectively.

Since the two populations are competing for the same resource, the density of each population has a restraining effect, proportional to this density, on the other. So the per-capita death rate for Y is proportional to X, and that for X is proportional to Y, which in symbols is

$$\left\{\begin{array}{c} \text{rate of} \\ \text{species-}X \\ \text{deaths} \end{array}\right\} = (c_1 Y)X, \qquad \left\{\begin{array}{c} \text{rate of} \\ \text{species-}Y \\ \text{deaths} \end{array}\right\} = (c_2 X)Y.$$

Here c_1 and c_2 are the constants of proportionality for this restraining effect.

Our model becomes

$$\frac{dX}{dt} = \beta_1 X - c_1 XY, \qquad \frac{dY}{dt} = \beta_2 Y - c_2 XY. \tag{5.24}$$

While β_1 and β_2 were per-capita birth rates, we may also consider them as overall per-capita growth rates that incorporate deaths (independent of the other species) as well as births. Thus they are per-capita growth rates, or per-capita reproduction rates, while the parameters c_1 and c_2 are the interaction parameters. These equations are known as *Gause's equations* and are a coupled pair of first-order, nonlinear differential equations.

Numerical solution

Again, although they appear reasonably simple, the equations cannot be solved analytically, and we use `MATLAB` or `Maple` to draw the time-dependent graphs, obtained using approximate numerical techniques as discussed in Chapter 4. Parameter values have been estimated for two competing species of microorganisms (see Renshaw (1991)) as

$$\beta_1 = 0.21827, \qquad \beta_2 = 0.06069,$$
$$c_1 = 0.05289, \qquad c_2 = 0.00459.$$

The `MATLAB` code is given in Listing 5.5, the `Maple` code is given in Listing 5.6 and the results are graphed in Figure 5.14.

Listing 5.5: MATLAB code: c_pe_compet.m

```
function c_pe_compet
global beta1 beta2 c1 c2;
beta1=0.22; beta2=0.06;
c1=0.053; c2=0.0046;
tend = 50; %the end time
u0 = [0.5; 1.5]; %set IC
[tsol, usol] = ode45(@rhs, [0, tend], u0);
Xsol = usol(:,1); Ysol = usol(:,2);
plot(tsol, Xsol, 'b'); hold on;
plot(tsol, Ysol, 'r:');
axis([0, tend, 0, 10]);

function udot = rhs(t, u)
global beta1 beta2 c1 c2;
X - u(1); Y-u(2),
Xdot = beta1*X - c1*X*Y;
Ydot = beta2*Y - c2*X*Y;
udot = [Xdot; Ydot];
```

Listing 5.6: Maple code: c_pe_compet.mpl

```
restart; with(plots): with(DEtools):
beta[1]:=0.22: beta[2]:=0.061: c[1]:=0.053: c[2]:=0.0046:
de1 := diff(X(t),t) = beta[1]*X(t)-c[1]*X(t)*Y(t);
de2 := diff(Y(t),t) = beta[2]*Y(t)-c[2]*X(t)*Y(t);
inits := [X(0)=0.5, Y(0)=1.5];
mydeopts := arrows=none, method=rkf45:
plot1 := DEplot([de1,de2], [X,Y], t=0..50,
            [inits],scene=[t,X],linecolor=black,mydeopts):
plot2 := DEplot([de1,de2], [X,Y], t=0..50,
            [inits],scene=[t,Y],linecolor=gray,mydeopts):
display(plot1, plot2, view = [0 .. 50, 0 .. 10]);
```

As usual, we are interested in what happens in the long run. From the time-dependent diagram of Figure 5.14 it appears that one of the species dies out over time. Varying the initial conditions illustrates that it is possible to choose values such that the other species dies out instead (e.g., with $X(0) = 2.6$, $Y(0) = 1.6$).

The results suggest that in the case of two species competing for the same resource, in the long run one species will survive and the other become extinct. This is known as

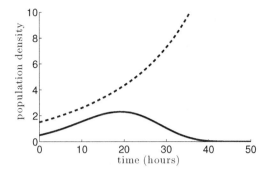

Figure 5.14: A sample numerical simulation of the competing species model with each species having unrestricted growth in the absence of the other species. The parameter values are $\beta_1 = 0.22$, $\beta_2 = 0.060$, $c_1 = 0.053$ and $c_2 = 0.0046$, with the initial condition $x_0 = 0.5$, $y_0 = 1.5$. Species X (black), species Y (grey dotted).

Gause's Principle of Competitive Exclusion, from a series of experiments conducted by the microbiologist on competing species of yeast cells. Furthermore, unlike in the predator-prey model, the populations do not appear to oscillate with time.

Interpretation of parameters

The system of equations developed so far for the model of competitive species has four parameters: the two per-capita growth rates β_1 and β_2, which are independent of any other species, and the two constants of proportionality c_1 and c_2, which describe the interaction between the species.

The competition interaction can be interpreted as stating that the more deaths there are in species Y, the more resources will be available to species X, and thus the fewer deaths there will be for X. To explain the interaction we consider a small time interval Δt, sufficiently small so that X and Y do not change significantly. Let a' be the rate of area 'used' by one individual in species Y (for food), hence making it unavailable to X (for food). Assume that no such areas can overlap. Then one individual 'uses' an area of $A_s \Delta t$, and Y individuals will 'use' $A_s Y(t) \Delta t$. Assuming that each area of resource 'used' by Y implies it is unavailable to X, the rate at which area is made unavailable to X by a single Y individual is the density (number per unit area) of X in an area divided by Δt, which is $A_s X(t)$. Hence the total effect of the removal of resources from X by Y becomes $A_s X(t) Y(t)$. So A_s, or c_1 in our model, can be interpreted as the rate at which X dies due to the removal of resources by Y.

If we let $f = c_2/c_1$, then f is a measure of the efficiency of the competitive interaction. It can be considered as the number of units of species Y required to reduce species X by one unit.

Limitations of the model

One immediate and obvious limitation of this model is that each population grows exponentially in the absence of the other. In the next section, we improve on this by including density-dependent growth.

In the time-dependent diagram, Figure 5.14, it appears that only one of the species will survive and the other will become extinct. (In the next chapter we show this to be the case for the competition model under nearly all conditions.) This idea was emphasised in the results of experimental work carried out by Gause.

Early this century, research tended to accept and support Gause's principle; however, more detailed research has led to a questioning of this model. How does the degree of competition modify the predictions? What actually constitutes a competitive model when some resources are shared and others are not? By the 1980s, there was a lively debate underway among the leaders in the field, some rejecting outright the credibility of such models, while others argued that there was a value in, and a place for, the years of research already accomplished on these competition models. Two articles, Lewin (1983a) and Lewin (1983b), provide an informative insight into this controversy and the positions adopted by the various players.

Density-dependent growth

In 1932, the Russian microbiologist Gause described an experiment with two strains of yeast, *Saccharomyces cervevisiae* and *Schizosaccharomyces kefir*, hereafter called Species X and Species Y. These are described in Renshaw (1991). Gause found that, grown on their own, each species exhibited a logistic growth curve, but when grown together, the growth pattern changed with Species X dying out.

We now extend the competition model to account for logistic growth in both species, in the absence of the other species. Let K_1 and K_2 be the carrying capacities for Species X and Y, respectively. Then including density-dependent growth in (5.24), we have

$$\frac{dX}{dt} = \beta_1 X \left(1 - \frac{X}{K_1}\right) - c_1 XY, \qquad \frac{dY}{dt} = \beta_2 Y \left(1 - \frac{Y}{K_2}\right) - c_2 XY. \tag{5.25}$$

With $d_1 = \beta_1/K_1$ and $d_2 = \beta_2/K_2$, system (5.25) becomes

$$\frac{dX}{dt} = \beta_1 X - d_1 X^2 - c_1 XY, \qquad \frac{dY}{dt} = \beta_2 Y - d_2 Y^2 - c_2 XY. \tag{5.26}$$

Using Gause's data (Renshaw, 1991), it is also possible to estimate the parameters d_1 and d_2 as in the model as

$$d_1 = 0.017, \qquad d_2 = 0.010.$$

Note that these parameters correspond to the carrying capacities (in the absence of the other species) of $K_1 = \beta_1/d_1 = 13.0$ and $K_2 = \beta_2/d_2 = 5.8$.

Re-running the `Maple` or `MATLAB` code used to produce Figure 5.14, with an adjustment made for the inclusion of the density-dependent growth for both X and Y, and the parameter values from Gause's experiment, produces Figure 5.15.

We find that after a certain time, the figure suggests that only one of the species survives, stabilising to a fixed density, which is the carrying capacity for that species. This is what Gause observed (see Renshaw (1991) for a comparison between data and numerical solutions).

To get a better understanding of how this model behaves as we change initial conditions, or the parameter values, we could generate further numerical graphs. However, this is a somewhat hit-and-miss approach. For example, does one population always die out, or can they sometimes coexist? We return to a more detailed analysis of the dynamics of this system in Section 6.5, and also in Chapter 7, as we develop some helpful analytical tools.

Renshaw (1991) gives a detailed discussion of the competing species of yeast. See Kormondy (1976) for a discussion of competition between species of microscopic protozoa (*Paramecium aurelia* and *Paramecium caudatum*).

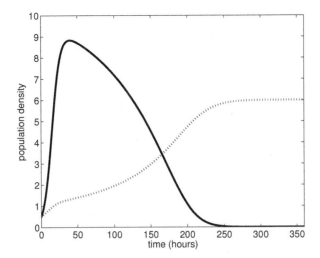

Figure 5.15: Graph of the population density of the two species of yeast populations, species $X(t)$ (black line), and species $Y(t)$ (grey dotted line)) as functions of time for the model with density dependence and with parameter values from Gause's experiment, as given in the text. Both populations have an initial density of 0.5.

> ***Summary of skills developed here:***
>
> - *Formulate differential equations for variations on the competition model presented here, including models combining three or more competing species.*
> - *Obtain numerical solutions to the competition model and extensions of it.*
> - *Understand what is meant by, and the implications of, the 'random' approach of generating numerical solutions.*
> - *Understand the competitive exclusion principle.*

5.8   Scenario: Aggressive protection of lerps and nymphs

The following scenario does not incorporate the above theory specifically, but describes the occurrence of species competing for territory with a colony of birds protecting its territory in order to maintain an adequate food supply to support itself. It is included for interest and to provide an example of the value of interspecific territoriality, a system for which the above theory would have relevance. Adapted from an article by Loyn et al. (1983).

In regions of southeastern Australia, infestations of psyllids have inflicted severe damage on the eucalypt forests causing deterioration of the foliage and, in extreme cases, completely destroying certain trees. This has prompted an examination of the conditions under which it may occur. Many species of birds live in the canopy of these eucalypts and compete to feed on these insects; the Bell Miners are one such species.

The Bell Miner (*Manorina melanophrys*) is a honeyeater living in large colonies that they defend aggressively against other competing birds, often substantially larger than themselves. They feed on the nymphs, sweet secretions and lerps (protective carbohydrate covers) of psyllids; however, where their colonies occur, the canopy foliage appears unhealthy and infested with these insects. The Miners aggressively protect this abundance of prey. In an experiment, a colony of Bell Miners was removed to join another colony 45 km away. They settled into this new location and did not return.

Other bird flocks soon moved into the site of the old colony and the foliage improved rapidly. The insect infestation declined, with the numbers of psyllids remaining very small thereafter.

It appears that the insect infestations can be controlled by the removal of the Bell Miner colonies. The interspecific territorial behaviour of the Miners prevents this control and introduces a control of its own. Miners effectively maintain an abundant and exclusive food supply, albeit detrimental to the eucalypts, enabling a colony to remain within the same territory for up to 40 years.

5.9   Model of a battle

We now consider a novel type of population interaction: a destructive competition or battle between two opposing groups. These may be battles between two hostile insect groups, athletic teams, or human armies. While the models we derive here apply to the last case, the principles can be generalised and would apply to many other examples. The model we develop turns out to be a system of two coupled, linear differential equations.

Background

Battles between armies have been fought since antiquity. In ancient times battles were primarily hand-to-hand combat. With the development of archery and then gunpowder, a crucial feature of battles has been aimed fire. Although many factors can affect the outcome of a battle, experience has shown that numerical superiority and superior military training are critical. The model we present was first developed in the 1920s by F. W. Lanchester who was also well known for his contributions to the theory of flight.

Our aim is to develop a simple model that predicts the number of soldiers in each army at any given time, provided we know the initial number of soldiers in each army. (As with epidemics, we consider the number, rather than the density, of individuals.)

Model assumptions

First we make some basic assumptions and then develop the model based on these:

- We assume the number of soldiers to be sufficiently large so that we can neglect random differences between them.

- We also assume that there are no reinforcements and no operational losses (i.e., due to desertion or disease).

These are assumptions that can easily be relaxed at a later stage if the model is inadequate.

General compartmental model

The first step is to develop two word equations that describe how the two populations change, based on the input-output principle of the balance law. Suppose the two opposing groups or populations are the red army and the blue army.

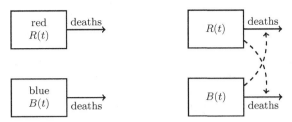

Figure 5.16: Compartmental diagram for the simple battle model.

Example 5.5: *Determine the appropriate input-output diagram and associated word equations for the number of soldiers in both the red and blue armies.*

Solution: *Since there are no reinforcements or operational losses, each population changes by the number of soldiers who are wounded by the other army. Thus we can set up the input-output diagram of Figure 5.16.*

Thus, in any given instant of time,

$$\left\{ \begin{array}{c} \text{rate of} \\ \text{change of} \\ \text{red soldiers} \end{array} \right\} = - \left\{ \begin{array}{c} \text{rate red soldiers} \\ \text{wounded by} \\ \text{blue army} \end{array} \right\}$$

$$\left\{ \begin{array}{c} \text{rate of} \\ \text{change of} \\ \text{blue soldiers} \end{array} \right\} = - \left\{ \begin{array}{c} \text{rate blue soldiers} \\ \text{wounded by} \\ \text{red army} \end{array} \right\}.$$

(5.27)

In a real battle there will be a mixture of shots: those fired directly at an enemy soldier and those fired into an area known to be occupied by an enemy, but where the enemy cannot be seen. Some battles may be dominated by one or the other firing method. We consider these two idealisations of shots fired as *aimed fire* and *random fire*. For the model we assume only aimed fire for both armies.

In the aimed fire idealisation, we assume all targets are visible to those firing at them. If the blue army uses aimed fire on the red army, then each time a blue soldier fires, he/she takes aim at an individual red soldier. The rate of loss of soldiers of the red army depends only on the number of blue soldiers firing at them and not on the number of red soldiers. We see later that this assumption is equivalent to assuming a constant probability of success (on average) for each bullet fired.

For random fire, a soldier firing a gun cannot see his/her target, but fires randomly into an area where enemy soldiers are known to be. The more enemy soldiers in that given area, the greater the rate of wounding. For random fire we thus assume that the rate of enemy soldiers wounded is proportional to both the number firing and the number being fired at.

In summary we make the following further assumptions:

- For aimed fire, the rate of soldiers wounded is proportional to the number of enemy soldiers only.

- For random fire, the rate at which soldiers are wounded is proportional to both numbers of soldiers.

Formulating the differential equations

Let $R(t)$ denote the number of soldiers of the red army and $B(t)$ the number of soldiers of the blue army. We assume aimed fire for both armies. This information is expressed mathematically by writing

$$\left\{\begin{array}{c} rate\ red\ soldiers \\ wounded\ by \\ blue\ army \end{array}\right\} = a_1 B(t), \qquad \left\{\begin{array}{c} rate\ blue\ soldiers \\ wounded\ by \\ red\ army \end{array}\right\} = a_2 R(t), \qquad (5.28)$$

where a_1 and a_2 are positive constants of proportionality. The constants a_1 and a_2 measure the effectiveness of the blue army and red army, respectively, and are called *attrition coefficients*.

We thus assume that attrition rates are dependent only on the firing rates and are a measure of the success of each firing.

We now substitute (5.28) into the basic word equation (5.27), where the rate of change in the number of red soldiers is dR/dt and for the blue soldiers it is dB/dt. The two simultaneous differential equations are thus

$$\frac{dR}{dt} = -a_1 B, \qquad \frac{dB}{dt} = -a_2 R. \qquad (5.29)$$

Numerical solution

During the Battle of Iwo Jima in the Pacific Ocean (1945), daily records were kept of all U.S. combat losses. These data are graphed and referenced in Braun (1979). The values of the attrition coefficients a_1 and a_2 have been estimated from the data[1] as $a_1 = 0.0544$ and $a_2 = 0.0106$, and the initial numbers in the red and blue armies, respectively, were $r_0 = 66,454$ and $b_0 = 18,274$.

Using MATLAB or Maple we can obtain accurate numerical solutions to the differential equations (5.29). The results are shown in Figure 5.17 with the red army as the U.S. forces and the blue army as the Japanese. The model shows remarkably good agreement with the data. The MATLAB and Maple code used to obtain the graphs are given in Listing 5.7 and Listing 5.8.

Listing 5.7: MATLAB code: c_pe_combat.m

```
function c_cp_combat
global a1 a2;
tend = 30;    %the endtime
a1=0.0544; a2=0.0106;
u0 = [66; 18]; %set initial conditions as a column vector
[tsol, usol] = ode45(@rhs, [0, tend], u0);
Rsol = usol(:, 1); Bsol = usol(:, 2);
plot(tsol, Rsol, 'r'); hold on; plot(tsol, Bsol, 'b');

function udot = rhs(t, u)
global a1 a2;
R=u(1); B=u(2);
Rdash = -a1*B;
Bdash = -a2*R;
udot = [Rdash; Bdash];
```

[1]The method of least squares was used to determine the parameter values. It varies the parameters to minimise a function that is the sum of the squares of the errors between the numerical solution and the data points.

Listing 5.8: Maple code: c_pe_combat.mpl

```
restart:with(plots):with(DEtools):
a[1]:=0.0544:a[2]:=0.0106:
de1:=diff(R(t),t)=-a[1]*B(t);
de2:=diff(B(t),t)=-a[2]*R(t);
inits:=[R(0)=66,B(0)=18]:
myopts:=stepsize=0.1,arrows=NONE:
plot1:=DEplot([de1,de2],[R,B],t=0..30,[inits],scene=[t,R],linecolour=red,myopts):
plot2:=DEplot([de1,de2],[R,B],t=0..30,[inits],scene=[t,B],linecolour=black,myopts):
display(plot1,plot2);
```

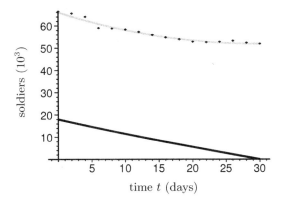

Figure 5.17: Numbers of U.S. soldiers in the battle of Iwo Jima measured in thousands. The solid lines (U.S.-grey, Japanese-black) are the model predictions from `Maple` and the dots are the measured data. From Braun (1979).

Despite the excellent fit to the data, there are still some reservations about using this model for the Battle of Iwo Jima. The values of the parameters indicate the Japanese soldiers were approximately five times more effective than the U.S. soldiers, since $a_1 \simeq 5a_2$. In this battle, the Japanese army already occupied the island and were battle hardened and knew the area. Perhaps a better model would be to assume that the U.S. soldiers used random fire while the Japanese soldiers used aimed fire. It turns out (see Exercise 5.21) that this alternative model gives an equally good fit to the data.

The differential equations (5.29) are linear, unlike the systems of differential equations in previous sections (Sections 5.4, 5.7 and 5.2) and it is possible here to obtain an exact solution. This is done in the exercises in Chapter 6; see Exercise 6.15.

Interpretation of parameters

We can further refine the model by trying to express the parameters a_1 and a_2 in terms of possible quantities that could be measured. The rate at which soldiers are wounded depends on both the firing rate and the probability of a shot hitting a target.

We return to equations (5.28). Consider a single blue soldier firing at the red army. We assume each blue soldier fires at a constant rate f_b. Then

$$\left\{ \begin{array}{c} \text{rate red soldiers} \\ \text{wounded by} \\ \text{single blue soldier} \end{array} \right\} = \left\{ \begin{array}{c} \text{rate bullets} \\ \text{fired in time} \\ \text{interval} \end{array} \right\} \times \left\{ \begin{array}{c} \text{probability of} \\ \text{a single bullet} \\ \text{hitting target} \end{array} \right\}$$

$$= f_b p_b,$$

where p_b is the probability (constant) that a single bullet from a blue soldier wounds a red soldier. Hence, for the entire blue army, we multiply by the number of blue soldiers, $B(t)$, to obtain the total rate of red soldiers wounded by the blue army (per unit time). This gives

$$\left\{ \begin{matrix} \text{rate red soldiers} \\ \text{wounded by} \\ \text{blue army} \end{matrix} \right\} = f_b p_b B(t). \qquad (5.30)$$

Equating this to (5.28) we obtain the attrition rates, or coefficients, a_1 and a_2 as

$$a_1 = f_b p_b, \qquad a_2 = f_r p_r, \qquad (5.31)$$

where f_r is the firing rate by a single red soldier and p_r is the probability that a single red bullet hits its target. We can think of f_b, f_r and p_b, p_r as factors influenced by morale, training and technology.

For random fire we do not assume the probability of a single bullet wounding a soldier to be constant. It will vary depending on the number of target soldiers within a given area. Thus, this probability will depend on both the number of target soldiers and the area into which fire is being directed.

Limitations and extensions of the model

The model we have developed here was based on the assumption of aimed fire. More generally, battles occur where one army uses aimed fire and the other uses random fire (for example, guerrilla warfare) or where both armies use random fire (for example, long-range artillery). These models lead to the differential equations

$$\frac{dR}{dt} = -a_1 B, \qquad \frac{dB}{dt} = -c_2 RB,$$

in the case of guerrilla warfare and

$$\frac{dR}{dt} = -c_1 RB, \qquad \frac{dB}{dt} = -c_2 RB,$$

for long-range artillery or trench warfare.

There are several other obvious extensions and variations of the basic model. These include incorporating both random and aimed fire, or modelling regular reinforcement and/or operational losses, such as from disease.

Other good references include Taylor (1980), Przemieniecki (1994) and Tung (2007) who discuss further extensions of the basic models to include reinforcements, range dependent firing and geometric mean fire. In Braun (1979) there is a clear discussion of this model and a similar model for guerrilla warfare, where one army is hidden (e.g., jungle warfare). Also discussed is the Battle of Iwo Jima. See Taylor (1980) for a substantial discussion of these types of models and ways of estimating attrition coefficients. Stochastic (random) effects may also be included in the basic models; see Przemieniecki (1994).

Summary of skills developed here:

- *Modify the model to account for one, or both, of the armies using random fire.*
- *Modify the model to account for loss due to disease and/or gains due to reinforcements.*
- *Obtain numerical solutions to the model developed and some extensions.*

5.10   Case Study: Rise and fall of civilisations

This case study is based on an article by Feichtinger et al. (1996). It illustrates the use of mathematical models to provide a possible explanation for the rise and fall of dynasties in ancient China.

China is one of the oldest human civilisations that can claim to have some degree of continuity. In ancient China there have been many dynasties, such as the Xia, Shang, and Zhou dynasties of ancient times and the Qin, Han and Ming dynasties. Between these dynasties there have been periods of rapid population decline corresponding to the fall of dynasties. During these times of anarchy, the ruling classes (which include the soldiers under their command) have been weak and unable to control the numbers of bandits and outlaws. However, over time this state does not persist and the ruling classes are able to suppress the outlaws and bandits and the general population of peasants and farmers increases.

Why do these periods of anarchy occur? There are probably many causes; however, one plausible explanation is that it could be due to the natural dynamics of various interacting groups in the population. It is the aim of a mathematical model to try and explain this behaviour with as few variables as possible. To try and help explain the fall of dynasties, a simple three-population model has been developed.

The three sub-populations are the farmers, $F(t)$, (i.e., the peasant class), the bandits, $B(t)$, and the ruling class, $R(t)$ (which includes the soldiers hired by the emperor). The differential equations are

$$\frac{dF}{dt} = rF\left(1 - \frac{F}{K}\right) - \frac{aFB}{b+F} - hFR, \tag{5.32}$$

$$\frac{dB}{dt} = \frac{eaFB}{b+F} - mB - \frac{cBR}{d+B}, \tag{5.33}$$

$$\frac{dR}{dt} = \frac{faFB}{b+F} - gR, \tag{5.34}$$

where r, K, a, e, h, b d, m, c, f, and g are all positive constants in the model. This model is like a predator-prey model where the bandits 'prey on' the farmers and the soldiers (ruling class) 'prey on' the bandits as well as the farmers. Rulers impose taxes on farmers and punish bandits.

In the absence of rulers and bandits, the farmers exhibit logistic growth,

$$\frac{dF}{dt} = rF\left(1 - \frac{F}{K}\right).$$

The term $aFB/(b+F)$ is a saturating predation rate of bandits upon farmers. The number of farmers killed by an individual bandit is $aF/(b+F)$ and as F becomes large, this rate tends to a constant rate (i.e., as $F \to \infty$, then $aF/(b+F) \to a$, a constant rate), but is approximately proportional to the number of farmers, F, for small numbers of farmers.

Both the bandits and the rulers (soldiers) have a natural mortality term proportional to their numbers, mB and gR. The bandits have an additional mortality term $cBR/(d+B)$ from the contact of bandits with soldiers. This term also incorporates saturation, reflecting the fact that soldiers have to search out bandits, so with increasing numbers of bandits the rate that each individual soldier can kill is $cB/(d+B)$, which tends to a constant as $B \to \infty$ and is proportional to B for small numbers of bandits.

The term hFR in the first differential equation represents an additional mortality rate for the farmers. This is due to excessive taxing of the farmers by the ruling class, which causes additional hardship and results in additional deaths due to starvation. Note that this is proportional to both the number of farmers and the number of the ruling class; the greater the number of soldiers, the more taxes are needed to pay for them, and the higher the mortality for each individual farmer.

The parameter values for this model have been listed by Feichtinger et al. (1996) as given in Table 5.2. The populations are measured as fractions of the farmer carrying capacity, so, in these units, the carrying capacity $K = 1$. Similarly, the time is scaled with respect to r^{-1}, where r is the farmer intrinsic growth rate, so $r = 1$ here. (See Appendix D.1 for a discussion of scaling.) Using initial conditions $F(0) = 0.7$, $B(0) = 0.1$ and $R(0) = 0.2$, we can easily use `Maple` or `MATLAB` to solve the differential equations numerically and to graph the populations, as in Figure 5.18 and Figure 5.19.

Table 5.2: Parameter values for the dynastic cycle model, from Feichtinger et al. (1996). These assume the populations are measured as a fraction of the maximum farmer population, so $K = 1$, and $r = 1$ in these units. These parameters were chosen to illustrate the possibility of cycles of population growth rather than an accurate estimate from historical data.

$r = 1$	$K = 1$	$a = 1$
$b = 0.17$	$h = 0.1$	$d = 0.42$
$m = 0.4$	$c = 0.4$	$f = 0.1$
$g = 0.009$	$e = 1.2$	

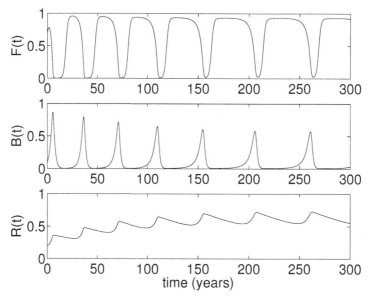

Figure 5.18: Numerical simulation of the farmer-bandit-rulers model for the rise and fall of dynasties in ancient China. Parameter values used are given in Table 5.2 and initial values are $F(0) = 0.7$, $B(0) = 0.1$ and $R(0) = 0.2$, where these values correspond to a fraction of the carrying capacity of the farmer population.

With this choice of parameters the populations of farmers and bandits oscillate, with the *farmer population near the carrying capacity most of the time, but where the population periodically undergoes a dramatic fall. This directly coincides with the time when the bandit population increases dramatically. The ruling population (the soldiers) peaks at the same time as the bandit population peaks, and then falls slightly until another farmer population collapse.*

It is interesting to explore the effect of changing some of the parameters. In Figure 5.19 the parameter h has been changed. This parameter corresponds to the coefficient of the additional mortality rate of farmers due to taxing by the rulers. If this value is increased from $h = 0.14$ to $h = 2.0$, this changes the dynamics of the populations. After some degree of oscillation all three populations settle into a steady state, and are in equilibrium. The surprising implication is that a higher degree of severity of taxation of farmers leads to a more stable society, in the long term. A possible interpretation of this change in behaviour is as follows. With a larger exploitation rate h, the ruling class is more able to maintain order by having sufficient resources (i.e., numbers of soldiers) to keep the bandit population from becoming too large.

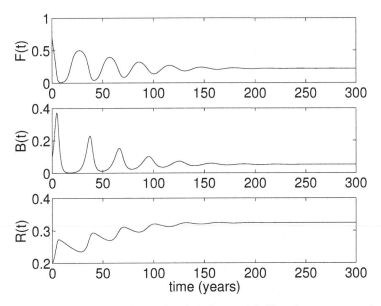

Figure 5.19: Numerical simulation of the farmer-bandit-rulers model. Here the parameter values in Table 5.2 are used, but the parameter h, corresponding to the hiring rate of soldiers, is increased from 0.1 to 2.0.

It is also interesting to determine the effect of changing other parameters in the model, and trying to interpret the results. For example, increasing the parameter $c = 0.4$ to $c = 0.8$ causes longer periods between periods of anarchy (see Exercise 5.28). Note that changing the initial conditions changes the graphs, but they eventually settle into the same long-term behaviour.

This is a fairly complex model (with 11 separate parameters) and there is much scope for exploring its behaviour. Some progress can be made by making the observation that the ruling class population generally changes slowly compared with the other populations. Feichtinger et al. (1996) exploit this by holding the variable R constant and analysing the resulting system of two simultaneous differential equations. They do this using bifurcation theory. This is beyond the scope of this book; however, some of the techniques developed in subsequent chapters on phase-plane analysis are used in their analysis.

Despite the apparent complexity of behaviour of this model, it is still of some interest to explore some simple extensions. Foremost is one identified by Feichtinger et al. (1996) who suggest adding an additional term corresponding to a loss of the soldiers from contacts with the bandits. This might be significant where there are large numbers of bandits or if the bandits were sufficiently good fighters compared with the soldiers. Another interesting extension would be to include an additional term in the farmer differential equation corresponding to recruitment of the farmers as soldiers into the ruling class, with a corresponding term in the differential equation for the rulers (soldiers).

5.11   Exercises for Chapter 5

5.1. Basic SIR numerical solution. In Section 5.2, a model for an epidemic was developed, which led to the system of differential equations in the form

$$\frac{dS}{dt} = -\beta SI, \quad \frac{dI}{dt} = \beta SI - \gamma I.$$

Use parameter values $\beta = 0.002$ and $\gamma = 0.4$, and assume that initially there is only one infective but there are 500 susceptibles. Use MATLAB or Maple to generate the time-dependent plot on the interval $t = [0, 20]$.

(a) How many susceptibles never get infected, and what is the maximum number of infectives at any time?

(b) What happens, as time progresses, if the initial number of susceptibles is doubled, $S(0) = 1,000$? How many people were infected in total?

(c) Return the initial number of susceptibles to 500. Suppose the transmission coefficient β is doubled. How does this affect the maximum number of infected individuals? Is this what you expect?

(d) Draw the compartment diagram for the SIR model with an additional dashed line that indicates which rates are also influenced by any other compartments.

5.2. The SI and SIS models. For each of the following, draw suitable compartment diagrams and hence obtain a model in the form of a system of differential equations.

(a) Consider a disease where all those infected remain contagious for life (they never recover). Ignore any births and deaths. Develop a pair of differential equations for the number of susceptibles S and number of infectives I. (Define any notation you introduce.)

(b) Consider a disease where all who recover then become susceptible to catching the disease again. Assume that these individuals have the same chance of catching the disease as those who have never been infected before. Assume the mean period of infection is γ^{-1}. Ignore any births and deaths.

5.3. SEIR model. *Many diseases have a latent period, which is when there is a period of time between infection and when an infected individual becomes infectious. One example is measles, where the latent period is approximately 5 days.*

Extend the SIR epidemic model to one with an additional population class $E(t)$, corresponding to individuals who have been **exposed** *to the disease, so they are no longer susceptibles, but are not yet infectious. You may assume a latent period σ^{-1}. Also, infectives recover in a mean time γ^{-1} and have lifelong immunity.*

5.4. Waning immunity. *Consider an infectious disease with immunity, but instead of assuming that immunity lasts for the individual's entire life, assume that it gradually decreases (wanes) and individuals eventually become susceptible again after an average time w^{-1}.*

(a) *Ignore births and deaths and assume individuals are infectious for an average time γ^{-1} and give differential equations for the numbers of susceptibles, infectives and recoverers.*

(b) *Include births and deaths with the same per-capita rate a.*

5.5. Continuous vaccination. *We formulate two models for continuous vaccination of the susceptible population; the first assumes the vaccine gives perfect life-long protection and the second assumes the vaccine is only temporary. Ignore any births or deaths.*

(a) *Formulate a model for the spread of a disease where lifelong immunity is attained after catching the disease. The susceptibles are continuously vaccinated against the disease at a rate where a proportion ν per unit time of susceptibles are vaccinated. We want to track the number vaccinated so you need to have an additional variable $V(t)$.*

(b) *Assume the vaccine is only partially effective, where vaccinated individuals return to the susceptible state after being protected by vaccination for an average time μ^{-1}.*

5.6. Cholera model compartment diagram. *Draw a suitable compartment diagram for the system of equations used in the cholera case study in Section 5.3,*

$$S'(t) = -c\frac{BS}{k_{50} + B},$$
$$I'(t) = c\frac{BS}{k_{50} + B} - \gamma I,$$
$$B'(t) = eI + (n_b - m_b)B.$$

5.7. Endemic cholera model. *The cholera model considered in Section 5.3 was*

$$S'(t) = -c\frac{BS}{k_{50} + B},$$
$$I'(t) = c\frac{BS}{k_{50} + B} - \gamma I,$$
$$B'(t) = eI + (n_b - m_b)B,$$

but this model can only predict outbreaks and not deal with the situation where cholera has become endemic.

(a) *To model endemic cholera, modify this model to account for susceptible births and also incorporate natural deaths of all populations. Give the differential equation for $R(t)$, the number of recovered. Assume per-capita birth rate b for the whole population. Assume the per-capita death rate a is the same for susceptibles, infected and recovered.*

(b) *When is the total population constant? Find the disease-free equilibrium ($I = 0$) for the case where the deaths are balanced exactly with the births.*

(c) What is the basic reproduction number R_0 for this new model? Assume the initial number of susceptibles is the disease-free equilibrium in (b).

5.8. Two prey and one predator. Develop a model with three differential equations describing a predator-prey interaction, where there are two different species of prey and one species of predator. (Base these on the assumptions used in Section 5.4.) Also, assume that prey are necessary for predator births and that prey are only killed by predators (ignore natural prey deaths).

5.9. Effect of DDT. Consider the predator-prey model developed in Section 5.4, which describes the threat to the American citrus industry posed by the accidental introduction of the Australian scale insect, and the later introduction of the ladybird to combat the pest.

(a) Use MATLAB or Maple to generate the time-dependent graphs of the populations over time with $\beta_1 = 1$, $\alpha_2 = 0.5$, $c_1 = 0.01$ and $c_2 = 0.005$. Initialise with $(x_0, y_0) = (200, 80)$ and $(x_0, y_0) = (80, 200)$. Comment on any differences in the long-term behaviour.

(b) Verify that $X' = 0$ and $Y' = 0$, if $X = 100$ and $Y = 100$ simultaneously. Plot these values (as horizontal lines) in the time-dependent diagrams.

(At these values there is no change in the populations over time, so they are equilibrium populations. We see how to find such equilibrium solutions in Chapter 6.)

(c) Now include the effect of DDT in the model, with $p_1 = 0.1$ and $p_2 = 0.1$. Generate the time-dependent plots as before. Verify that $X' = 0$ and $Y' = 0$ when $X = 120$, $Y = 80$, and plot these values on the time-dependent diagrams.

(d) Comment on your observations from the above.

5.10. Predator-prey with density dependence. Starting with the Lotka–Volterra model, we wish to include the effect of logistic growth for the prey and DDT pesticide acting on both species. The differential equations are

$$\frac{dX}{dt} = \beta_1 X \left(1 - \frac{X}{K}\right) - c_1 XY - p_1 X, \qquad \frac{dY}{dt} = c_2 XY - \alpha_2 Y - p_2 Y.$$

(a) With parameter values as in Figure 5.9, and also Figure 5.10, use Maple or MATLAB to generate time-dependent graphs for the populations over time. Use $x_0 = 100$ and $y_0 = 100$ as initial conditions and set $K = 800$. Comment on your observations, particularly with respect to Figure 5.10.

(b) What happens if $K = 100$?

5.11. Symbiosis. Symbiosis is where two species interact with each other in a mutually beneficial way. Starting with a compartmental diagram, formulate a differential equation model describing this process, based on the following.

Assume the per-capita death rate for each species to be constant, but the per-capita birth rate to be proportional to the density of the other species. In other words, the presence of the other species is necessary for continued existence. (Define all parameters and variables of the model.)

5.12. Competing species with no density dependence. In Section 5.7, a simple model describing the interaction between two competing species was developed, where each population exhibits exponential growth.

(a) Use MATLAB or Maple to generate time-dependent graphs for the populations over time, with parameter values as given in Figure 5.14, but using a variety of initial conditions.

(b) Show that $X' = 0$ and $Y' = 0$ simultaneously, when $X = 2.5$ and $Y = 1.5$. Include them in the time-dependent plots as horizontal lines.

(c) *In this case it is not so easy to make some 'general conclusions' about the population be-haviour over long periods of time. Why?*

5.13. Competing species with density dependence. *Consider the following model for two competing species, with densities, $X(t)$ and $Y(t)$, given by the differential equations*

$$\frac{dX}{dt} = \beta_1 X - c_1 XY - d_1 X^2, \qquad \frac{dY}{dt} = \beta_2 Y - c_2 XX - d_2 Y^2,$$

with parameter values $\beta_1 = 3$, $\beta_2 = 3$, $c_1 = 2$, $c_2 = 1$, $d_1 = 2$ and $d_2 = 2.5$.

(a) *What is the carrying capacity for each of the species, evaluated for the given parameter values? (Hint: Compare this form with the form given by (5.25) in Section 5.7 and find K_1 and K_2 in terms of the parameters here.)*

(b) *With the above parameter values, and the initial values $X = 2$ and $Y = 2$, use* MATLAB *or* Maple *to draw time-dependent plots for these populations. Over a period of time, what population densities do you estimate they will approach?*

5.14. Simple age-based model. *Consider a population split into two groups: adults and juveniles, where the adults give birth to juveniles but juveniles are not yet fertile. Eventually juveniles mature into adults. You may assume constant per-capita birth and death rates for the population, and also assume that the young mature into adults at a constant per-capita rate σ.*

Starting from suitable word equations or a compartment diagram, formulate a pair of differential equations describing the density of adults, $A(t)$, and the density of juveniles, $J(t)$. Define all variables and parameters used.

5.15. Beetle populations. *A population of beetles has three different age stages: larvae (grub), pupae (cocoon), and adult. Assume constant per-capita death rates for each population class of a_1 for larvae, a_2 for pupae and a_3 for adults. Also assume adults produce larvae at a constant per-capita birth rate of larvae b_1. The larvae turn into pupae at a constant per-capita rate σ_1 and pupae turn into adults at a constant per-capita rate σ_2.*

Let $A(t)$ denote the number of adults, $L(t)$ the number of larvae and $P(t)$ the number of pupae at time t and formulate a mathematical model in the form of three differential equations.

5.16. Wine fermentation. *In the fermentation of wine, yeast cells digest sugar from the grapes and produce alcohol as a waste product, which is toxic to the yeast cells. Develop a model consisting of three coupled differential equations for the density of yeast cells, the amount of alcohol and the amount of sugar. In the model assume the yeast cells have a per-capita birth rate proportional to the amount of sugar, and a per-capita death rate proportional to the amount of alcohol present. Assume the rate of alcohol produced is proportional to the density of yeast cells, and the rate of sugar consumed is proportional to the density of yeast cells.*

5.17. Density-dependent contact rate. *For a fatal disease, if the basic epidemic model of Section 5.2 is modified to include density-dependent disease transmission, the resulting differential equations are*

$$\frac{dS}{dt} = -p\frac{c(N)}{N}SI, \qquad \frac{dI}{dt} = p\frac{c(N)}{N}SI - \gamma I,$$

where $N = S + I$, p is a constant (the probability of infection) and the contact rate function $c(N)$ is given by

$$c(N) = \frac{c_m N}{K(1 - \epsilon) + \epsilon N},$$

where ϵ is a positive constant between 0 and 1, and K is a positive constant.

Consider $\epsilon = 0.5$, and using Maple *or* MATLAB *with $pc_m = 1.62$, $r = 0.44$, $K = 1,000$, and initial values $i_0 = 1$, $s_0 = 762$ (time is measured in days), graph the number of susceptibles over time and determine when the number of infectives is at a maximum. Compare this model to those with $\epsilon = 1$ and $\epsilon = 0$, and discuss the differences.*

5.18. **Control of stray dogs.** *Read the case study (Section 5.6) for a model that compares control strategies for stray dogs. Not all mathematical details for the equations derived are given in that case study, and thus these details are established here.*

(a) *Derive (5.21) from equations (5.18), (5.19) and (5.20).*

(b) *Find the equilibrium points for (5.21) and by considering the sign of dQ/dt, establish when Q is increasing or decreasing over time as σ varies. These solutions should agree with those illustrated in Figure 5.11.*

5.19. **Crime.** *A model for the movement of petty criminals in and out of prisons assumes that all new criminals arise from contacts of existing criminals with law abiding citizens. Assume there is a mean time σ^{-1} before a criminal is caught and sent to prison. The mean time for a prison sentence is μ^{-1}. Upon release from prison, a small fraction f of ex-prisoners become law abiding citizens with the remainder returning to being criminals. Formulate a model to describe the movement of criminals into and out of prison. Your model should consist of three differential equations: one for $L(t)$ the number of law abiding citizens, a second for $C(t)$ the number of criminals, and a third for $P(t)$ the number of prisoners.*

5.20. **Battle loss due to disease.** *Develop a model (a pair of differential equations) for a battle between two armies where both groups use aimed fire. Assume that the red army has a significant loss due to disease, where the associated death rate (from disease) is proportional to the number of soldiers in that army.*

5.21. **Jungle warfare.** *In Section 5.9 we developed a simple model for a battle between two armies. We assumed that the probability of a single bullet hitting its target is constant. This is not a good assumption in jungle warfare or guerrilla warfare where one, or both, of the soldiers may be hidden from view of the other.*

Suppose that soldiers from the red army are visible to the blue army, but soldiers from the blue army are hidden. Thus, all the red army can do is fire randomly into an area and hope they hit something. The blue army uses aimed fire.

(a) *Write appropriate word equations describing the rate of change of the number of soldiers in each of the armies.*

(b) *By making appropriate assumptions, obtain two coupled differential equations describing this system.*

(c) *Extend the model to include reinforcements if both of the armies receive reinforcements at constant rates.*

5.22. **Jungle warfare (continued).** *This question refers to the differential equations model, developed in Exercise 5.21, and gives estimates of the parameters. Suppose the blue army defends an area of $A = 10^5\,\mathrm{m}^2$, with an initial number of 150 soldiers. The red army has 500 soldiers initially, all of whom are exposed to fire from the blue army. Each soldier, in either army, fires at the same rate of $f_r = f_b = 400$ bullets per day. Field data have shown that each single bullet fired from the blue army has the constant probability 1/100 of killing or removing a red soldier.*

(a) *Write a formula for the probability of a single bullet fired from a single red soldier wounding a blue soldier in terms of the total area A and the area exposed by a single blue soldier $A_b = 0.1\,\mathrm{m}^2$.*

(b) *Hence, estimate the coefficients in your model (i.e., write the rate of wounding in terms of the probability in (a) and the firing rate.)*

(c) *With your estimates in (b), use* `MATLAB` *or* `Maple` *to calculate the number of soldiers left in each army after 5 days.*

5.23. Spread of malaria by mosquitoes. With the disease malaria in humans, the disease is carried by mosquitoes that cannot infect each other. Infectious mosquitoes can only infect susceptible humans and infected humans can only infect susceptible mosquitoes when they are bitten by a susceptible mosquito.

Assume the rate of transmission is proportional to both numbers of mosquitoes and number of humans for transmission in both directions, and assume (for this model) once infected, both humans and mosquitoes never recover.

Ignoring any births and deaths, develop a mathematical model for susceptible and infected humans $S_h(t)$, $I_h(t)$, and susceptible and infected mosquitoes $S_m(t)$, $I_m(t)$.

5.24. Two infectious stages. Consider an infectious disease model based on the standard SIR model, where an individual passes through two infectious states with different degrees of infectivity, with transmission coefficients β_1 and β_2 (this could be due to different rates of virus shedding or to different contact rates, for example).

Ignoring any births or deaths, formulate a model as three differential equations, for S the number of susceptible, I_1 the first infectious state, and I_2 the second infectious state. Recovery (with lifelong immunity) can only occur when the individual has passed through the second infectious state.

5.25. Spread of a religion. A new religion is spreading through a community in a remote country by missionaries recruited from the local population. The community is made up of unbelievers (with numbers denoted by $U(t)$), converts (numbers $C(t)$) and missionaries (numbers $M(t)$)). Assume that only contacts between missionaries and unbelievers result in an unbeliever becoming a convert. A constant proportion of converts each year decide to become missionaries.

Formulate a system of differential equations for these populations. Your model should have the property that the total population remains constant over time. Births and deaths may be ignored and relapses to unconverted of either converts or missionaries may be neglected.

5.26. Predator-prey with child-care. Formulate a mathematical model for a predator-prey system where the prey protect their young from the predators. The model should have three dependent variables: $X_1(t)$, the juvenile prey numbers; $X_2(t)$, the adult prey numbers; and $Y(t)$, the predator numbers. In your model, assume the juvenile prey are completely protected from the predators.

5.27. Basic reproduction number. The basic reproduction number, R_0, is the number of new infections produced by a single infective, over the the duration of time they are infectious. For the SIR model,

$$\frac{dS}{dt} = -\beta SI, \qquad \frac{dI}{dt} = \beta SI - \gamma I,$$

we have $R_0 = \beta S_0 \gamma^{-1}$.

Consider the following model, which includes births and deaths,

$$\frac{dS}{dt} = -\beta S \frac{I}{N} + aN - aS, \quad \frac{dI}{dt} = \beta S \frac{I}{N} - \gamma I - aI, \quad \frac{dR}{dt} = \gamma I - aR,$$

where $N = S + I + R$.

5.28. Farmers, bandits and soldiers. Read over the case study in Section 5.10 for a model of farmer, bandit and soldier population in ancient China.

(a) Write a `Maple` or `MATLAB` program to reproduce the graph in Figure 5.18.

(b) Change the parameter c from $c = 0.4$ to $c = 0.8$ and plot the graph. Describe the changes and interpret.

5.29. Frequency-dependent transmission. *The equations for the SIR epidemic with frequency dependent transmission,*

$$\frac{dS}{dt} = -\beta_f S \frac{I}{N}, \qquad \frac{dI}{dt} = \beta_f S \frac{I}{N} - \gamma I$$

were derived for the numbers of suscetibles and infectives.

(a) *What are the equations for the density of susceptibles and infectives? [Hint: define $s(t) = S(t)/A$ and $i(t) = I(t)/A$, where A is a constant area.]*

(b) *Also deal with the density-dependent transmission case.*

5.30. Diseases with carriers. *Formulate a model for an infectious disease where there is immunity for only some of those who recover; others 'recover' to become permanent carriers, who can still cause infections. Thus susceptibles, $S(t)$, may be infected by either infectious individuals, $I(t)$, or carriers, $C(t)$. A carrier can infect others at a reduced rate compared to infectious individuals but shows no symptoms.*

(a) *Assume there is a fixed proportion q of those recovering from the infection become carriers. Assume transmission rates β_1 for normal infectives and β_2 for carriers and assume that individuals remain infective for a mean time γ^{-1}. Give equations for the number of susceptibles $S(t)$, the number of infectives $I(t)$, the number of carriers $C(t)$, and the number of recovered who are immune $R(t)$.*

(c) *Give at least one example of an infectious disease that could be modelled by the equation you have developed.*

Chapter 6

Phase-plane analysis

For pairs of coupled first-order differential equations it can be useful to use the chain rule to eliminate time and reduce the pair of equations to a single first-order differential equation. The graph of all the solutions to this single equation is called the phase-plane. We apply this method to the basic models from Chapter 5. Consequently, we are able to draw some general conclusions about the models and confirm the observations that were made using only the numerical solutions.

6.1 Introduction

In the previous chapter, we developed models for interacting populations that resulted in pairs of nonlinear, coupled differential equations and we then used `Maple` to solve the equations numerically. However, each solution was for a particular pair of initial conditions and a particular combination of parameters. Clearly there are infinitely many such combinations, and we need some other tools to understand the system behaviour as a whole. What would be valuable is to understand how any changes in the initial conditions or values of the parameters might affect these solutions and the subsequent system dynamics.

We see how to gain insight into the behaviour of the solutions to these systems by finding the equilibrium points (points corresponding to where the derivatives are zero) and eliminating time from the differential equations (by converting the coupled differential equations to a single first-order differential equation). Further, we are interested in the long-term behaviour of the systems. We ask questions such as: Does a population die out or settle down to some fixed size, or, can a fatal disease wipe out a population completely or recur periodically? The techniques we develop to deal with such questions are designed to investigate the behaviour of solutions to the systems and require an understanding of equilibrium points and the phase-plane. First, we introduce these techniques using a very simple example system and then apply them to the systems developed in the previous chapter.

Consider the coupled pair of first-order (linear) differential equations

$$\frac{dX}{dt} = Y, \qquad \frac{dY}{dt} = -X. \tag{6.1}$$

Equilibrium points

Equilibrium points correspond to solutions of a coupled system of differential equations where the solutions are constant, that is, where $dX/dt = 0$ and $dY/dt = 0$, simultaneously. Hence, for the differential equations (6.1), we obtain

$$Y = 0, \qquad X = 0$$

and so $(X, Y) = (0, 0)$ is the only equilibrium solution.

Trajectories and the phase-plane diagram

Let us consider the (X, Y)-plane: this is called the *phase-plane*. Dividing the plane into four quadrants in the manner illustrated in Figure 6.1, we have, in the first quadrant where $X > 0$ and $Y > 0$, that $dX/dt = Y > 0$ and $dY/dt = -X < 0$. Thus $X(t)$ is increasing and $Y(t)$ is decreasing, and we obtain a direction vector for any solution in that quadrant, given by the arrow in Figure 6.1. Each quadrant can be considered in the same manner. We can thus infer that the solutions, that is, the *phase-plane trajectories*, move in a clockwise direction.

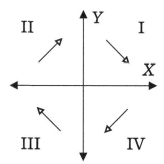

Figure 6.1: Direction vectors for the trajectories in the phase-plane of example equations (6.1).

Using the chain rule

We note that neither of the differential equations involves the time variable t explicitly, that is, t does not appear on the RHS of either equation. (Of course, the solutions will be time dependent, since the derivatives on the LHS are with respect to time.) This means that we should be able to eliminate the time variable and find an expression, independent of t, which relates X and Y. Alternatively stated, we express Y as a function of X. That is, we are making X the independent variable where it was previously a variable dependent on t.

An expression for the chain rule is

$$\frac{dY}{dt} = \frac{dY}{dX}\frac{dX}{dt},$$

which gives the derivative of Y with respect to t in terms of the derivative of Y with respect to X and the derivative of X with respect to t. Dividing by the derivative of X with respect to t gives

$$\frac{dY}{dX} = \frac{dY/dt}{dX/dt}. \tag{6.2}$$

We substitute from the coupled pair of differential equations (6.1) into (6.2) to give

$$\frac{dY}{dX} = -\frac{X}{Y}. \tag{6.3}$$

Here we have a first-order differential equation with Y a function of X.

It is not always a simple matter to solve the differential equation; however, in this case the differential equation (6.3) is separable since we can write it in the form

$$Y\frac{dY}{dX} = -X. \tag{6.4}$$

Example 6.1: *Solve the differential equation* (6.4) *using the separation of variables technique.*

Solution: *Integrate both sides with respect to the independent variable* X *to obtain*

$$\int Y \frac{dY}{dX} \, dX = \int -X \, dX.$$

By the substitution rule for integration, this simplifies to

$$\int Y \, dY = \int -X \, dX.$$

Carrying out both integrations gives

$$\frac{1}{2}Y^2 = -\frac{1}{2}X^2 + C,$$

where C *is the constant of integration. Multiplying throughout by 2, and rearranging the terms, we have that*

$$X^2 + Y^2 = K$$

with $K = 2C$. *The value of* K *will be determined by the initial conditions.*

This solution equation is that of a circle. It describes the paths traced out by the (X, Y) pair over time, depending on the initial values or starting conditions. These are the exact solutions to the phase-plane trajectories.

Interpretation of the phase-plane

As an interpretation of these trajectories, if a system has initial values x_0 and y_0 then the system starts at the point (x_0, y_0) in the phase-plane and, as time evolves, it traces out the trajectory curve (in this case a circle) in a clockwise direction as was established in Figure 6.1. At any subsequent time the values of $X(t)$ and $Y(t)$ will be the coordinates of this trajectory. Since, for this example, the trajectory is a circle, the motion is repeated continuously in time. In fact, this would be the case for any closed trajectory.

To see how the system evolves in time we would normally need an exact solution of the original coupled equations giving both Y and X as functions of time. Often this is not possible. Nevertheless, we can still use the chain rule to infer useful information about the system. We see how to do this in the following sections, which examine the systems of differential equations for populations developed in the previous chapter.

In essence, the idea of phase-plane analysis is to draw a phase-plane diagram together with the phase-plane trajectories in order to understand some general features of the system. (The terminology arises from the use of these ideas to calculate planetary orbits.) If the differential equations are sufficiently simple, we can go further and use the chain rule to eliminate time. In this way, we may obtain an exact expression relating the two dependent variables that describe the trajectory path.

The phase-plane diagram is useful for determining the behaviour of solutions for a variety of initial conditions. In the above example, we saw that all solutions of the differential equations have phase-trajectories that are circles. This means that the plots for both variables as functions of time must be oscillations, which follows because X and Y must always return to their original values as we move along the trajectory. Furthermore, as the initial point approaches the equilibrium point, the amplitude of the oscillation is reduced with the equilibrium point itself corresponding to a solution that is constant in time.

Note, however, that in this procedure we have lost information about time. This was the price for reducing the coupled, first-order differential equations to a single, first-order equation.

Summary of skills developed here:

- *Understand the concept of equilibrium solutions.*
- *Establish the directions of trajectories.*
- *Use the information on equilibrium points and trajectory directions to draw a phase-plane.*
- *Understand how the chain rule can eliminate time and reduce a coupled pair of differential equations to a single differential equation.*

6.2   Phase-plane analysis of epidemic model

Previously, in Section 5.2, we developed a model for an epidemic of an infectious disease. We now use the chain rule and some analysis to prove that the disease, described by this model, can never infect the entire population.

Review of the model

The epidemic model, developed in Section 5.2, assumed the population N was divided into susceptibles, denoted by $S(t)$, infectives, denoted by $I(t)$ and removals $N - S(t) - I(t)$. We assumed the disease to confer life-long immunity and we neglected to include natural births and deaths. The model pair of differential equations we obtained was

$$\frac{dS}{dt} = -\beta SI, \qquad \frac{dI}{dt} = \beta SI - \gamma I. \tag{6.5}$$

The parameter β is the transmission coefficient and γ is the recovery (or removal) rate.

For the parameter values we considered in the example of Section 5.2, our results indicated that the number of infectives always tended to zero while the number of susceptibles approached some finite number. It seemed there were always certain individuals who never contracted the disease. It is interesting to speculate whether, according to our model, this is always the case. We use the chain rule below to show this is indeed so.

A picture of the phase-plane plot can be obtained by solving the differential equations numerically. This allows us to determine the behaviour of the model for a range of initial conditions, but for a fixed set of parameter values. Some **Maple** code to give the phase-plane diagram in Figure 6.2 is given in Listing 6.1. Similarly, some **MATLAB** code for a similar diagram is given in Listing 6.2.

Listing 6.1: Maple code: c_ps_epidemic.mpl

```
restart:with(plots):with(DEtools):
unprotect(gamma): gamma :='gamma';
interface(imaginaryunit=i); I='i';
beta:=2.18*10^(-3): gamma:=0.44:
de1:=diff(S(t),t)=-b*S(t)*I(t);
de2:=diff(I(t),t)=beta*S(t)*I(t)-gamma*I(t);
inits:=[0,762,1],[0,600,20],[0,400,50]:
plot1:=DEplot([de1,de2],[S(t),I(t)],t=0..30,
    [inits],scene=[S,I],stepsize=0.1,dirgrid=[10,10],
```

```
    arrows=medium,linecolour=black):
display(plot1);
```

Listing 6.2: MATLAB code: c_ps_epidemic.m

```
function c_ps_epidemic
global beta gamma;

tend = 15; %the end time
beta=2.0*10^(-3); gamma=0.44;
u0vec = [762, 600, 400; % make a matrix of 3 ICs
           1,  20,  50];
u0size = size(u0vec); % size of the matrix of ICs
numICs = u0size(2); % number of ICs

for k = 1:numICs %loop over each case of ICs
    u0 = u0vec(:, k); %extract the kth column of matrix
    [tsol, usol] = ode45(@rhs, [0, tend], u0); %solve the DE
    Ssol = usol(:, 1); Isol = usol(:, 2);
    plot(Ssol, Isol); hold on; %plot each trajectory
end
%produce arrows (see appendix for code for this function)
c_dirplot(@rhs, 0, 800, 0, 300, 10);
axis([0,800, 0, 300]);

function udot = rhs(t, u)
global beta gamma;
S=u(1); I=u(2);
Sdot = -beta*S*I;
Idot = beta*S*I - gamma*I;
udot = [Sdot; Idot];
```

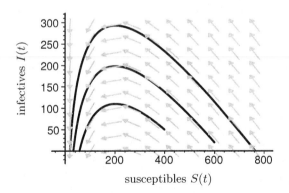

Figure 6.2: Maple generated phase-plane diagram for the influenza in a school example, with transmission coefficient $\beta = 2.18 \times 10^{-3}$ and recovery rate $\gamma = 0.44$ with a variety of initial values. These parameter values are the same as those used in Section 5.2. This shows the behaviour for the given parameter values but for a range of different initial conditions.

In Figure 6.2 we follow the curve (a trajectory) from the initial values in the direction of the arrows. The arrows come from the direction field as determined by the differential equation for each value of S and I. We see from this diagram that as $I \to 0$, so S does not tend to zero — there are some susceptibles left uninfected. Of course this conclusion is only valid for the given parameter values for β and α. We can form conclusions for general values of parameter values, in this model, to obtain analytic expressions for the trajectories.

Applying the chain rule

For the simple differential equations comprising this model it is possible to obtain an analytic expression for the trajectories. We can do this by eliminating time (using the chain rule), which results in a differential equation involving the variables I and S. The solution to the differential equation establishes a relation between these two variables, as derived in the following example.

Example 6.2:   *Use the chain rule to find I in terms of S, given that the initial number of susceptibles is s_0 and the initial number of contagious infectives is i_0.*

Solution: *Using the chain rule,*

$$\frac{dI}{dS} = \frac{dI/dt}{dS/dt}.$$

Substituting from the differential equations (6.5), we eliminate the time variable, obtaining the single differential equation

$$\frac{dI}{dS} = \frac{\beta SI - \gamma I}{-\beta SI},$$

which simplifies to

$$\frac{dI}{dS} = -1 + \frac{\gamma}{\beta S}. \tag{6.6}$$

This differential equation relates the number of infectives I to the number of susceptibles S but does not retain explicit information about time.

The differential equation (6.6) is a first-order separable differential equation of a trivial kind and to solve it we have only to integrate both sides with respect to S. The solution is

$$I = -S + \frac{\gamma}{\beta} \ell n(S) + K, \tag{6.7}$$

where K is an arbitrary constant of integration.

The initial numbers of susceptibles and infectives are s_0 and i_0, respectively, and hence the initial condition for the differential equation (6.6) is

$$I(s_0) = i_0.$$

Applying this to the general solution (6.7) gives an equation for K from which we deduce that

$$K = i_0 + s_0 - \frac{\gamma}{\beta} \ell n(s_0).$$

Sketching the phase-plane trajectories

The solution to the differential equation (6.6), given by (6.7), is not very complicated so it is worthwhile to try and obtain a general sketch of I versus S.

Example 6.3: *Determine and sketch the family of phase-plane curves given by (6.7).*

Solution: *To find turning points we set $dI/dS = 0$. That is,*

$$-1 + \frac{\gamma}{\beta S} = 0.$$

Solving for S gives

$$S = \frac{\gamma}{\beta},$$

corresponding to a potential turning point. Since there is only one potential turning point, we do not need to find a second derivative to determine whether it is a maximum, minimum or inflection point. Instead, we can use the values of the function at $S = 0$ and $S \to \infty$.

As $S \to 0$, we see that $I \to -\infty$ since $\ell n(S) \to -\infty$. As $S \to \infty$, the limit is harder to work out. Note that $\ell n(S) \to \infty$ and $-S \to -\infty$ as $S \to \infty$. To determine the limit, let us consider the derivative dI/dS. Note that from (6.6), $dI/dS \to -1$ as $S \to \infty$. This implies that $I \to -\infty$ since S grows faster than $\ell n\, S$ for large S. (A formal method for showing this is to use l'Hôpital's rule.)

Since $I \to -\infty$ as $S \to 0$ and $S \to \infty$, thus the point corresponding to $S = \gamma/\beta$ must be a local maximum. Hence the curve must cross the positive S-axis twice, for sufficiently large K. This is sketched in Figure 6.4.

Direction of trajectories

The direction of trajectories in the phase-plane is determined from the differential equations, as below.

Example 6.4: Determine the directions of trajectories in the phase-plane.

Solution: From $dS/dt = -\beta SI$ we see that for $S > 0$ and $I > 0$, dS/dt is always negative. This means that $S(t)$ is always decreasing. Similarly, from the other differential equation, dI/dt is positive provided $I(\beta S - \gamma) > 0$. Thus $dI/dt > 0$ if $S > \gamma/\beta$ and negative if $S < \gamma/\beta$. The phase-plane, for positive S and I, is therefore divided into two regions, as shown in Figure 6.3.

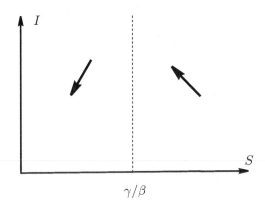

Figure 6.3: Diagram showing directions of trajectories in the phase-plane for the epidemic model.

The resulting sketch of Figure 6.4 agrees with the earlier theory, which indicated (see Figure 6.3) that from any initial condition (s_0, i_0), the trajectory of $S(t)$ decreases continuously.

Interpretations

A useful conclusion from Figure 6.4 concerns the number of susceptibles left after the disease has run its course. We see that it is impossible for the disease to infect all of the susceptibles. (However, the number of susceptibles may become sufficiently small for it to be regarded as effectively zero.)

We can interpret some interesting results from Figure 6.4. Note that if s_0 is greater than the critical value γ/β, then the number of infectives must increase for a time before it decreases, whereas if $s_0 < \gamma/\beta$, then the disease dies out. Thus for an epidemic to occur, in which the number of infectives increases from an initial number i_0, there needs to be at least some threshold number of susceptibles present prior to the outbreak, that is,

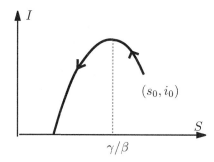

Figure 6.4: General sketch of the phase-plane diagram, illustrating a typical phase-plane curve with the direction in which the curve is traced indicated by the arrow.

$s_0 > \gamma/\beta$. The combination $\beta s_0/\gamma$ is known as the *basic reproduction ratio*, often denoted by the symbol R_0, as introduced in Section 5.2. It is a measure of the number of infections caused by a single infective in a fully susceptible population. This is an important quantity in epidemiology: a basic reproduction ratio of less than one indicates the propensity of the disease to die out. This can be exploited to determine control strategies to reduce the impact and even eradicate an infectious disease.

Chapter 8 includes a case study where different strategies are tested in a bid to eradicate bovine tuberculosis in possums in New Zealand. However, the analysis in this case is a little different from models of diseases in humans; for the possums (an introduced and destructive feral pest) we are unconcerned if the population is eradicated in the process!

Summary of skills developed here:

- *Use the chain rule to eliminate time from any epidemic model.*
- *Further practice at graph sketching.*
- *Use this model (and extensions) to examine vaccination or other disease eradication strategies.*

6.3   Analysis of a battle model

We apply the chain rule to the coupled system of differential equations for the Lanchester battle model developed in Section 5.9. This reduces the coupled equations to a single first-order differential equation. By solving this we obtain an expression that relates the number of soldiers in one army to the number of those in the other. This in turn provides a more general understanding of the system dynamics.

Review of the model

In the previous chapter we developed a model describing a battle between two armies. The model resulted in a pair of coupled differential equations, where $R(t)$ and $B(t)$ denote the number of soldiers in the red army and blue army, respectively. We assumed both armies to use only aimed fire. We derived the pair of differential equations

$$\frac{dR}{dt} = -a_1 B, \qquad \frac{dB}{dt} = -a_2 R, \qquad (6.8)$$

where a_1 and a_2 are positive constants (attrition coefficients). Recall that these attrition coefficients can also be expressed in terms of firing rates and probabilities of hitting a target.

The phase-plane diagram from Maple

We can use Maple to plot some typical phase-plane trajectories, see Listing 6.3, or MATLAB, see Listing 6.4. We let $a_1 = 0.0544$ and $a_2 = 0.0106$. The phase-plane trajectories corresponding to three different initial conditions are shown in Figure 6.5.

Listing 6.3: Maple code: c_ps_combat.mpl

```
restart:with(plots):with(DEtools):
a[1]:=0.0544:a[2]:=0.0106:
de1:=diff(R(t),t)=-a[1]*B(t);
de2:=diff(B(t),t)=-a[2]*R(t);
inits:=[R(0)=66,B(0)=18],[R(0)=45,B(0)=18],[R(0)=30,B(0)=18]:
plot1:=DEplot([de1,de2],[R,B],t=0..30,[inits],scene=[R,B],linecolour=black):
display(plot1);
```

Listing 6.4: MATLAB code: c_ps_combat.m

```
function c_ps_combat
global a1 a2;

tend = 30;%set the endtime to run simulation
a1=0.0544; a2=0.0106;
u0vec = [66 45 30; %make a matrix of 3 ICs
         18 18 18];
u0size = size(u0vec);
numICs = u0size(2);

% plot phase-plane curve for each Init Cond
for k = 1:numICs
    u0 = u0vec(:,k); %choose kth col for init cond
    [tsol, usol] = ode45(@rhs, [0, tend], u0);
    Rsol = usol(:, 1); Bsol = usol(:, 2);
    plot(Rsol, Bsol); hold on;
end
% makearrows, see Appendix for this function
c_dirplot(@rhs, 0, 70, 0, 20, 10);
axis([0,70, 0, 20]);

function udot = rhs(t, u)
global a1 a2;
R=u(1); B=u(2);
Rdash = -a1*B;
Bdash = -a2*R;
udot = [Rdash; Bdash];
```

Equilibrium points

Typically we set the two rates of change to zero and solve the equations simultaneously. In this case, the exercise is trivial and the only equilibrium solution is $(R, B) = (0, 0)$.

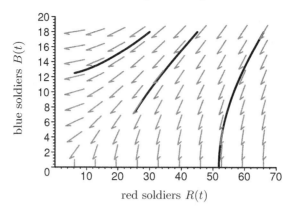

Figure 6.5: Phase-plane trajectories generated by `Maple` using parameter values $a_1 = 0.0544$ and $a_2 = 0.0106$ with various initial conditions.

Direction of trajectories

The direction of the trajectories can be obtained from the original differential equations (6.8). This is done in the following example.

Example 6.5: *Determine possible directions of phase-plane trajectories in the phase-plane.*

Solution: *From these equations, since R and B are both positive, and since the constants a_1 and a_2 are positive, then dX/dt is always negative and dY/dt is always negative. Hence $R(t)$ and $B(t)$ are always decreasing with time.*

Because of the directions in the phase-plane, it is clear that the trajectory must head towards one of the axes. Since the only equilibrium point is the origin, the trajectory (unless it passes through the origin) must cross the R or B axis in finite time, as the rate of change is not zero there. This means that the battle would be over in finite time. To determine where the trajectory intersects the axes, we need to determine the explicit form of the trajectories, and fortunately we can do this using the chain rule.

Applying the chain rule

We use the chain rule as before to eliminate the time variable t. Substituting from the differential equations (6.8) we obtain

$$\frac{dB}{dR} = \frac{dB/dt}{dR/dt} = \frac{a_2}{a_1} \frac{R}{B}. \tag{6.9}$$

Thus we obtain a single, first-order differential equation that relates B and R, but does not involve t. The solution is given by the following example.

Example 6.6: *Solve the differential equation (6.9).*

Solution: *The solution of the first-order differential equation (6.9) is obtained by separating the variables and then integrating with respect to the independent variable R. This gives*

$$\int B \frac{dB}{dR} \, dR = \int \frac{a_2}{a_1} R \, dR.$$

Using the substitution rule for integration, the LHS can be converted to an integral involving the variable B,

$$\int B \, dB = \int \frac{a_2}{a_1} R \, dR.$$

Carrying out the integrations gives the equation

$$\frac{1}{2}B^2 = \frac{a_2}{2a_1}R^2 + C,$$

where C is an arbitrary constant of integration. Multiplying both sides of the equation by 2, we obtain

$$B^2 = \frac{a_2}{a_1}R^2 + K, \tag{6.10}$$

where $K = 2C$ is also an arbitrary constant.

Applying the initial conditions $R(0) = r_0$ and $B(0) = b_0$, we obtain

$$b_0^2 = \frac{a_2}{a_1}r_0^2 + K,$$

so that

$$K = b_0^2 - \frac{a_2}{a_1}r_0^2. \tag{6.11}$$

If we suppose the battle is fought until one of the sides is wiped out, then we can use (6.10) to determine who wins the battle. We can also see this graphically by plotting the trajectories described by (6.10).

Returning to the phase-plane

Let us now examine the solution (6.10) graphically. From (6.10) we have

$$B = \sqrt{\frac{a_2}{a_1}R^2 + K}, \tag{6.12}$$

where the positive square root is the only valid one (as it only makes sense to have a positive number of soldiers), and K is a constant that is determined by the initial conditions.

To sketch the family of phase-plane trajectories satisfying (6.12), we use standard graph sketching techniques. First we look for possible turning points, then we examine the asymptotic behaviour for large R and finally, we look for intercepts on the axes.

Example 6.7: *Sketch the phase-plane trajectories.*

Solution: *For turning points we require $dB/dR = 0$. Rather than differentiate (6.12) directly, we can use the differential equation (6.9). For R and B both positive, dB/dR is positive and so B is always an increasing function of R and there are no turning points.*

We can also check the behaviour of B as R becomes large. For sufficiently large R the term R^2 will always be large compared with K. So for large R the trajectory asymptotes to $B = \sqrt{a_2/a_1}\,R$, which is the equation of a straight line.

We also calculate where the curves cross the axes. For B-intercepts, we set $R = 0$ and then solve for B to obtain the intercept $B = \sqrt{K}$ for $K > 0$. This only makes sense if K is positive. Similarly, for R-intercepts, we set $B = 0$ and then solve for R to obtain the intercept $R = \sqrt{-a_1K/a_2}$ for $K < 0$. There is a real solution only if K is negative. We thus get a different family of curves, depending on whether K is positive or negative.

Putting together the information that B is an increasing function of R, the fact that the curves all approach a straight line asymptotically, and the known values of the intercepts, allows us to infer the general form of the phase-plane curves. A sketch of these is given in Figure 6.6.

For a given initial condition (r_0, b_0), we start at that point on one of the curves. As time increases, we move along the curve in a direction towards one of the axes (as the number of soldiers can only decrease in this model).

From Figure 6.6 we see that if $K > 0$ then $R \to 0$ and the blue army wins. However, if $K < 0$ then $B \to 0$ and the red army wins.

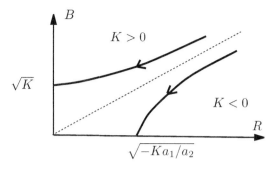

Figure 6.6: Typical phase-plane trajectories for the combat model with two armies exposed to each other's fire. The directions along the trajectories are indicated.

Discussion

An interesting use of this battle model is for the analysis of tactics. Although the model studied here is probably too simple to represent most real battles, it is still useful in understanding why some battle tactics work. One such example is the 'divide and conquer' strategy. In this strategy you divide your enemy's force, while not dividing your own, so that you fight two battles against two smaller forces. It has been used by many famous generals, including Napoleon. Why should the outcome of this be any different from fighting a single battle against the total force? The following example illustrates the reason.

Example 6.8:   *For simplicity let us assume that both armies have equal attrition coefficients* $a_1 = a_2$. *Let us suppose the red army has 10,000 soldiers initially and the blue army has 8,000. Determine who wins if (1) there is one battle between the two armies, and (2) there are two battles: the first with half the red army against the entire blue army and the second with the other half of the red army against the blue army survivors of the first battle.*

Solution: *If there is only one battle, the red army wins since* $a_1 = a_2$ *and* $r_0 > b_0$. *In other words,* $a_1 K = a_1 \times (8 \times 10^3)^2 - a_2 \times (10 \times 10^3)^2 = -36 \times 10^6 a_1$ *so* $K < 0$, *which is the condition for the red army to win.*

Instead, suppose that half the red army meets the entire blue army in a first battle and then the remaining half of the red army meets the remaining blue army soldiers in a second battle. For the first battle, we can calculate the number of blue soldiers left. For this calculation we use equation (6.10) with $r_0 = 5,000$, $b_0 = 8,000$ *and obtain the surviving number of blue soldiers (as* $R \to 0$),

$$B \to (b_0^2 - r_0^2)^{1/2} = (64 \times 10^6 - 25 \times 10^6)^{1/2} \approx 6.25 \times 10^3.$$

In the second battle, we now have $r_0 = 5 \times 10^3$ *and* $b_0 = 6.25 \times 10^3$. *Clearly, the blue army now wins the second battle since it has the greater numbers (and* $a_1 = a_2$).

This example demonstrates how an initially inferior army is able to defeat a superior army by forcing the superior army to engage in two separate battles. It illustrates the important military concept of concentrating forces and provides an indication of the role of this type of theory in the design of battle tactics and strategies.

> **Summary of skills developed here:**
> - *Apply the chain rule method to coupled systems.*
> - *Draw and interpret phase-plane diagrams for systems of differential equations.*
> - *Extend the results of this and more complicated battle models to give tactical advice.*

6.4 Analysis of a predator-prey model

We explore the Lotka–Volterra predator-prey equations with the aim of gaining a more general understanding of the dynamics using phase-plane analysis. We first obtain the phase-plane diagram for a specific set of parameters, using `Maple` to generate a numerical solution, and thereafter we infer what the phase-plane looks like for a general parameter set.

Review of model

In Section 5.4, we formulated a simple model for predator-prey interactions. The resulting differential equations were

$$\frac{dX}{dt} = \beta_1 X - c_1 XY, \qquad \frac{dY}{dt} = -\alpha_2 Y + c_2 XY. \qquad (6.13)$$

Recall that X denotes the prey population and Y denotes the predator population. For the positive constant parameters we had c_1 and c_2 as the interaction parameters, β_1 as the prey per-capita birth rate and α_2 as the predator per-capita death rate.

We found these equations produced oscillations in time for the various initial conditions and parameter combinations considered. Here we address the question of whether this model always predicts oscillations for any (positive) values of the parameters.

The phase-plane diagram from `Maple`

We let $\beta_1 = 1$, $\alpha_2 = 0.5$, $c_1 = 0.01$ and $c_2 = 0.005$, which are the same values for the parameters we used in Section 5.4. We can use `Maple` or `MATLAB` to solve the differential equations numerically and plot some phase-plane trajectories for this set of parameter values for some different initial conditions; see Listing 6.5 and Listing 6.6.

Listing 6.5: Maple code: c_ps_predprey.mpl

```
restart:with(plots):with(DEtools):
beta[1]:=1.0:alpha[2]:=0.5:c[1]:=0.01:c[2]:=0.005:
de1:=diff(X(t),t)=beta[1]*X(t)-c[1]*X(t)*Y(t);
de2:=diff(Y(t),t)=-alpha[2]*Y(t)+c[2]*X(t)*Y(t);
inits:=[0,100,80],[0,50,50],[0,100,170];
DEplot([de1,de2],[X,Y],t=0..50,[inits],scene=[X,Y],linecolour=black,stepsize=0.1);
```

Listing 6.6: MATLAB code: c_ps_predprey.m

```matlab
function c_ps_predprey
global beta1 alpha2 c1 c2;

beta1=1.0; alpha2=0.5; c1=0.01; c2=0.005;
tend = 20;%the end time to run the simulation
%  as a matrix
u0vec = [100, 50, 100; %make a matrix of ICs
         80, 50, 170];
u0size = size(u0vec); %size of the matrix
numICs = u0size(2); % extract number of ICs

for k = 1:numICs %loop over each IC
    u0 = u0vec(:, k); %extract the kth column
    [tsol, usol] = ode45(@rhs, [0, tend], u0);
    Xsol = usol(:, 1); Ysol = usol(:, 2);
    plot(Xsol, Ysol);
    hold on;
end
% makearrows, see Appendix for this function
c_dirplot(@rhs, 0, 800, 0, 300, 11);
axis([0,300, 0, 200]);

function udot = rhs(t, u)
global beta1 alpha2 c1 c2;
X = u(1); Y=u(2);
Xdot = beta1*X - c1*X*Y;
Ydot = -alpha2*Y + c2*X*Y;
udot = [Xdot; Ydot];
```

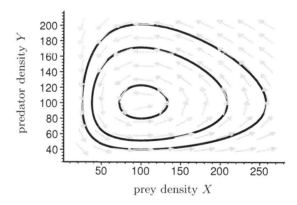

Figure 6.7: Maple generated phase-plane plot for the Lotka–Volterra equations. The parameter values $\beta_1 = 1$, $c_1 = 0.01$, $\alpha_2 = 0.5$ and $c_2 = 0.005$ have been used with various initial conditions.

The fact that the curves which appear in the phase-plane diagram of Figure 6.7 are closed is very significant. (Closed trajectories do not occur in all phase-plane diagrams.) Recall that as time evolves we travel along the curve. From the arrows in Figure 6.7 (the direction field) the direction of travel is in the anticlockwise direction. Thus, if initially the populations are x_0 and y_0, then as we travel along the curve we eventually return to the initial values x_0 and y_0. In fact, every point on the curve will be repeated each time we make a complete circuit. These closed curves imply that both populations must follow *periodic cycles*. This is consistent with the time-dependent graphs obtained in Section 5.4.

We now see what we can deduce about the model for general parameter values. We begin by finding a general expression for the equilibrium points. Then we analyse the directions followed by the trajectories in the phase-plane.

Equilibrium populations

As we saw from Figure 6.7, for different initial values we get different phase-plane trajectories enclosed within one another. We can imagine the trajectories shrinking to a single point. This point would correspond to equilibrium populations, where the populations do not change with time. The equilibrium solutions correspond to those values where the rates of change are zero. These are the constant solutions of the differential equations (6.13).

We can find these values analytically, as the following example shows, by solving a pair of simultaneous equations. When solving simultaneous nonlinear equations it is important to ensure you have found all the possible solutions and to this end it is often useful to factorise, as in the following example.

Example 6.9: *Find the equilibrium solutions of the differential equations (6.13).*

Solution: *We set $dX/dt = 0$ and $dY/dt = 0$ in equations (6.13) and obtain the equations*

$$\beta_1 X - c_1 XY = 0, \qquad -\alpha_2 Y + c_2 XY = 0.$$

To find solutions we first write the two simultaneous equations in factored form as

$$X(\beta_1 - c_1 Y) = 0, \tag{6.14}$$
$$Y(-\alpha_2 + c_2 X) = 0. \tag{6.15}$$

From (6.14) there are two possible solutions: $X = 0$ or $\beta_1 - c_1 Y = 0$. We need to look at each case. Recall that the parameters β_1, α_2, c_1 and c_2 are positive (non-zero) constants.

If $X = 0$, then substituting this into (6.15) gives $-\alpha_2 Y = 0$ so that $Y = 0$. This gives one possible solution of both equations at $(X, Y) = (0, 0)$.

Taking the other case, $\beta_1 - c_1 Y = 0$, then $Y = \beta_1/c_1$. Substituting this into (6.15) gives $-\alpha_2 + c_2 X = 0$, for which the only solution is $X = \alpha_2/c_2$. Hence we obtain a second solution to both equations: $(X, Y) = (\alpha_2/c_2, \beta_1/c_1)$.

To ensure we have all the solutions, we should also solve the second equation and substitute the solution into the first, but this yields exactly the same two solutions as we obtained above.

Summarising, we have obtained two equilibrium solutions,

$$(X, Y) = (0, 0) \quad \text{and} \quad (X, Y) = \left(\frac{\alpha_2}{c_2}, \frac{\beta_1}{c_1} \right).$$

Another way of being sure we have all the solutions is to think of the problem of finding the solutions geometrically. Solving the equation $dX/dt = 0$ gave $X = 0$ and $Y = \beta_1/c_1$, which represent the equations of two lines that we denote as L_0 and L_1 in the (X, Y)-plane. Similarly, $dY/dt = 0$ yields two lines, L_2 and L_3, given by $X = \alpha_2/c_2$ and $Y = 0$, respectively. The equilibrium points occur where the pair L_0 and L_1 intersect with L_2 and L_3 (not L_0 with L_1, or L_2 with L_3). These lines (or curves) are known as *nullclines* and their points of intersection are illustrated in Figure 6.8.

We can also use `Maple` or `MATLAB` (with symbolic toolbox) to solve the equations for the equilibrium solutions. The `Maple` code is given in Listing 6.7.

Listing 6.7: Maple code: c_ps_predprey_eqmpts.mpl

```
restart:
eq1 := beta[1]*X(t) - c[1]*X(t)*Y(t);
eq2 := -alpha[2]*Y(t) - c[2]*X(t)*Y(t);
solve({eq1,eq2}, {X(t),Y(t)});
```

Consider the phase-plane trajectories, which are equivalent to the oscillations in the time-dependent diagrams. We may regard the equilibrium populations in this case as the average density of the prey and predator populations, about which they oscillate.

Direction of trajectories

In order to examine the direction vectors in the phase-plane, we first rewrite the original differential equations (6.13) in the factored form

$$\frac{dX}{dt} = X\left(\beta_1 - c_1 Y\right), \qquad \frac{dY}{dt} = Y\left(-\alpha_2 + c_2 X\right). \tag{6.16}$$

We draw the nullcline curves corresponding to $dX/dt = 0$ and $dY/dt = 0$. They divide the phase-plane into regions where the trajectories have different directions.

The nullcline curves here are the lines $X = 0$, $Y = \beta_1/c_1$ (for $dX/dt = 0$) and the lines $Y = 0$, $X = \alpha_1/c_2$ (for $dY/dt = 0$). Effectively, they divide the phase-plane into the four regions labelled I, II, III and IV in Figure 6.8.

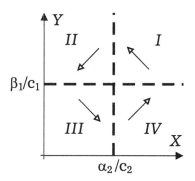

Figure 6.8: Regions in the phase-plane illustrating the trajectory directions. The dashed lines are the non-zero nullclines.

From the differential equations (6.16) we can establish whether $X(t)$ and $Y(t)$ are increasing or decreasing in each of the regions. The following example illustrates how this can be done.

Example 6.10: *Determine the trajectory directions for region II in Figure 6.8.*

Solution: *Everywhere in region II, $X < \alpha_2/c_2$ and $Y > \beta_1/c_1$.*

From (6.16), $dX/dt < 0$ everywhere in region II, since $\beta_1 - c_1 Y < 0$ if $c_1 Y > \beta_1$.

Similarly, $dY/dt < 0$ everywhere in region II, since $-\alpha_2 + c_2 X < 0$ when $c_2 X < \alpha_2$.

Thus both X and Y are decreasing in region II. This gives a vector with negative x-component and negative y-component, as shown in Figure 6.8.

By a similar argument,

- In region I, X is decreasing and Y is increasing.

- In region III, X is increasing and Y is decreasing.
- In region IV, X is increasing and Y is increasing.

This tells us that as we move along a phase-trajectory we must move in an anticlockwise direction. However, it does not tell us whether this is a closed curve or a spiral. In Section 6.6, we use the chain rule to prove that all trajectories for this system are closed.

Furthermore, for the equilibrium point $(0,0)$, we note from Figure 6.8 that the trajectories move towards it in one direction and away from it in another. It is thus an unstable equilibrium point in the sense that trajectories move away from it with time and we show in Section 6.6 that this is always the case, regardless of the parameter values. It is called a saddle point, which we discuss in detail in Chapter 7.

Density-dependent growth

In Chapter 5, the basic predator-prey model was extended to include density-dependent growth (logistic) for the prey species in order to curb the unrealistic exponential growth. It is left as an exercise (Exercise 6.6) to generate the associated phase-plane diagram, and examine how this change in growth alters the equilibrium points and/or the direction vectors of the system.

Summary of skills developed here:
- *Find equilibrium solutions.*
- *Establish and draw the nullclines for a system.*
- *Find directions of trajectories in the phase-plane.*
- *Deduce how populations behave in time from their phase-plane diagram.*

6.5 Analysis of competing species models

We now apply the phase-plane theory to the model we developed to describe the interaction between competing species. Again, we first look at the phase-plane obtained from numerical solutions (using `Maple`) for a specific set of parameter values and then extend the analysis to consider general values of the parameters, finding equilibrium points and trajectory directions.

Review of model

In Section 5.7, we formulated two models for competing species. The first model assumed exponential growth in the absence of a competitor, and later we introduced an improved model that assumed density-dependent growth for both species in the absence of a competitor. We consider the second improved model here, leaving analysis of the first model as an exercise (Exercise 6.9).

From Section 5.7 our model is

$$\frac{dX}{dt} = \beta_1 X - d_1 X^2 - c_1 XY, \qquad \frac{dY}{dt} = \beta_2 Y - d_2 Y^2 - c_2 XY, \qquad (6.17)$$

where β_1 and β_2 are per-capita birth rates or growth rates, d_1 and d_2 are density-dependent coefficients (independent of the other species), and c_1 and c_2 are interaction coefficients (the effect of species interaction on the death rates). In the absence of species Y, species X has logistic growth with carrying capacity $K_1 = \beta_1/d_1$. Similarly, if $X = 0$, then species Y has logistic growth with carrying capacity $K_2 = \beta_2/d_2$.

The phase-plane diagram from Maple

We can use Maple to draw the phase-plane, but only for a given set of parameter values. Take, for example, per-capita birth rates $\beta_1 = 0.22$, $\beta_2 = 0.061$, interspecies interaction parameters $c_1 = 0.053$, $c_2 = 0.0046$, and intraspecies interaction parameters $d_1 = 0.017$, $d_2 = 0.010$, which are those from Gause's experiments with yeast (see Section 5.7 and Edelstein–Keshet (1988)).

The results in Figure 6.9 show clearly that the trajectories approach equilibrium populations on the X and Y axes, for these values of the parameters. See Listing 6.8 and Listing 6.9 for sample Maple and MATLAB code.

Listing 6.8: Maple code: c_ps_competlogistic.mpl

```
restart:with(plots):with(DEtools):
tend:=350:
beta[1]:=0.22: beta[2]:=0.061:
d[1]:=0.017: d[2]:=0.010:
c[1]:=0.053: c[2]:=0.0046:
de1:=diff(X(t),t)=beta[1]*X(t)-c[1]*X(t)*Y(t)-d[1]*X(t)*X(t);
de2:=diff(Y(t),t)=beta[2]*Y(t)-c[2]*X(t)*Y(t)-d[2]*Y(t)*Y(t);
inits:=[0,0.5,0.5],[0,20,4],[0,20,8],[0,20,10];
plot1:=DEplot([de1,de2],[X,Y],t=0..tend,[inits],
      X=0..20,Y=0..10,scene=[X,Y],stepsize=0.05,dirgrid=[10,10],
      linecolour=black,arrows=medium):
display(plot1);
```

Listing 6.9: MATLAB code: c_ps_competlogistic.m

```
function c_ps_compet
global beta1 beta2 c1 c2 d1 d2;

beta1=0.22; beta2=0.06;
c1=0.053; c2=0.0046;
d1=0.017; d2=0.010;
tend = 360; %the end time
u0vec = [0.5, 20, 20, 20; % make matrix of ICs
        0.5,  4,  8, 10];
u0size = size(u0vec); %size of the matrix
numICs = u0size(2); % extract number of ICs

for k = 1:numICs
    u0 = u0vec(:, k); %extract the kth column
    [tsol, usol] = ode45(@rhs, [0, tend], u0);
    Xsol = usol(:, 1); Ysol = usol(:, 2);
    plot(Xsol, Ysol, 'b'); hold on;
end
% makearrows, see Appendix for this function
c_dirplot(@rhs, 0, 20, 0, 10, 10);
axis([0, 22, 0, 10]);

function udot = rhs(t, u)
global beta1 beta2 c1 c2 d1 d2;
X = u(1); Y=u(2);
Xdot = beta1*X - c1*X*Y - d1*X^2;
Ydot = beta2*Y - c2*X*Y - d2*Y^2;
udot = [Xdot; Ydot];
```

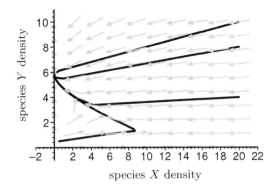

Figure 6.9: `Maple` produced phase-plane diagram for the competing species model including density-dependent growth, with parameter values as given for Gause's experiment.

Note that the trajectory that starts at $(0.5, 0.5)$ moves to the right before turning around and heading towards an equilibrium point at approximately $(0, 6)$. The other trajectories, from a variety of initial conditions, tend to the same equilibrium point. This is consistent with the time-dependent graph in Section 5.7 where one species died out and the other tended to an equilibrium point.

Equilibrium populations

As in Section 6.1 and Section 6.4, we find the equilibrium points for arbitrary parameter values by setting the rates of change to 0. This is done in the following example.

Example 6.11: *Find all the equilibrium points for the system of differential equations (6.17).*

Solution: *We need to solve (simultaneously) $dX/dt = 0$ and $dY/dt = 0$. This implies (after factorising)*

$$X(\beta_1 - c_1 Y - d_1 X) = 0, \tag{6.18}$$

$$Y(\beta_2 - c_2 X - d_2 Y) = 0. \tag{6.19}$$

From (6.18) we have two possibilities: $X = 0$ or $\beta_1 - c_1 Y - d_1 X = 0$. If $X = 0$, then (6.19) simplifies to $Y(\beta_2 - d_2 Y)$, which has two possible solutions: $Y = 0$ or $Y = \beta_2/d_2$. Thus we obtain two solutions to the equations, $(X, Y) = (0, 0)$ and $(0, \beta_2/d_2)$.

From (6.19) we have two possibilities: $Y = 0$ or $\beta_2 - c_2 X - d_2 Y = 0$. If $Y = 0$, then (6.18) simplifies to $X(\beta_1 - d_1 X) = 0$. This has two solutions $X = 0$ and $X = \beta_1/d_1$. Thus, simultaneous solutions (X, Y) of both equations are $(0, 0)$ (already obtained above) and $(\beta_1/d_1, 0)$.

One further possibility not yet considered is where (6.18) is satisfied by letting $\beta_1 - d_1 X - c_1 Y = 0$ and where (6.19) is satisfied by letting $\beta_2 - c_2 X - d_2 Y = 0$. By substituting one equation into the other it can be shown that these two equations are satisfied when

$$X = \frac{c_1 \beta_2 - d_2 \beta_1}{c_1 c_2 - d_1 d_2}, \qquad Y = \frac{c_2 \beta_1 - d_1 \beta_2}{c_1 c_2 - d_1 d_2}.$$

Summarising, we have found (at most) four equilibrium solutions,

$$(0, 0), \quad \left(0, \frac{\beta_2}{d_2}\right), \quad \left(\frac{\beta_1}{d_1}, 0\right), \quad \left(\frac{c_1 \beta_2 - d_2 \beta_1}{c_1 c_2 - d_1 d_2}, \frac{c_2 \beta_1 - d_1 \beta_2}{c_1 c_2 - d_1 d_2}\right),$$

where the fourth equilibrium point is relevant only if both components are positive. Two of the equilibrium points suggest the extinction of one species and the survival and stabilisation of the other.

We can also obtain the equilibrium points using `Maple`. The code for this is given below in Listing 6.10. Similar code for calculating the equilibrium points using `MATLAB` (with symbolic toolbox) is given in Listing 6.11.

Listing 6.10: Maple code: c_ps_compete_eqm.mpl

```
restart:
eq1 := beta[1]*X(t) - c[1]*X(t)*Y(t) -d[1]*X(t)*X(t);
eq2 := beta[2]*Y(t) - c[2]*X(t)*Y(t) -d[2]*Y(t)*Y(t);
solve( {eq1,eq2}, {X(t),Y(t)} );
```

Listing 6.11: MATLAB code: c_ps_compete_eqm.m

```
de1 = 'beta*X - c1*X*Y - d1*X^2= 0';
de2 = 'c2*X*Y - alpha*Y -d2*Y^2= 0';
soln = solve(de1, de2, 'X', 'Y');
disp(soln.X); disp(soln.Y);
```

As before, define the nullcline lines L_0, L_1, L_2, L_3 as

$$L_0: \quad X = 0, \qquad L_1: \quad \beta_1 - d_1X - c_1Y = 0,$$
$$L_3: \quad Y = 0. \qquad L_2: \quad \beta_2 - c_2X - d_2Y = 0,$$

There is an equilibrium point corresponding to the intersection of the pair of lines L_0 and L_1 with the pair of lines L_2 and L_3 (but not L_0 with L_1 or L_2 with L_3). This is shown in Figure 6.10. The diagrams confirm that there are four equilibrium points if the lines L_1 and L_2 intersect in the positive quadrant of the (X, Y)-plane and only three equilibrium points otherwise.

Direction of trajectories

As usual, we are interested in the predictions of the model over time. We establish the trajectory behaviour by determining trajectory directions and considering only positive values of X and Y. The linear nullclines L_0, L_1, L_2 and L_3 divide the phase-plane (with X and Y both positive) into different regions in four different ways. These four cases are illustrated in Figure 6.10.

We determine the directions of the trajectories (seen in Figure 6.10) by looking at a typical point in each of the regions bounded by the nullclines. The example below illustrates this process for one of the cases in Figure 6.10.

Example 6.12: *Find the directions of the trajectories in each of the regions bounded by the nullclines for Case 3 of Figure 6.10.*

Solution: *We examine each of the four regions of Case 3 separately. The differential equations (6.17) are written as*

$$X' = X(\beta_1 - d_1X - c_1Y), \qquad Y' = Y(\beta_2 - d_2Y - c_2X),$$

where X' denotes dX/dt and Y' denotes dY/dt.

Consider the region below both L_1 and L_2. From the equations for L_1 and L_2 ($\beta_1 - d_1X - c_1Y = 0$ and $\beta_2 - d_2Y - c_2X = 0$) a point in the region below L_1 has $Y < (\beta_1 - d_1X)/c_1$ and a point in the region below L_2 has $Y < (\beta_2 - c_2X)/d_2$. Thus,

$$c_1Y < \beta_1 - d_1X \quad \Rightarrow \quad 0 < \beta_1 - d_1X - c_1Y \quad \Rightarrow \quad X' > 0,$$
$$d_2Y < \beta_2 - c_2X \quad \Rightarrow \quad 0 < \beta_2 - c_2X - d_2Y \quad \Rightarrow \quad Y' > 0.$$

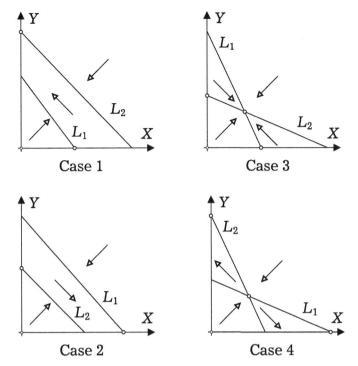

Figure 6.10: The four cases of the phase-plane diagram for competing species including density-dependent growth. The direction vectors for the trajectories are illustrated with equilibrium points marked with open circles.

Consider the region below L_1 and above L_2. Here,

$$c_1 Y < \beta_1 - d_1 X \quad \Rightarrow \quad 0 < \beta_1 - d_1 X - c_1 Y \quad \Rightarrow \quad X' > 0,$$
$$d_2 Y > \beta_2 - c_2 X \quad \Rightarrow \quad 0 > \beta_2 - c_2 X - d_2 Y \quad \Rightarrow \quad Y' < 0.$$

Similarly, for the region above L_1 and below L_2,

$$c_1 Y > \beta_1 - d_1 X \quad \Rightarrow \quad 0 > \beta_1 - d_1 X - c_1 Y \quad \Rightarrow \quad X' < 0,$$
$$d_2 Y < \beta_2 - c_2 X \quad \Rightarrow \quad 0 < \beta_2 - c_2 X - d_2 Y \quad \Rightarrow \quad Y' > 0.$$

Finally, for the region above both L_1 and L_2,

$$c_1 Y > \beta_1 - d_1 X \quad \Rightarrow \quad 0 > \beta_1 - d_1 X - c_1 Y \quad \Rightarrow \quad X' < 0,$$
$$d_2 Y > \beta_2 - c_2 X \quad \Rightarrow \quad 0 > \beta_2 - c_2 X - d_2 Y \quad \Rightarrow \quad Y' < 0.$$

This provides all the information required to establish the direction vectors of Case 3.

In Figure 6.10 we show the directions for each of the four different ways of sectioning the phase-plane with nullclines. In Case 1, the trajectories all move towards the equilibrium point on the Y-axis. This corresponds to extinction of the X-species, in the long term. In Case 2, the trajectories all move towards the equilibrium point on the X-axis. This corresponds to extinction of the Y-species, in the long term. In Case 3, the trajectories move to the equilibrium point where the nullclines L_1 and L_2 intersect. This corresponds to a coexistence of the two species, in the long term. In Case 4, the trajectories may end up at either of the equilibrium points on the axes, but not at the equilibrium point where the nullclines L_1 and L_2 intersect. Which equilibrium point is approached, or which species survives, will depend on the initial conditions.

We can use the differences between the intercepts on the X and Y axes to determine which of the above cases provides the solution. Let

$$\text{Int}_X = \frac{\beta_1}{d_1} - \frac{\beta_2}{c_2} \quad \text{and} \quad \text{Int}_Y = \frac{\beta_2}{d_2} - \frac{\beta_1}{c_1},$$

the signs of which establish the equilibrium point towards which the trajectories move. The results are displayed in Table 6.1. (In Chapter 7, we develop further theory to establish which of these equilibrium points will attract or repel the trajectories.)

Table 6.1: Predicting the equilibrium point from the sign of Int_X and Int_Y.

Int_X	Int_Y	Equilibrium point approached
$+$	$-$	$\left(\dfrac{\beta_1}{d_1}, 0\right)$
$-$	$+$	$\left(0, \dfrac{\beta_2}{d_2}\right)$
$-$	$-$	$\left(\dfrac{c_1\beta_2 - d_2\beta_1}{c_1c_2 - d_1d_2}, \dfrac{c_2\beta_1 - d_1\beta_2}{c_1c_2 - d_1d_2}\right)$
$+$	$+$	$\left(\dfrac{\beta_1}{d_1}, 0\right)$ or $\left(0, \dfrac{\beta_2}{d_2}\right)$ depending on starting conditions

Interpretation of parameters

Above we established that β_1/d_1 was the carrying capacity K_1 for species X and that β_2/d_2 was the carrying capacity K_2 for population Y. Thus we can interpret the trajectory approach to $(\beta_1/d_1, 0) = (K_1, 0)$ as the survival of X and the extinction (or migration) of Y. Likewise, an approach to $(0, \beta_2/d_2) = (0, K_2)$ can be seen as the survival of Y and the extinction of X.

Initially, when our model allowed for only exponential growth (that is, infinite carrying capacity in the absence of the other species), it can be shown that there is an equilibrium point at $(\beta_2/c_2, \beta_1/c_1)$, the only point for which both species can possibly exist simultaneously. Now Int_X is defined as the difference between the X coordinate β_2/c_2 and the carrying capacity of the species, as was introduced with logistic growth. (Similarly for Int_Y.) Thus Int_X and Int_Y can be considered as *interference factors*, or a measure of how much one species interferes with the other.

Discussion

Our competition model predicts, in almost all cases, the extinction of one species and the survival of the other. However, observations indicate that there is substantial diversity in many ecosystems with competing species coexisting. As we have discussed previously, these ecosystems do not exist in isolation but are integral components of some much larger system.

In Chapter 8, we present a system that uses the competition model developed here, and embeds it within another system, a predator-prey model, establishing that in this more complicated system the two competing species can coexist. This lends credence to the earlier assertion of larger systems having greater stability than isolated ecosystems.

> **Summary of skills developed here:**
> - *Reinforcement of methods for finding equilibrium points by solving simultaneous equations.*
> - *Reinforcement of methods for finding direction vectors for trajectories and establishing the general behaviour of a system from the phase-plane.*

6.6 Closed trajectories for the predator-prey

We return to the analysis of the basic predator-prey model, the Lotka–Volterra system, introduced in Section 6.4. We show by applying the chain rule that, for all parameter values in this model, the phase-plane curves are always closed curves. This allows us to infer that, according to this model, both populations always oscillate over time.

Review of model

The Lotka–Volterra predator-prey equations are

$$\frac{dX}{dt} = \beta_1 X - c_1 XY, \qquad \frac{dY}{dt} = -\alpha_2 Y + c_2 XY, \tag{6.20}$$

where the parameters β_1, α_2, c_1 and c_2 are all positive constants. We also found in Section 6.4 that the equilibrium points were given by

$$(X, Y) = (0, 0) \quad \text{and} \quad (X, Y) = \left(\frac{\alpha_2}{c_2}, \frac{\beta_1}{c_1} \right).$$

Furthermore, we deduced the directions of the phase-plane trajectories from which we could infer that the trajectories were either closed curves or spirals, traversed in an anti-clockwise direction. We show, by applying the chain rule to find a relation between X and Y, that we can prove there is never a spiral, regardless of the parameters we may choose.

Applying the chain rule

We use the chain rule to eliminate time from the differential equations, thus obtaining a differential equation relating the predator population Y to the prey population X.

Example 6.13: Use the chain rule to find an equation relating X and Y.

Solution: By the chain rule,

$$\frac{dY}{dX} = \frac{dY/dt}{dX/dt}.$$

Hence, using (6.20) *we obtain the first-order differential equation*

$$\frac{dY}{dX} = \frac{Y(-\alpha_2 + c_2 X)}{X(\beta_1 - c_1 Y)}. \tag{6.21}$$

This differential equation is separable. Separating the variables and integrating both sides with respect to X, we obtain

$$\int \left(\frac{\beta_1}{Y} - c_1 \right) dY = \int \left(-\frac{\alpha_2}{X} + c_2 \right) dX.$$

Carrying out both integrations yields

$$\beta_1 \ell n(Y) - c_1 Y = -\alpha_2 \ell n(X) + c_2 X + K, \qquad (6.22)$$

where K is an arbitrary constant of integration.

Proving all trajectories are closed

From the direction vector diagram of the earlier section (Figure 6.8) it is clear that the trajectories must be closed curves or spirals. We now provide an argument that the phase-plane trajectories are closed curves for all positive values of the constant parameters.

To get further information, we return to formula (6.22). Suppose we hold the variable X constant, $X = X_1$. Then we have the equation

$$\ell n(Y) = \frac{c_1}{\beta_1} Y + K_1 \qquad (6.23)$$

where K_1 is the constant defined by $K_1 = (-\alpha_2 \ell n(X_1) + c_2 X_1 + K)/\beta_1$. If the phase-plane curve were a closed curve, it could have, at most, two intersection points with any vertical line, in particular $X = X_1$. On the other hand, if it were a spiral, it would have an infinite number of intersection points. By sketching the curves $Z = \ell n(Y)$ and $Z = (c_1/\beta_1)Y + K_1$ we see there are, at most, two solutions. This is illustrated in Figure 6.11.

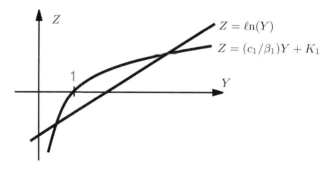

Figure 6.11: A general sketch of the intersection between the two curves $Z = \ell n(Y)$ and $Z = (c_1/\beta_1)Y + K_1$ showing, at most, two intersection points.

Returning to the phase-plane

We have concluded that all the phase-plane curves are closed curves. Using this, together with Figure 6.8 and the fact that X and Y are positive, we can sketch the general form of a typical phase-plane trajectory. Recall, from Section 6.4, that the trajectory is traversed in an anticlockwise direction. This is illustrated in Figure 6.12.

All the phase-plane curves are traversed in the anticlockwise direction along closed trajectories enclosing an equilibrium point, which in this case represents average population densities for the two populations. We note that as we trace out the trajectory, the prey (X) reach their maximum population before the predator population (Y) reaches its maximum

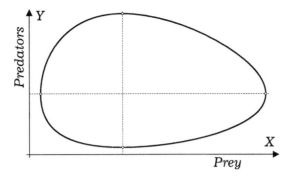

Figure 6.12: Phase-plane diagram showing a typical trajectory for the predator-prey problem enclosing the equilibrium point.

value. Thus the predator oscillation cycle always lags behind the prey oscillation, no matter what parameter values or initial conditions we use.

A feature of this model is that if you suddenly perturb the populations, they begin to move on a different phase-plane trajectory. This is one undesirable feature of the Lotka–Volterra equations from a biological perspective. Biologists would prefer a more realistic model where the populations tend to return to the original phase-plane curve after they have been perturbed slightly. This can be achieved using a modified version of the Lotka–Volterra equations, which we consider in Chapter 8.

Summary of skills developed here:

- *Use the chain rule to prove that the trajectories of the the Lotka–Volterra equations are always closed.*
- *Consider the points of intersection between two simple graphs to infer information about solutions to a more complicated equation.*

6.7 Case Study: Bacteria battle in the gut

Simple models can often be applied to complex systems and successfully predict phenomena that have been observed. In the following case study we model, using the mechanisms discussed in this book, the interaction between different strains of *Escherichia coli* in the gut of animals. With this simple approach we show that a gut with a slow turnover rate favours different strains from one with a fast turnover rate, where these rates of turnover relate to diet. Adapted from Barnes et al. (2007).

Colicins are a class of protein antibiotics known as bacteriocins that are produced by certain strains of Escherichia coli (E. coli). They are produced through a process called cell lysis, where a single cell of bacteria produces many colicins. The advantage of producing colicins is to kill off other competing strains of E. coli in the same environment, although in doing so the colicin itself is destroyed. Thus colicins are important in mediating intraspecific interactions.

E. coli are found in the gut of animals, and colicin production varies markedly between populations. The process of cell lysis has been studied in detail by Levin (1988), Frank (1994), Gordon and Riley (1999), Kerr et al. (2002) and Kirkup and Riley (2004); however, in this case study we investigate a different phenomenon. We examine whether the turnover rate in the gastro-intestinal tract determines which strains of E. coli dominate when colicin producing strains interact with colicin sensitive strains. It has been observed that in hosts with fast gut turnover rates, such as carnivores, non-colicin producing strains dominate, while in hosts with slow turnover rates, such as herbivores, colicin producing strains dominate.

To examine the dynamics, we consider a gastro-intestinal tract of fixed volume V and consider the interaction between two strains of E. coli: one a colicin producing strain x with density X in the gut, and one a colicin sensitive strain y with density Y. We assume a constant flow rate of food F into and from the gut, and the densities with which the strains enter with the flow are X_{in} and Y_{in}. Within the gut we assume exponential growth for each strain with growth rates of β_x and β_y, which is a reasonable assumption during the initial stages of the dynamics. It should be noted that the qualitative results were unchanged when logistic growth was considered (see Exercise 6.22).

We now introduce the process of cell lysis. We assume that cells of strain x lyse at the per-capita rate α_x, with each lysed cell producing 10^6 colicin molecules, and each molecule capable of killing a cell of the opposing strain, destroying itself in the process. A model for this process is

$$\begin{aligned}
\frac{dX}{dt} &= \beta_x X + (X_{in} - X)\frac{F}{V} - \alpha_x X, \\
\frac{dY}{dt} &= \beta_y Y + (Y_{in} - Y)\frac{F}{V} - c_2 XY.
\end{aligned} \tag{6.24}$$

Note that the last term of dY/dt, $c_2 XY$ is a mortality rate of Y cells due to the rate of contact between the two different strains of cells. From Gordon and Riley (1999), the probability of an encounter of a colicin cell with a sensitive cell is of the order 10^{-11}. We can therefore determine the form of the $c_2 XY$ terms and estimate the c_2 parameter. We can calculate the rate of of Y cells dying from a colicin molecule produced by all the lysed X cells as

$$\begin{Bmatrix} \text{rate of} \\ Y \text{ cells} \\ \text{dying} \end{Bmatrix} = \begin{Bmatrix} \text{rate of} \\ X \text{ cells} \\ \text{lysed} \end{Bmatrix} \times \begin{Bmatrix} \text{no. of colcin} \\ \text{molecules produced} \\ \text{per } X \text{ cell} \end{Bmatrix} \times \begin{Bmatrix} \text{probability of} \\ \text{contact} \end{Bmatrix} \times \begin{Bmatrix} \text{no. of} \\ Y \text{ cells} \end{Bmatrix}$$

$$\simeq (\alpha_x X) \times 10^6 \times 10^{-11} \times Y$$

$$= 10^{-5} \alpha_X XY.$$

Since typical cell densities are of the order 10^6, we rescale both X and Y in the equations (6.24) to be in units of 10^6 cells. This now gives $c_2 \simeq 10\alpha_x$ (i.e., a factor of 10^6 cancels out from all terms on both sides of each equation except for the XY term).

Our aim is to investigate whether the mean residence time $\tau = V/F$ (where turnover rate is $F/V = 1/\tau$) determines which strain dominates. To do so we examine the nullclines, by setting $dX/dt = 0$ and $dY/dt = 0$, and consider any stable equilibrium points in the positive phase-plane. Our results are illustrated in Figure 6.13, where the stable node is plotted against F (recalling that F/V is turnover rate that increases with F). We note that the 'change' between the dominance of colicin producing cells (X) and non-colicin producing cells (Y), according to this model, occurs for $F \approx 6$. It is evident that for low turnover rates (herbivores), the colicin producing strain dominates, while for high turnover rates (carnivores), the non-colicin producing strain dominates. This is in agreement with *observations. Thus, in spite of the relative simplicity of this model, the results seem to encapsulate the observed phenomena.*

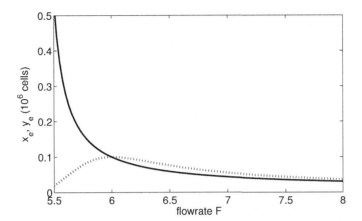

Figure 6.13: The equilibrium values for X (solid line) and Y (dashed line) are plotted for increasing values of F, and hence increasing turnover rate. (Parameter values were $\beta_x = \beta_y = 0.3$, $X_{\text{in}} = Y_{\text{in}} = 0.01 \times 10^6$, $V = 20$ units and $\alpha_x = 0.03$.)

We note that in reality bacteria enter at variable and random rates, not a constant rate as applied here. Barnes et al. (2007) considered this scenario using random values for X_{in} and Y_{in}. They found that the qualitative results were unchanged in that, over hundreds of simulations, colicin producing strains dominated in the majority of cases for low turnover rates, and vice versa.

Interest in such bacteria and their dynamics continues to increase since they have the potential for use as biocontrol agents for the management of fungal and bacterial plant pathogens, and also as the active agent in probiotic formulations. Probiotic therapy is a disease prevention strategy used in humans and domestic animals, as well as being a means of enhancing growth in livestock. The ultimate aim is to ensure the presence of 'good' bacteria in the gut that knock out bacterial pathogens. However, effective use of bacteria will require a sound understanding of microbial ecology and, from the results presented here, diet has a significant impact.

6.8 Exercises for Chapter 6

6.1. Simple example. *Find the only equilibrium point for the system*

$$x' = x - 5y, \qquad y' = x - y.$$

Then use `Maple` *or* `MATLAB` *to numerically plot phase-plane trajectories for this system, showing the trajectories are closed and enclose this equilibrium point.*

6.2. Finding equilibrium points. For the following, find all the equilibrium solutions.

(a) $\dfrac{dX}{dt} = 3X - 2XY, \qquad \dfrac{dY}{dt} = XY - Y,$

(b) $\dfrac{dX}{dt} = 2X - XY, \qquad \dfrac{dY}{dt} = Y - XY,$

(c) $\dfrac{dX}{dt} = Y - 2XY, \qquad \dfrac{dY}{dt} = XY - Y^2.$

6.3. Using the chain rule. For the differential equations

$$\frac{dX}{dt} = -XY, \qquad \frac{dY}{dt} = -2Y,$$

use the chain rule to find a relation between Y and X.

6.4. Contagious for life. Consider a disease where all those infected remain contagious for life. A model describing this is given by the differential equations

$$\frac{dS}{dt} = -\beta SI, \qquad \frac{dI}{dt} = \beta SI,$$

where β is a positive constant.

(a) Use the chain rule to find a relation between S and I.

(b) Obtain and sketch the phase-plane curves. Determine the direction of travel along the trajectories.

(c) Using this model, is it possible for all the susceptibles to be infected?

6.5. Disease with reinfection. A model for the spread of a disease, where all infectives recover from the disease and become susceptibles again, is given by the pair of differential equations

$$\frac{dS}{dt} = -\beta SI + \gamma I, \qquad \frac{dI}{dt} = \beta SI - \gamma I,$$

where β and γ are positive constants, $S(t)$ denotes the number of susceptibles and $I(t)$ denotes the number of infectives at time t.

(a) Use the chain rule to find a relationship between the number of susceptibles and the number of infectives given the initial number of susceptibles is s_0 and there was initially only one infective.

(b) Draw a sketch of typical phase-plane trajectories. Deduce the direction of travel along the trajectories, providing reasons.

(c) Using the phase-plane diagram, describe how the number of infectives changes with time.

6.6. Predator-prey with density-dependent growth of prey. Consider the system

$$\frac{dX}{dt} = \beta_1 X \left(1 - \frac{X}{K}\right) - c_1 XY, \qquad \frac{dY}{dt} = c_2 XY - \alpha_2 Y,$$

for the dynamics of a predator-prey model, with density-dependent growth of the prey, and all parameters positive constants.

(a) Find all the equilibrium points. How do they differ from those of the standard Lotka–Volterra system in Section 6.4?

(b) Use `Maple` or `MATLAB` to explore the phase-plane trajectories for different parameter values and initial conditions.

6.7. One prey and two predators. *A model of a three species interaction with two predators that compete for a single prey food-source is*

$$\frac{dX}{dt} = a_1 X - b_1 XY - c_1 XZ, \quad \frac{dY}{dt} = a_2 XY - b_2 Y, \quad \frac{dZ}{dt} = a_3 XZ - b_3 Z,$$

where a_i, b_i, c_i, for $i = 1, 2, 3$, are all positive constants. Here $X(t)$ is the prey density and $Y(t)$ and $Z(t)$ are the two predator densities.

(a) *Find all possible equilibrium populations. Is it possible for all three populations to coexist in equilibrium?*

(b) *What does this suggest about introducing an additional predator into an ecosystem?*

6.8. Predator-prey with DDT. *The predator-prey equations with additional deaths by DDT are*

$$\frac{dX}{dt} = \beta_1 X - c_1 XY - p_1 X, \quad \frac{dY}{dt} = -\alpha_2 Y + c_2 XY - p_2 Y,$$

where all parameters are positive constants.

(a) *Find all the equilibrium points.*

(b) *What effect does the DDT have on the non-zero equilibrium populations compared with the case when there is no DDT? Compare this with Figure 5.9 in Section 5.4, which was for a specific set of parameter values. What general conclusions can you draw?*

(c) *Show that the predator fraction of the total average population is given by*

$$f = \frac{1}{1 + \left(\frac{c_1}{c_2} \frac{(\alpha_2 + p_2)}{(\beta_1 - p_1)} \right)}.$$

What happens to this proportion f as the DDT kill rates, p_1 and p_2, increase?

6.9. Competing species without density dependence. *For the competing species model developed in Section 5.7, exhibiting exponential growth in the absence of a competitor, find (a) the equilibrium points and trajectory directions in different parts of the (X, Y)-plane. (b) Hence sketch the phase-plane diagram.*

(Note that this diagram is consistent with the numerical results generated in Section 5.7, and illustrated in Figure 5.14.)

6.10. Competing species with density dependence. *Consider the competition population model with density-dependent growth*

$$\frac{dX}{dt} = X(\beta_1 - c_1 Y - d_1 X), \quad \frac{dY}{dt} = Y(\beta_2 - c_2 X - d_2 Y).$$

Find all four equilibrium points for the system (either using `Maple` *or* `MATLAB`, *or, if you like algebra, by hand).*

6.11. Predator-prey with density dependence and DDT. *For the predator-prey model with density dependence for the prey and DDT acting on both species,*

$$\frac{dX}{dt} = \beta_1 X \left(1 - \frac{X}{K} \right) - c_1 XY - p_1 X,$$

$$\frac{dY}{dt} = c_2 XY - \alpha_2 Y - p_2 Y,$$

show that (X, Y) is one equilibrium point, where

$$X = \frac{\alpha_2 + p_2}{c_2}, \quad Y = \frac{\beta_1 \left(1 - \frac{X_e}{K} \right) - p_1}{c_1},$$

and determine if there are any other equilibrium points.

6.12. Control of stray dogs. *Read the case study from the previous chapter (Section 5.6) for a model that compares control strategies for stray dogs.*

(a) *Show how to obtain system (5.22) starting from equations (5.18) and (5.19).*

(b) *Find the equilibrium points for system (5.22). (These solutions should agree, with substitution, with those in Exercise 5.18 in Chapter 5.)*

6.13. Rabbits and foxes. *A population of sterile rabbits $X(t)$ is preyed upon by a population of foxes $Y(t)$. A model for this population interaction is the pair of differential equations*

$$\frac{dX}{dt} = -aXY, \qquad \frac{dY}{dt} = bXY - cY,$$

where a, b and c are positive constants.

(a) *Use the chain rule to obtain a relationship between the density of foxes and the density of rabbits.*

(b) *Sketch typical phase-plane trajectories, indicating the direction of movement along the trajectories.*

(c) *According to the model, is it possible for the foxes to completely wipe out the rabbit population? Give reasons.*

6.14. Microorganisms and toxins. *The pair of differential equations*

$$\frac{dP}{dt} = rP - \gamma PT, \qquad \frac{dT}{dt} = qP,$$

where r, γ and q are positive constants, is a model for a population of microorganisms P, which produces toxins T that kill the microorganisms.

(a) *Given that initially there are no toxins and p_0 microorganisms, obtain an expression relating the population density and the amount of toxins. (Hint: Use the chain rule.)*

(b) *Hence, give a sketch of a typical phase-plane trajectory. Using this, describe what happens to the microorganisms over time.*

6.15. Exact solution for battle model. *Consider the aimed fire battle model developed in the text:*

$$\frac{dR}{dt} = -a_1 B, \qquad \frac{dB}{dt} = -a_2 R.$$

The exact solution can be found using theoretical techniques as follows:

(a) *Take the derivative of the first equation to get a second-order differential equation, and then eliminate dB/dt from this equation by substituting the second equation (given above) into this second-order equation.*

(b) *Now assume the solution to be an exponential of the form $e^{\lambda t}$. Substitute it into the second-order equation and solve for the two possible values of λ. The general solution for R will be of the form*

$$R(t) = c_1 e^{\lambda_1} + c_2 e^{\lambda_2},$$

where c_1 and c_2 are the arbitrary constants of integration. The solution for B is then found using the equation $dR/dt = -a_1 B$.

Write the solutions in terms of hyperbolic functions cosh and sinh (this makes it more convenient to solve for the arbitrary constants).

(c) *Now find the arbitrary constants by applying the initial conditions $R(0) = r_0$ and $B(0) = b_0$, when $t = 0$.*

(d) Using `Maple` or `MATLAB` (with symbolic toolbox), check the solution above. Note: The software may give the solution in terms of exponential functions, in which case you will need to convert to check.

(Note: Further details about methods for solving second-order differential equations, in particular for differential equations with constant coefficients, as used here, can be found in Appendix A.5.)

6.16. Battle model with desertion. The following battle model represents two armies where both are exposed to aimed fire, and for one of the armies (red) there is significant loss due to desertion (at a constant rate c). The numbers of soldiers, R and B, satisfy the differential equations

$$\frac{dR}{dt} = -a_1 B - c, \qquad \frac{dB}{dt} = -a_2 R,$$

where a_1, a_2 and c are positive constants.

(a) If the initial number of red soldiers is r_0 and the initial number of blue soldiers is b_0, use the chain rule to find a relationship between B and R.

(b) For $a_1 = a_2 = c = 0.01$, give a sketch of typical phase-plane trajectories and deduce the direction of travel along the trajectories.

6.17. Jungle warfare. A simple mathematical model describing jungle warfare, with one army exposed to random fire and the other to aimed fire, is given by the coupled differential equations

$$\frac{dR}{dt} = -c_1 RB, \qquad \frac{dB}{dt} = -a_2 R,$$

where c_1 and a_2 are positive constants.

(a) Use the chain rule to find a relation between R and B, given initial numbers of soldiers r_0 and b_0. Hence sketch some typical phase-plane trajectories. Give directions of travel along the trajectories, providing reasons for your choice.

(b) Given that, initially, both the red and blue armies have 1,000 soldiers, and the constants c_1 and a_2 are 10^{-4} and 10^{-1}, respectively, determine how many soldiers are left if the battle is fought so that all the soldiers of one army are killed.

(c) In this model, one of the armies is hidden whereas the other is visible to their enemy. Which is the hidden army? Give reasons for your answer.

6.18. Battle with long-range weapons. In a long-range battle, neither army can see the other soldiers, but fires into a known area. A simple mathematical model describing this battle is given by the coupled differential equations

$$\frac{dR}{dt} = -c_1 RB, \qquad \frac{dB}{dt} = -c_2 RB,$$

where c_1 and c_2 are positive constants.

(a) Use the chain rule to find a relationship between R and B, given the initial numbers of soldiers for the two armies are r_0 and b_0, respectively.

(b) Draw a sketch of typical phase-plane trajectories.

(c) Explain how to estimate the parameter c_1 given that the blue army fires into a region of area A.

6.19. Fatal disease. The following model is for a fatal disease, where all infectives die from the disease with death rate α, and where the transmission is assumed to be frequency dependent so the contact rate is constant. The differential equations are

$$\frac{dS}{dt} = -pcS\frac{I}{N}, \qquad \frac{dI}{dt} = pcS\frac{I}{N} - \alpha I,$$

where $N = S + I$ is the total population size and c is the (constant) per-capita contact rate and p is the constant probability of a contact resulting in an infection.

(a) Write the system in terms of the variables N and I (i.e., obtain a differential equation for N) and hence show that

$$\frac{dI}{dN} = \frac{pc}{\alpha}\frac{I}{N} + \left(1 - \frac{pc}{\alpha}\right).$$

(b) Solve the differential equation in (a) to obtain

$$I = N + KN^{pc/\alpha},$$

where K is the arbitrary constant of integration.

(c) Discuss any qualitative differences between this model and the one studied in Section 6.2. (Hint: Substitute $S = N - I$ for a meaningful comparison, and graph I against S.)

6.20. SIR model, estimating the transmission coefficient. Adapted from Brauer and Castillo-Chàvez (2001).

For the standard SIR epidemic model,

$$S' = -\beta SI, \qquad I' = \beta SI - \gamma I.$$

From the equation for the phase-plane solution,

$$I = -S + \frac{\gamma}{\beta}\ell n(S) + K.$$

(a) If s_f denotes the remaining number of susceptibles when there are no remaining infectives, show that

$$\frac{\beta}{\gamma} = \frac{\ell n(s_0/S_f)}{S_0 + i_0 - s_f},$$

where s_0 and i_0 are the initial numbers of susceptibles and infectives, respectively.

(b) A study at Yale university in 1982 described an influenza epidemic with initial proportions of susceptibles of the student population as 91.1% and final proportion of susceptibles as 51.3%. (Assume, initially, that no one had recovered).

Given that the mean infectious period for influenza γ^{-1} is approximately 3 days, estimate the combination βN, where N is the total population size, and hence estimate R_0, the basic reproduction number.

6.21. Endemic disease model. Consider the endemic infectious disease model from Section 5.2, with natural per-capita birth rate equal to natural per-capita death rate, $b = a$,

$$\frac{dS}{dt} = aN - \beta SI - aS,$$

$$\frac{dI}{dt} = \beta SI - \gamma I - aI,$$

$$\frac{dR}{dt} = \gamma I - aR,$$

where $N(t) = S(t) + I(t) + R(t)$.

(a) Show there are only two equilibrium points, one corresponding to a disease-free equilibrium (where $I = 0$) and one corresponding to an endemic equilibrium (where $I > 0$).

(b) Write the endemic equilibrium solution in terms of $R_0 = \beta N/(\gamma + a)$, the basic reproduction number for this system. Discuss the cases $R_0 > 1$ and $R_0 < 1$.

6.22. Bacteria in the gut. *Read over the case study on bacteria in the gut in Section 6.7.*

Using `Maple` *or* `MATLAB`, *or by hand, determine the equilibrium points of the system (6.24),*

$$\frac{dX}{dt} = \beta_x X + (X_{\text{in}} - X)\frac{F}{V} - \alpha_x X$$
$$\frac{dY}{dt} = \beta_y Y + (Y_{\text{in}} - Y)\frac{F}{V} - c_2 XY.$$

$$(6.25)$$

Chapter 7

Linearisation analysis

In this chapter we develop some powerful theory, which often allows us to predict the dynamics of a system in general terms. It provides the means by which we can establish the phase-plane behaviour of a system and predict the outcome for any possible parameter combination. The theory is developed for both linear and nonlinear systems, and is applied to some of the nonlinear systems studied in Chapters 5 and 6.

7.1 Introduction

From the small sample of examples we have examined in the previous chapters, we have seen that systems of equations can result in many different types of behaviour. Depending on the initial conditions or the chosen parameter values, the outcome may be stable or unstable, cyclic or divergent.

In this chapter we develop some powerful theory that allows us to predict the dynamical behaviour of a system. In the first case, we consider only linear systems of equations. However, since many of the interacting population models we have met are nonlinear (and in fact most natural systems are nonlinear), we then show how this theory for the linear case can be extended to nonlinear systems as well. In the final section we apply the process to the nonlinear models of population interactions, which we developed in Chapters 5 and 6, and extend in Chapter 8.

7.2 Linear theory

So far, in our phase-plane analysis of systems of equations, we have encountered a variety of behaviours of trajectories near equilibrium points. For example, we saw trajectories approaching some of these points, trajectories being repelled by others, as well as spiralling trajectories and closed loops.

In the following theory we develop techniques that allow us to predict, for each equilibrium point, the behaviour of the trajectories close to that point. From this we can establish a complete picture of the system phase-plane. Initially, we consider the linear case (a pair of coupled linear equations in two unknowns) and then show how this can be extended to the nonlinear case. (While we restrict our analysis to two equations in two unknowns, the theory is applicable to larger systems with many unknowns.)

The general linear system

We start by considering the following general form of a pair of coupled linear equations:

$$X' = a_1 X + b_1 Y,$$
$$Y' = a_2 X + b_2 Y,$$

where differentiation is with respect to time t (i.e., $X' = dX/dt$, $Y' = dY/dt$) and a_1, a_2, b_1 and b_2 are constant.

We denote an equilibrium point (critical point or steady state) for the system by (x_e, y_e). Thus $a_1 x_e + b_1 y_e = 0$ and $a_2 x_e + b_2 y_e = 0$.

Linear algebra notation

The system above can be written in terms of matrices and vectors in the following way. Let

$$\boldsymbol{x}(t) = \begin{bmatrix} X(t) \\ Y(t) \end{bmatrix} \quad \text{and} \quad \boldsymbol{x}'(t) = \begin{bmatrix} X'(t) \\ Y'(t) \end{bmatrix},$$

where $\boldsymbol{x}(t)$ is a vector function of time. Let

$$\mathbf{A} = \begin{bmatrix} a_{11} & a_{12} \\ a_{21} & a_{22} \end{bmatrix},$$

where $\mathbf{A}$ is a matrix. Then the above system of two equations can be written as

$$\boldsymbol{x}' = \mathbf{A}\boldsymbol{x}.$$

This means that

$$\begin{bmatrix} X' \\ Y' \end{bmatrix} = \begin{bmatrix} a_{11} & a_{12} \\ a_{21} & a_{22} \end{bmatrix} \begin{bmatrix} X \\ Y \end{bmatrix} = \begin{bmatrix} a_{11}X + a_{12}Y \\ a_{21}X + a_{22}Y \end{bmatrix} \tag{7.1}$$

using normal matrix multiplication.

What needs to be understood in general is the effect of multiplying a vector by a matrix $\mathbf{A}$. We look at an example to illustrate this.

Example 7.1:   Carry out the multiplication $\mathbf{A}\boldsymbol{x}_1$ and $\mathbf{A}\boldsymbol{x}_2$, where

$$\mathbf{A} = \begin{bmatrix} 3 & -2 \\ 1 & 0 \end{bmatrix}, \qquad \boldsymbol{x}_1 = \begin{bmatrix} -1 \\ 1 \end{bmatrix}, \qquad \boldsymbol{x}_2 = \begin{bmatrix} 2 \\ 1 \end{bmatrix}.$$

Solution:

$$\mathbf{A}\boldsymbol{x}_1 = \begin{bmatrix} -3 - 2 \\ -1 \end{bmatrix} = \begin{bmatrix} -5 \\ -1 \end{bmatrix}$$

and

$$\mathbf{A}\boldsymbol{x}_2 = \begin{bmatrix} 6 - 2 \\ 2 \end{bmatrix} = \begin{bmatrix} 4 \\ 2 \end{bmatrix} = 2 \begin{bmatrix} 2 \\ 1 \end{bmatrix}.$$

These results are illustrated in Figure 7.1. So multiplication by a matrix $\mathbf{A}$ maps a vector onto another vector. In the case of $\boldsymbol{x}_2$, we have that

$$\mathbf{A}\boldsymbol{x}_2 = 2\boldsymbol{x}_2,$$

so that the effect of multiplying by A is the same as multiplying by a scalar or number (which is 2 in this case). We use this notion of the 'equivalence' of multiplication by a matrix and a scalar (number) in the process of finding eigenvalues and eigenvectors: the latter will be the vectors for which there is a nontrivial solution to the equation, and the former will be the associated scalars. These values turn out to be essential in predicting the behaviour of trajectories in the phase-plane associated with the system. For further details of matrix algebra, see Appendix B.1.

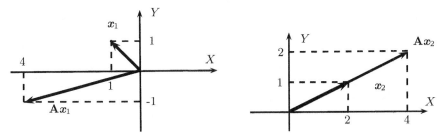

Figure 7.1: The effect of multiplying vectors x_1 and x_2 by a matrix $\mathbf{A}$ as defined in Example 7.1.

Outline of method to solve equations

In order to solve a general system of equations, as in (7.1), we can transform the equations onto a different system of axes. In choosing the new axes carefully, the equations transform to simple differential equations of exponential growth or decay, and thus are simple to solve. Finally, these solutions can be transformed back to the original system of axes to provide a solution in the required form.

In order to transform a general pair of differential equations, which may be hard to solve, into a system that is easy to solve, we find and use the eigenvalues and eigenvectors. (These are described fully in Appendix B.1.) Essentially, *eigenvectors* x are the non-zero solutions of the matrix equation

$$\mathbf{A}x = \lambda x,$$

where the *eigenvalues* λ are the values for which these non-zero solutions exist.

Rewriting the equation as $Ax - \lambda x = 0$ and then expanding using matrix multiplication, we obtain

$$\mathbf{A}x - \lambda x = (\mathbf{A} - \lambda \mathbf{I})x = \begin{bmatrix} a_1 - \lambda & b_1 \\ a_2 & b_2 - \lambda \end{bmatrix} x,$$

where $\mathbf{I}$ is the identity matrix

$$\mathbf{I} = \begin{bmatrix} 1 & 0 \\ 0 & 1 \end{bmatrix}.$$

From the theory of linear algebra (see Appendix B.1) it follows that for non-zero solutions of this equation to exist, the determinant of the expression must be zero, and thus we get the *characteristic equation* of matrix $\mathbf{A}$

$$|\mathbf{A} - \lambda \mathbf{I}| = 0 \quad \Rightarrow \quad \lambda^2 - \lambda(a_1 + b_2) + (a_1 b_2 - a_2 b_1) = 0.$$

This equation is central to the theory that is developed here. Note that the coefficient of λ is the sum of the diagonal elements of the matrix $\mathbf{A}$, namely the *trace* of $\mathbf{A}$, and also that the last term is the *determinant* of matrix $\mathbf{A}$. We use these values extensively when applying the theory that we now develop.

We can establish the eigenvalues as the solutions to this characteristic equation. (We could also then solve for the associated pair of eigenvectors from the vector equation, but do not need these values in the applications.) From here the trajectory behaviour can be determined since, as will become apparent, it is dependent entirely on the eigenvalues.

Establishing the trajectory behaviour

We return to the general pair of linear first-order equations:

$$X' = a_1 X + b_1 Y,$$
$$Y' = a_2 X + b_2 Y,$$

which has an equilibrium point at the origin, $(x_e, y_e) = (0, 0)$. In vector notation,

$$x' = \mathbf{A}x.$$

Suppose we have found the eigenvalues, λ_1 and λ_2, as well as the associated eigenvectors for A, namely

$$u = \begin{bmatrix} u_1 \\ u_2 \end{bmatrix} \quad \text{and} \quad v = \begin{bmatrix} v_1 \\ v_2 \end{bmatrix}.$$

We define $\mathbf{U}$ to be the matrix whose columns are the eigenvectors; thus

$$\mathbf{U} = \begin{bmatrix} u & v \end{bmatrix} = \begin{bmatrix} u_1 & v_1 \\ u_2 & v_2 \end{bmatrix}.$$

From the definition of eigenvectors and eigenvalues, we have

$$\mathbf{A}u = \lambda_1 u \quad \text{and} \quad \mathbf{A}v = \lambda_2 v,$$

which implies that

$$\mathbf{A}\begin{bmatrix} u & v \end{bmatrix} = \begin{bmatrix} \lambda_1 u & \lambda_2 v \end{bmatrix} = \begin{bmatrix} u & v \end{bmatrix} \begin{bmatrix} \lambda_1 & 0 \\ 0 & \lambda_2 \end{bmatrix} \quad \text{or} \quad \mathbf{AU} = \mathbf{UD}$$

with

$$\mathbf{D} = \begin{bmatrix} \lambda_1 & 0 \\ 0 & \lambda_2 \end{bmatrix}.$$

Assuming that U is invertible, we can write

$$\mathbf{U}^{-1}\mathbf{AU} = \mathbf{D}. \tag{7.2}$$

We use this equation below.

First we express x as a linear combination of the eigenvectors and, assuming this is possible, we have

$$x = z_1 u + z_2 v.$$

Letting

$$z = \begin{bmatrix} z_1 \\ z_2 \end{bmatrix}, \quad \text{then} \quad x = \mathbf{U}z.$$

Since X and Y are functions of time, and the eigenvectors are not (since $\mathbf{A}$ is not a function of time), thus z_1 and z_2 must also be functions of time. We now establish two expressions for x':

$$x = \mathbf{U}z \quad \text{so} \quad x' = \mathbf{U}z',$$

and also

$$x' = \mathbf{A}x \quad \text{so} \quad x' = \mathbf{AU}z.$$

Equating these two expressions for x' and then using (7.2) gives

$$\mathbf{U}z' = \mathbf{AU}z,$$

and then

$$z' = \mathbf{U}^{-1}\mathbf{AU}z$$
$$= \mathbf{D}z.$$

We are now in a position to solve the differential equations easily. Expanding $z' = \mathbf{D}z$,

$$z_1' = \lambda_1 z_1,$$
$$z_2' = \lambda_2 z_2,$$

we obtain two equations that are easy to solve. They are the equations for exponential growth and decay with which, by now, we are familiar. We have as solutions $z_1 = k_1 e^{\lambda_1 t}$ and $z_2 = k_2 e^{\lambda_2 t}$, where k_1 and k_2 are arbitrary constants.

Using these we can find solutions for X and Y by retracing our steps through this process and carrying the solutions with us. We have

$$x = k_1 e^{\lambda_1 t} u + k_2 e^{\lambda_2 t} v$$
$$= e^{\lambda_1 t} \hat{u} + e^{\lambda_2 t} \hat{v},$$

where $\hat{u} = k_1 u$ and $\hat{v} = k_2 v$ are two eigenvectors (as any scalar multiple of an eigenvector is again an eigenvector) and so

$$X = e^{\lambda_1 t} \hat{u}_1 + e^{\lambda_2 t} \hat{v}_1,$$
$$Y = e^{\lambda_1 t} \hat{u}_2 + e^{\lambda_2 t} \hat{v}_2.$$

Geometric interpretation
What have we done?

- We start with a set of axes, X and Y.

- We find eigenvectors u and v, which give us the directions of a new system of axes z_1 and z_2 along which the effect of multiplying by $\mathbf{A}$ is the same as multiplying by a scalar λ. (It is clear from the diagram that any scalar multiple of the eigenvectors will suffice as an eigenvector.)

- We consider the plane described by the axes z_1 and z_2 and examine the behaviour of the solutions (which are now easy to find) $z_1 = k_1 e^{\lambda_1 t}$ and $z_2 = k_2 e^{\lambda_2 t}$ in this phase-plane. How the trajectories behave depends entirely on λ_1 and λ_2, the eigenvalues. Suppose they are as illustrated in the left-hand diagram of Figure 7.2.

- Next we return to the (X, Y) phase-plane with the transformation of this solution, which is distorted (stretched or contracted) in some way since we have changed axes again: note that it retains the main features, or dynamics, of the simpler system.

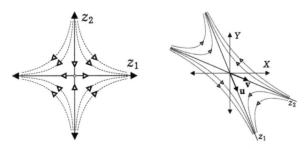

Figure 7.2: Sample saddle solution in the eigenvector phase-plane (left figure) and translated to the original phase-plane (right figure).

Equilibrium point classifications

For the systems described above, we had the origin $(0,0)$ as the equilibrium (or critical) point. What we have found, using the techniques of eigenvalues and eigenvectors, is the behaviour of the trajectories in the phase-plane close to this point. The behaviour depends on the eigenvalues (λ_1 and λ_2) since the trajectories can be described by $z_1 = k_1 e^{\lambda_1 t}$ and $z_2 = k_2 e^{\lambda_2 t}$. Clearly different values of λ_1 and λ_2 may result in very different behaviours, and thus each case is dealt with separately below.

The following summarises the relationship between the eigenvalues and the forms of the trajectories:

- Case $\lambda_1 < 0$ and $\lambda_2 < 0$ (eigenvalues real and negative): We have

$$\lim_{t \to \infty} k_1 e^{\lambda_1 t} = 0 \quad \text{and} \quad \lim_{t \to \infty} k_2 e^{\lambda_2 t} = 0,$$

  and thus all trajectories approach the equilibrium point at the origin. Such a point is called a *stable node* and is illustrated in Figure 7.3.

- Case $\lambda_1 > 0$ and $\lambda_2 > 0$ (eigenvalues real and positive): We have both z_1 and z_2 approaching ∞ (diverging) as t increases and thus all trajectories diverge from the equilibrium point. Such a point is called an *unstable node* (see Figure 7.3).

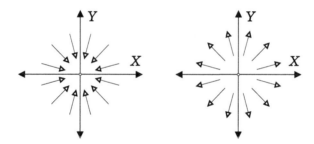

Figure 7.3: Trajectory behaviour close to a stable node (left) and an unstable node (right).

- Case $\lambda_1 > 0$ and $\lambda_2 < 0$ (eigenvalues real and of different sign): We have that $z_1 = k_1 e^{\lambda_1 t}$ and $z_2 = k_1 e^{\lambda_2 t}$, so $z_2 \to 0$ and $z_1 \to \infty$ as time increases. The trajectories approach zero along one axis and approach ∞ along the other. Such a point is called a *saddle* or an unstable saddle point and is illustrated in Figure 7.4.

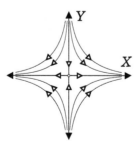

Figure 7.4: Trajectory behaviour close to an unstable saddle point.

- Case $\lambda_1 = \alpha + i\beta$ and $\lambda_2 = \alpha - i\beta$ (complex conjugate eigenvalues with $\alpha \neq 0$ and $\beta \neq 0$): In this case, the solutions can be written in the form $z_1 = e^{\alpha t} \cos \beta t$, $z_2 = e^{\alpha t} \sin \beta t$ and the trajectories spiral around the equilibrium point. If $\alpha < 0$, then they spiral inwards towards the equilibrium point. Such a point is called a *stable focus*. If $\alpha > 0$, then they spiral outwards and away from the equilibrium point. Such a point is called an *unstable focus*. A stable and unstable focus are illustrated in Figure 7.5.

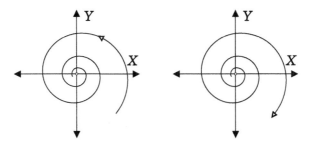

Figure 7.5: Trajectory behaviour close to a stable focus (left) and an unstable focus (right).

- Case λ_1 and λ_2 purely imaginary. In this case, the solutions can be written in the form $z_1 = \cos \beta t$ and $z_2 = \sin \beta t$ and the trajectories form closed loops enclosing the equilibrium point. Such a point is called a *centre* and the solutions are called periodic. A centre is illustrated in Figure 7.6.

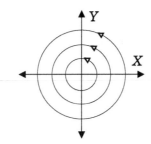

Figure 7.6: Trajectory behaviour close to a centre.

We have already come across examples of most of these types of equilibrium points in Chapter 6, where we considered the behaviour of the trajectories in the phase-plane for some basic population models. The power of the above results is that, having located an equilibrium point, we now have the means to predict its type, once we have found the associated eigenvalues.

Summary

- We start with the characteristic equation

$$\lambda^2 - \lambda(a_1 + b_2) + (a_1 b_2 - a_2 b_1) = 0$$

and solve for λ. This provides us with the eigenvalues.

- Solving the quadratic characteristic equation gives

$$\lambda_1 = \frac{1}{2}p + \frac{1}{2}\sqrt{\Delta}, \qquad \lambda_2 = \frac{1}{2}p - \frac{1}{2}\sqrt{\Delta},$$

where $p = a_1 + b_2$ is the trace of matrix $\mathbf{A}$, $q = a_1 b_2 - a_2 b_1$ is the determinant of A and $\Delta = p^2 - 4q$ the discriminant of the characteristic equation. The different possible classifications of the equilibrium points are given in Table 7.1.

- This can all be displayed in a diagram, Figure 7.7, that illustrates the general classifications.

Table 7.1: Table showing different classifications of equilibrium points

Δ	p	q	Equilibrium point
$\Delta > 0$	$p < 0$		$\Rightarrow$ stable node
$\Delta > 0$	$p > 0$		$\Rightarrow$ unstable node
		$q < 0$	$\Rightarrow$ saddle point
$\Delta < 0$	$p < 0$		$\Rightarrow$ stable spiral
$\Delta < 0$	$p = 0$		$\Rightarrow$ centre
$\Delta < 0$	$p > 0$		$\Rightarrow$ unstable spiral

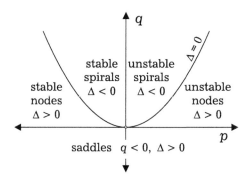

Figure 7.7: General classification diagram for equilibrium points using p, q and Δ from the characteristic equation.

Discussion

Note that we have not included the cases where $\Delta = 0$ or $q = 0$. When $\Delta = 0$, the roots are equal and the equilibrium point is a stable/unstable inflected node. Such equilibrium points are called degenerate. When $q = 0$ the system 'equilibrium point' consists of a line and the trajectories are parallel. These special cases are illustrated in Figure 7.8.

Stable nodes and spirals are known as attractors, while unstable nodes and spirals are known as repellers. It is important to note that changes in the parameters a_1, a_2, b_1, b_2 can result in very different dynamics. Thus, it is possible to choose the dynamics through controlled variation of the parameters.

We have considered only the case of an equilibrium point at the origin, that is, $(x_e, y_e) = (0, 0)$, but in general this is not so. Suppose we have a system as before, but with $(x_e, y_e) \neq$

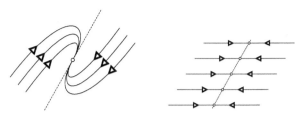

Figure 7.8: Trajectory behaviour when $\Delta = 0$ (left) and $q = 0$ (right), illustrated for the stable case when $p < 0$.

$(0, 0)$,

$$X' = a_1 X + b_1 Y + c_1,$$
$$Y' = a_2 X + b_2 Y + c_2.$$

Let $\xi = X - x_e$ and $\eta = Y - y_e$ (so that $X = x_e$ implies that $\xi = 0$ and $Y = y_e$ implies that $\eta = 0$). Then,

$$\begin{aligned}
\xi' = X' &= a_1(\xi + x_e) + b_1(\eta + y_e) + c_1 \\
&= (a_1 x_e + b_1 y_e + c_1) + a_1 \xi + b_1 \eta \\
&= a_1 \xi + b_1 \eta
\end{aligned}$$

since $a_1 x_e + b_1 y_e + c_1 = 0$. Similarly,

$$\eta' = Y' = a_2(\xi + x_e) + b_2(\eta + y_e) + c_2 = a_2 \xi + b_2 \eta.$$

This process is called a change of variable and allows us to transform the original system in X and Y with an equilibrium at (x_e, y_e), to a system with variables ξ and η that has its equilibrium point at the origin. We can now apply the above theory to this system, which is equivalent to the original system of equations.

Another assumption we made was that we could express $\boldsymbol{x}$ as a linear combination of the eigenvectors and the new axes z_1 and z_2. This is a result of the linear algebra presented in Appendix B.1.

Furthermore, we assumed that $\mathbf{U} = \begin{bmatrix} \boldsymbol{u} & \boldsymbol{v} \end{bmatrix}$ was invertible, that is, there exists a matrix U^{-1} such that $\mathbf{U}^{-1}\mathbf{U} = \mathbf{I} = \mathbf{U}\mathbf{U}^{-1}$. This follows as a consequence of Theorem 5 in Appendix B.1.

An alternative approach

The above approach requires some understanding of linear algebra theory, and this can be avoided if we make the assumption that the solution is an exponential of the form $ke^{\lambda t}$, with λ possibly imaginary. In this case we need to write the coupled linear system

$$X' = a_1 X + b_1 Y,$$
$$Y' = a_2 X + b_2 Y,$$

as a single second-order equation:

$$X'' - (a_1 + b_2)X' + (a_1 b_2 - a_2 b_1)X = 0.$$

(This can be done by differentiating the first equation and then eliminating Y using substitution from the second equation. For further details, see Appendix A.5.)

Assuming that there is a solution of the form $X = ke^{\lambda t}$, we can calculate X'' and X' and substitute them into the second-order equation to give

$$k\lambda^2 e^{\lambda t} - (a_1 + b_2)k\lambda e^{\lambda t} + (a_1 b_2 - a_2 b_1)ke^{\lambda t} = 0,$$

and then

$$\lambda^2 - (a_1 + b_2)\lambda + (a_1 b_2 - a_2 b_1) = 0.$$

Note that this is the characteristic equation once again. The solutions of this equation are the eigenvalues (λ_1 and λ_2) and from here the general solution is a linear combination of $e^{\lambda_1 t}$ and $e^{\lambda_2 t}$, that is,

$$X(t) = C_1 e^{\lambda_1 t} + C_2 e^{\lambda_2 t}.$$

We are now in a position to classify the equilibrium points, as before.

Summary of skills developed here:

- *Write a system of equations in vector notation, or convert a vector equation into a system of equations.*
- *Find the trace and determinant of a matrix.*
- *Calculate the eigenvalues of a matrix.*
- *From the eigenvalues of a matrix, and for a given equilibrium point, classify the equilibrium point and sketch the trajectory behaviour in the phase-plane close to this point.*

7.3   Applications of linear theory

What we are most interested in is how this theory can be applied to the models we have been studying. However, to start with, we apply the linear theory developed above to the example of a simple harmonic oscillator. This consists of a mass attached to a spring, which oscillates back and forth. (For further details see, for example, Fulford et al. (1997).) One example is a regularly vibrating (oscillating) object such as an idealised skateboard on a ramp, equipped with perfect bearings to ensure that no friction damps the motion. Such a skateboard would continue oscillating indefinitely between the edges of the ramp.

Simple harmonic motion

The following example examines an equation for simple harmonic motion.

Example 7.2: *Find all equilibrium points associated with the system*

$$X' = Y \qquad Y' = -\omega^2 X \tag{7.3}$$

and determine their classification(s).

Solution: *In matrix form, the system is*

$$\boldsymbol{x}' = \boldsymbol{A}\boldsymbol{x} \qquad \text{where} \qquad \boldsymbol{A} = \begin{bmatrix} 0 & 1 \\ -\omega^2 & 0 \end{bmatrix}.$$

The only equilibrium point is $(x_e, y_e) = (0, 0)$.

To classify this point we need to examine the characteristic equation

$$\lambda^2 - \lambda \operatorname{trace}(\boldsymbol{A}) + \det(\boldsymbol{A}) = 0, \qquad \text{whence} \quad \lambda^2 - \lambda \times 0 + \omega^2 = 0.$$

Solving this equation yields imaginary eigenvalues

$$\lambda = \pm i\omega.$$

Now, with the notation as in Section 7.2,

$$p = \operatorname{trace}(\boldsymbol{A}) = 0,$$

$$\Delta = p^2 - 4q = 0 - 4\omega^2 < 0.$$

This implies that the equilibrium point at $(0, 0)$ *is a centre, and all solutions are periodic, as could be expected.*

By eliminating the variable Y, the system (7.3) can be written as a single second-order differential equation $X'' = -\omega^2 X$. This is the well-known simple harmonic oscillator, and periodic solutions are exactly what we would expect.

We will now include damping, that is, some friction. In this case our idealised skateboard will eventually come to rest if the skateboarder does no work to keep it going. We consider what the changes will be. The differential equation is

$$X'' = -cX' - \omega^2 X$$

where c is a positive constant. Setting $Y = X'$, this can be written as the system

$$X' = Y, \qquad Y' = -cY - \omega^2 X. \tag{7.4}$$

The term $-cY$ describes the damping. In the following example we classify the equilibrium point.

Example 7.3: *Find and classify the equilibrium point(s) of the system (7.4).*

Solution: *The system can be written in matrix form as*

$$\boldsymbol{x}' = \boldsymbol{A}\boldsymbol{x} \quad \text{where} \quad \boldsymbol{A} = \begin{bmatrix} 0 & 1 \\ -\omega^2 & -c \end{bmatrix}.$$

We still have $(0, 0)$ *as the only equilibrium point.*

For its classification we consider the characteristic equation

$$\lambda^2 - \lambda \operatorname{trace}(\boldsymbol{A}) + \det(\boldsymbol{A}) = 0 \quad \text{or} \quad \lambda^2 - \lambda(-c) + \omega^2 = 0.$$

Hence,

$$\lambda = \frac{-c}{2} \pm \frac{1}{2}\sqrt{c^2 - 4\omega^2}.$$

Again, with the notation of Section 7.2, we have $p = \operatorname{tr}\boldsymbol{A} = -c < 0$, *as* $c > 0$, *and* $\Delta = c^2 - 4\omega^2$. *There are three cases to consider:*

- *If* $\Delta > 0$ *(i.e.,* $c > 2\omega$*), then* $(0, 0)$ *is a stable node. This case is known as strong damping.*
- *If* $\Delta < 0$ *(i.e.,* $c < 2\omega$*), then this implies that* $(0, 0)$ *is a stable focus. This case is known as weak damping.*
- *If* $\Delta = 0$ *(i.e.,* $c = 2\omega$*), then* $(0, 0)$ *is a stable inflected node. This case is known as critical damping.*

Battle model

In Section 5.9 we introduced a simple model of a battle between two armies using aimed fire against each other. The resulting coupled system of differential equations for the numbers of soldiers of the two armies were the linear Lanchester equations

$$\frac{dR}{dt} = -a_1 B, \qquad \frac{dB}{dt} = -a_2 R, \tag{7.5}$$

where the attrition coefficients a_1 and a_2 are positive constants.

Example 7.4: *Classify the equilibrium points for the system (7.5).*

Solution: *The system (7.5) can be written in matrix form as*

$$x' = Ax, \qquad A = \begin{bmatrix} 0 & -a_1 \\ -a_2 & 0 \end{bmatrix},$$

and the only equilibrium point is $(0,0)$.

The characteristic equation is

$$\lambda^2 - \mathrm{trace}(A) + \det(A) = 0 \quad or \quad \lambda^2 + (-a_1 a_2) = 0.$$

With the notation of Section 7.2,

$$q = \det(A) < 0,$$

which implies the equilibrium point is a saddle point. This is consistent with what we found in Section 6.3.

7.4   Nonlinear theory

In Chapters 5 and 6, we examined coupled systems of nonlinear equations to describe interacting populations. We found equilibrium points and, at times, it was possible to establish the long-term behaviour of the system using graphical techniques such as the phase-plane and time-dependent diagrams. However, it would be extremely useful if we could extend the system of classification of the equilibrium points for linear systems, to nonlinear systems as well.

We can do this by approximating the nonlinear system with a linear system close to the equilibrium point: this process is called *linearisation*. We then examine the dynamics of the linearised system close to an equilibrium point, and use the classifications from linear theory to predict the dynamics in the nonlinear system. It can be shown that in all classification cases (except that of a centre) the predictions of linear theory extend to the nonlinear system. (For proof see, for example, Hurewicz (1990).)

Linearisation

Our first task is to establish a linear approximation to the nonlinear equations. Consider a general system of two nonlinear differential equations

$$\frac{dX}{dt} = F(X, Y),$$

$$\frac{dY}{dt} = G(X, Y).$$

Let (x_e, y_e) be any equilibrium point for this system, not necessarily at $(0,0)$, and then

$$F(x_e, y_e) = 0 \quad \text{and} \quad G(x_e, y_e) = 0.$$

Consider solutions close to the steady-state (equilibrium) solutions

$$X(t) = x_e + \xi(t), \qquad Y(t) = y_e + \eta(t),$$

where $\xi(t)$ and $\eta(t)$ are small and approach zero when X and Y approach the equilibrium point. (ξ and η are called *perturbations* of the steady state.)

We now change the variables in the system from X and Y to ξ and η. Then,

$$\frac{d(x_e + \xi)}{dt} = F(x_e + \xi, y_e + \eta),$$

$$\frac{d(y_e + \eta)}{dt} = G(x_e + \xi, y_e + \eta),$$

where ξ and η are functions of t. But we have, since x_e and y_e are constant,

$$\frac{dX}{dt} = \frac{d(x_e + \xi)}{dt} = \frac{d\xi}{dt},$$

$$\frac{dY}{dt} = \frac{d(y_e + \eta)}{dt} = \frac{d\eta}{dt}.$$

So,

$$\frac{dX}{dt} = \frac{d\xi}{dt} = F(x_e + \xi, y_e + \eta),$$

$$\frac{dY}{dt} = \frac{d\eta}{dt} = G(x_e + \xi, y_e + \eta).$$

We now apply the Taylor series expansion in two variables to expand $F(x_e + \xi, y_e + \eta)$ and $G(x_e + \xi, y_e + \eta)$ and then take a linear approximation for each: a *tangent-plane approximation* for the system. (For details of partial derivatives and the Taylor series expansion, refer to Appendix B.2.) Applying the Taylor series expansion in two variables we find

$$\xi' = F(x_e, y_e) + F_\xi(x_e, y_e)\, \xi + F_\eta(x_e, y_e)\, \eta$$
$$+ \text{(terms of higher order)},$$

$$\eta' = G(x_e, y_e) + G_\xi(x_e, y_e)\, \xi + G_\eta(x_e, y_e)\eta$$
$$+ \text{(terms of higher order)},$$

where $F_\xi = \partial F / \partial \xi$, $F_\eta = \partial F / \partial \eta$ and likewise for G. Now $F(x_e, y_e) = 0$, since (x_e, y_e) is an equilibrium point, and similarly $G(x_e, y_e) = 0$. Then, taking the linear approximation of each Taylor series expansion (i.e., ignoring all terms of higher order) we are left with the tangent-plane approximation

$$\xi' = F_\xi(x_e, y_e)\, \xi + F_\eta(x_e, y_e)\, \eta,$$
$$\eta' = G_\xi(x_e, y_e)\, \xi + G_\eta(x_e, y_e)\, \eta.$$

In vector notation, the linearised system is

$$\begin{bmatrix} \xi' \\ \eta' \end{bmatrix} = \begin{bmatrix} F_\xi & F_\eta \\ G_\xi & G_\eta \end{bmatrix} \begin{bmatrix} \xi \\ \eta \end{bmatrix}. \tag{7.6}$$

Equilibrium point classification

Recall that we can predict the behaviour (the dynamics) of a linear system $\boldsymbol{x}' = A\boldsymbol{x}$ merely by considering the eigenvalues of A. Further, these eigenvalues are the solutions to the characteristic equation $\lambda^2 - \lambda(\text{trace}\mathbf{A}) + |\mathbf{A}| = 0$. Clearly, from (7.6)

$$\text{trace}(\mathbf{A}) = F_\xi + G_\eta,$$
$$|\mathbf{A}| = \det(A) = F_\xi\, G_\eta - F_\eta\, G_\xi,$$

and λ_1, λ_2 can be found for any equilibrium point.

Recall that ξ and η were not variables of the original equation. However, $X = x_e + \xi$ and $Y = y_e + \eta$ so that

$$\frac{\partial F}{\partial \xi} = \frac{\partial F}{\partial X}\frac{dX}{d\xi} = \frac{\partial F}{\partial X}\frac{d(x_e + \xi)}{d\xi} = \frac{\partial F}{\partial X},$$

and similarly

$$\frac{\partial F}{\partial \eta} = \frac{\partial F}{\partial Y}, \qquad \frac{\partial G}{\partial \xi} = \frac{\partial G}{\partial X} \qquad \text{and} \qquad \frac{\partial G}{\partial \eta} = \frac{\partial G}{\partial Y}.$$

This means that we have

$$\begin{bmatrix} F_\xi & F_\eta \\ G_\xi & G_\eta \end{bmatrix} = \begin{bmatrix} F_X & F_Y \\ G_X & G_Y \end{bmatrix}$$

and in vector notation the linearised system is

$$\begin{bmatrix} \xi' \\ \eta' \end{bmatrix} = \begin{bmatrix} F_\xi & F_\eta \\ G_\xi & G_\eta \end{bmatrix}\begin{bmatrix} \xi \\ \eta \end{bmatrix} \qquad \text{or} \qquad \begin{bmatrix} X' \\ Y' \end{bmatrix} = \begin{bmatrix} F_X & F_Y \\ G_X & G_Y \end{bmatrix}\begin{bmatrix} X - x_e \\ Y - y_e \end{bmatrix}. \qquad (7.7)$$

This is in the form $\boldsymbol{x}' = \mathbf{J}\boldsymbol{x}$. Here $\mathbf{J}$ is the matrix of first-order partial derivatives and is called the *Jacobian* matrix of the system above. Clearly, from this Jacobian matrix, which follows directly from the original system, the characteristic equation and eigenvalues can be calculated immediately.

Summary

In summary, if we start with a general nonlinear system

$$\frac{dX}{dt} = F(X, Y),$$
$$\frac{dY}{dt} = G(X, Y),$$

with any equilibrium point (x_e, y_e) we can establish the dynamics of the system close to this point in the following way:

- First, linearise the system to get

$$\begin{bmatrix} \xi' \\ \eta' \end{bmatrix} = \begin{bmatrix} F_X & F_Y \\ G_X & G_Y \end{bmatrix}\begin{bmatrix} \xi \\ \eta \end{bmatrix}.$$

- Evaluate the Jacobian matrix at the equilibrium point of interest. Each equilibrium point will produce different values of the Jacobian matrix.

- Then apply the linear theory of classification to this system.

It can be proved (see, for example, Hurewicz (1990)) that, with the exception of the prediction of a centre, the classification established for the linearised system can be extended to the nonlinear system. Furthermore, if an equilibrium point is a centre in the linearised system, then it is either a focus or a centre in the nonlinear system. In this case it is necessary to consider the nonlinear terms to establish the exact dynamics. However, such analysis is beyond the scope of this book and we refer the interested reader to, for example, Hurewicz (1990).

Summary of skills developed here:

- *Find the Jacobian matrix.*
- *Linearise a system of nonlinear equations and derive the linearised system.*
- *From the eigenvalues of the Jacobian matrix, classify any equilibrium point and sketch the trajectory behaviour in the phase-plane close to this point.*

7.5 Applications of nonlinear theory

We now apply this linearisation technique to some of the models of population dynamics we developed in Chapter 5.

Predator-prey model

Recall the model of the predator-prey interaction from Chapter 5:

$$X' = \beta_1 X - c_1 XY,$$
$$Y' = -\alpha_2 Y + c_2 XY, \tag{7.8}$$

for which we found two equilibrium points $(0,0)$ and $(\alpha_2/c_2, \beta_1/c_1)$ (see Section 6.4).

Example 7.5: *Find the linearised system for model (7.8), and hence classify all equilibrium points of the basic predator-prey model.*

Solution: *From (7.8) we label* $F(X,Y) = \beta_1 X - c_1 XY$ *and* $G(X,Y) = -\alpha_2 Y + c_2 XY$. *The Jacobian matrix is*

$$\boldsymbol{J} = \begin{bmatrix} F_X & F_Y \\ G_X & G_Y \end{bmatrix} = \begin{bmatrix} \beta_1 - c_1 Y & -c_1 X \\ c_2 Y & -\alpha_2 + c_2 X \end{bmatrix}.$$

There were two equilibrium points for this system, and we consider each in turn.

For the case where the equilibrium point is $(x_e, y_e) = (0,0)$, *we have that*

$$\boldsymbol{J} = \begin{bmatrix} \beta_1 & 0 \\ 0 & -\alpha_2 \end{bmatrix}.$$

The trace is then $\beta_1 - \alpha_2$ *and the determinant is given by* $\det(J) = -\beta_1\alpha_2$. *Using the notation of Section 7.2, the characteristic equation follows as* $\lambda^2 - (\beta_1 - \alpha_2)\lambda + (-\beta_1\alpha_2) = 0$ *and we have* $q = -\beta_1\alpha_2 < 0$, *which implies that the equilibrium point is a saddle point.*

In the second case where the equilibrium point is

$$(x_e, y_e) = \left(\frac{\alpha_2}{c_2}, \frac{\beta_1}{c_1} \right),$$

we have that

$$\boldsymbol{J} = \begin{bmatrix} 0 & -\dfrac{c_1\alpha_2}{c_2} \\ \dfrac{\beta_1 c_2}{c_1} & 0 \end{bmatrix}.$$

The trace of this matrix is 0 and its determinant is given by $\det(J) = \beta_1 \alpha_2$. As above, with the notation of Section 7.2, the characteristic equation in this case is $\lambda^2 - 0\lambda + \beta_1 \alpha_2 = 0$ and we have $p = 0$ and $\Delta < 0$, which implies that this equilibrium point is a centre.

The results from this example are those we might have expected from the numerical solutions obtained earlier using `Maple`. However, the latter critical point is predicted as a centre and, as we have stated, linear theory is not adequate in this case where it is necessary to consider the nonlinear terms to establish the true dynamics. On the other hand, we were able, for this example, to solve for the phase-trajectories exactly in Section 6.6. The solution obtained there shows that the critical point is indeed a centre.

Epidemic model

Recall the basic model for the spread of a disease in a population from Section 5.2. The differential equations were

$$\frac{dS}{dt} = -\beta SI, \qquad \frac{dI}{dt} = \beta SI - \gamma I. \qquad (7.9)$$

Here $S(t)$ describes the susceptible population and $I(t)$ describes contagious infectives, as functions of time. Parameter β is the disease transmission rate (transmission coefficient) and γ is the removal rate.

Example 7.6:   Classify the equilibrium points of the epidemic model (7.9).

Solution: We established in Section 6.2 that (7.9) has an infinite set of equilibrium points, $(S, I) = (s_e, 0)$, where s_e is an arbitrary positive real number.

From (7.9) we label $F(S, I) = -\beta SI$ and $G(S, I) = \beta SI - \gamma I$. The Jacobian matrix is

$$\mathbf{J} - \begin{bmatrix} \dfrac{\partial F}{\partial S} & \dfrac{\partial F}{\partial I} \\ \dfrac{\partial G}{\partial S} & \dfrac{\partial G}{\partial I} \end{bmatrix} - \begin{bmatrix} -\beta I & -\beta S \\ \beta I & \beta S - \gamma \end{bmatrix}.$$

At any given equilibrium point, $(S, I) = (s_e, 0)$, we have the linearised system

$$\begin{bmatrix} S' \\ I' \end{bmatrix} = \begin{bmatrix} 0 & -\beta s_e \\ 0 & \beta s_e - \gamma \end{bmatrix} \begin{bmatrix} S - s_e \\ I - 0 \end{bmatrix}.$$

At these equilibrium points the trace $\text{trace}(\mathbf{J}) = \beta s_e - \gamma$ is negative if $s_e < \gamma/\beta$ and positive if $s_e > \gamma/\beta$. This implies stable points for $S < \gamma/\beta$ and unstable points for $S > \gamma/\beta$. With the notation of Section 7.2, we have that $q = \det(\mathbf{J}) = -\beta I(\beta S - \gamma) + \beta^2 SI = 0$ at $(S, I) = (s_e, 0)$. This is a case of parallel trajectories as discussed in Section 7.2.

We have a line of stable nodes for $S < \gamma/\beta$ and a line of unstable nodes for $S > \gamma/\beta$. The same is true for the original nonlinear system (7.9).

These results are consistent with the phase-plane analysis of Section 6.2.

Computer algebra analysis

Most symbolic mathematics software packages, such as `Maple`, will have built-in routines to find equilibrium solutions, Jacobian matrices, characteristic polynomials and eigenvalues. What follows, in Listing 7.1, is some `Maple` code that has been used to solve the basic predator-prey example at the beginning of Section 7.5.

Listing 7.1: Maple code: c_pl_predprey.mpl

```
restart: with(VectorCalculus): with(LinearAlgebra):
eqn1 := beta[1]*X-c[1]*X*Y;
eqn2 := -alpha[2]*Y+c[2]*X*Y;
## equilibrium points
critpts := solve({eqn1 = 0, eqn2 = 0}, {X, Y});
## Jacobian matrix
sys := Vector([eqn1, eqn2]);
J := Jacobian(sys, [X, Y]);
## characteristic equation
ch := CharacteristicPolynomial(J, lambda);
## eigenvalues
ch1 := subs(critpts[1], ch);
ev1 := solve({ch1}, {lambda});
ch2 := subs(critpts[2], ch);
ev2 := solve({ch2}, {lambda});
## using p, q and delta for classification
p := Trace(J);
q := Determinant(J);
delta := p^2-4*q;
critpts[1];
p1 := subs(critpts[1], p);
q1 := subs(critpts[1], q);
delta1 := subs(critpts[1], delta);
critpts[2];
p2 := subs(critpts[2], p);
q2 := subs(critpts[2], q);
delta2 := subs(critpts[2], delta);
## calculating eigenvalues from p, q and delta
lambda1a := (1/2)*p1+(1/2)*sqrt(delta1);
lambda1b := (1/2)*p1-(1/2)*sqrt(delta1);
lambda2a := (1/2)*p2+(1/2)*sqrt(delta2);
lambda2b := (1/2)*p2-(1/2)*sqrt(delta2);
## calculating eigenvalues directly from Maple
eigs := Eigenvalues(J); eigs[1];
eig1a := subs(critpts[1], eigs[1]);
eig1b := subs(critpts[1], eigs[2]);
eig2a := subs(critpts[2], eigs[1]);
eig2b := subs(critpts[2], eigs[2]);
simplify(eig2b);
```

Further `Maple` commands, such as `simplify(X)` and `factor(X)`, can reduce complex expressions of X to a more manageable form.

Discussion

The above applications give some indication of the power of linearisation theory and the equilibrium point classification process developed. It should be noted that changes in the parameter values in each case may alter the equilibrium point classification and predict different dynamics. In this way it is not only possible to predict the system dynamics, but also to control the outcome in certain dynamical systems where such control parameters determine changes in the classification. This is an extremely important aspect in the study of dynamical systems.

> **Summary of skills developed here:**
> - *For a nonlinear system of equations, find all the equilibrium points and establish the trajectory behaviour close to each of them.*
> - *Sketch a picture of the entire phase-plane indicating the direction of the trajectories in any part of the plane.*
> - *Identify how a parameter might be changed to alter the dynamics in the phase-plane.*

7.6 Exercises for Chapter 7

7.1. Defining the trajectories. Consider the system

$$x' = x - 5y, \qquad y' = x - y,$$

where differentiation is with respect to time.

(a) Confirm there is only one equilibrium point which is a centre.

(b) Confirm that the family of ellipses given by

$$x^2 - 2xy + 5y^2 = K,$$

with K some constant, describes the solution trajectories in the (x, y)-plane. (Hint: One way to do this is to use direct substitution.)

7.2. Constructing a phase-plane. Consider the linear system

$$x' = x - y, \qquad y' = x + y,$$

where differentiation is with respect to time.

(a) Find the equilibrium point(s) and establish what type of equilibrium it is (they are).

(b) Using the above results draw a sketch of how you would expect the phase-plane trajectories to behave. (Note that, along the line $x = y$ (or equivalently $x' = 0$) there is no change in x, and along the line $x = -y$ there is no change in y.)

(c) Use `Maple` or `MATLAB` to draw the phase-plane to confirm your results.

7.3. Linearisation example. Consider the nonlinear system of equations

$$x' = x - y, \qquad y' = 1 - xy,$$

where differentiation is with respect to time.

(a) Find all equilibrium points for the system.

(b) Linearise the system and establish the classification of each equilibrium point.

(c) From the above results, sketch the trajectory behaviour in the phase-plane. (Note that along $x = y$ we have $x' = 0$, and along $y = 1/x$ we have $y' = 0$.)

(d) Use `Maple` or `MATLAB` (with symbolic toolbox) to check your results.

7.4. Linearisation example. *Given the system of equations*

$$x' = 3 + q(x - y), \qquad y' = 4 - xy,$$

where differentiation is with respect to time.

(a) *Assume $q = 1$. Give the linearised system of equations.*

(b) *Assume $q = 1$. Find all equilibrium points for the system and classify each point. Hence sketch the phase-plane.*

(c) *Consider q to be a variable parameter. Prove that one equilibrium point will always be a saddle point. Give full reasons for your answer.*

   (Hint: Show graphically, or otherwise, that for each such equilibrium point (x_e, y_e), the x and y coordinates will have the same sign.)

7.5. Linearisation of competition model. *Consider a competing species model, species X competing with species Y, including logistic growth for both species:*

$$X' = \beta_1 X - d_1 X^2 - c_1 XY, \qquad Y' = \beta_2 Y - d_2 Y^2 - c_2 XY.$$

Suppose that the parameter values are $\beta_1 = 3$, $\beta_2 = 2.5$, $d_1 = 2.8$, $d_2 = 2$, $c_1 = 2$ and $c_2 = 1$. The equilibrium values are

$$(0,0), \quad \left(0, \frac{\beta_2}{d_2}\right), \quad \left(\frac{\beta_1}{d_1}, 0\right), \quad \left(\frac{c_1\beta_2 - d_2\beta_1}{c_1 c_2 - d_1 d_2}, \frac{c_2\beta_1 - d_1\beta_2}{c_1 c_2 - d_1 d_2}\right),$$

and the phase-plane indicates that, with time, the populations will settle to coexistence of both species.

(a) *Linearise the system, giving the general form of the linearised equations and substituting in the parameter values as given.*

(b) *Use the linearised system to classify each equilibrium point. Sketch the phase-plane and illustrate the trajectory behaviour close to each equilibrium point from your classifications. Verify that these classifications agree with coexistence of the species.*

(c) *How, by varying the interaction parameters, could you change the phase-plane predictions to ensure that X will always die out?*

7.6. Comparison of nonlinear and linearised systems. *In order to compare the nonlinear system with the linearised system, we can draw them both on the same system of axes and compare them directly.*

(a) *For the predator-prey model*

$$X' = X(\beta_1 - c_1 Y), \qquad Y' = Y(-\alpha_2 + c_2 X),$$

   taking $\beta_1 = 1.3$, $c_1 = 0.01$, $\alpha_2 = 1$ and $c_2 = 0.01$, and considering the trajectories close to the non-zero equilibrium, draw on the same system of axes the time-dependent plots for the nonlinear and the linearised systems. (Use Maple *or* MATLAB *to solve the systems numerically.)*

(b) *Also compare the trajectories in the phase-plane of the nonlinear and the linearised systems.*

   (Hint: Set up the equations for the linearised system and then write them in terms of the X and Y coordinate system of the given nonlinear equations. For example, $X = x_e + \xi$ or $\xi = X - x_e$, where x_e is the equilibrium value. Then solve both systems of equations using Maple *or* MATLAB *and display the results in the same diagram.)*

7.7. Linearisation of an epidemic model. *Consider the following model for the spread of a disease, with susceptibles constantly renewed by births at a constant overall birth rate of μ:*

$$S' = -\beta S I + \mu, \qquad I' = \beta S I - \gamma I.$$

(a) *Show that the only equilibrium point is $(\gamma/\beta, \mu/\gamma)$.*

(b) *Linearise the system and determine the behaviour of the trajectories near the equilibrium point.*

(c) *Discuss, briefly, the implications for the time-dependent behaviour of both $S(t)$ and $I(t)$.*

7.8. Bacteria in the gut. *This question refers to the case study in Section 6.7.*

(a) *Using* `Maple` *or* `MATLAB` *obtain expressions for the equilibrium point(s) for the system of differential equations (6.24).*

(b) *Determine the type and stability of all points and establish, approximately, when the points change stability and classification as the turnover rate increases. (Use the parameter values as given in Figure 6.13.)*

7.9. Linearisation of jungle warfare model. *A model for a battle between red soldiers and blue soldiers, where one army uses random fire and the other uses aimed fire, is*

$$R' = -a_1 R B, \qquad B' = -a_2 R.$$

Linearise the system about the only equilibrium point, and hence determine the behaviour of trajectories near that point.

Chapter 8

Some extended population models

In this chapter, we present a variety of case studies based on research papers or articles. Each study significantly extends one or more of the simple models we have developed in Chapter 5 and applies the techniques of analysis developed in Chapter 6 and particularly Chapter 7. It serves to illustrate the relevance of the theory presented in this text and its current use in practice.

8.1 Introduction

There are many interesting problems requiring the use of population models, and we present a sample of them here with the aim of illustrating the versatility, diversity and applicability of these models to current issues.

In previous chapters, we developed the simplest type of population models. Each of those can be extended or combined with others to reflect the particular features of the population under study. As we see in the case studies that follow, the models examined so far form the basis on which the further models proposed are built. The concept of the balance law and the compartmental structure are integral in each case.

The references from which the following case studies are adapted are provided, and all figures presented have been generated using `Maple`. (The code has not been included here, although all figures can be generated from code integrated into the text of earlier chapters.) Code building and model exploration are encouraged in the exercises.

8.2 Case Study: Competition, predation and diversity

Mathematical models can be used to make predictions but also mathematical models are useful to investigate different hypotheses about nature. In Section 5.7, we developed simple models for predator-prey interactions and for competing species. For the latter we found that the model predicted, in most cases, the survival of one species and the extinction of all others. We noted that this was a limitation of the model, as in many instances observations in nature confirm the stability and coexistence of a number of competing species.

One notion that would explain coexistence as well as diversity is that of cooperation, as opposed to competition. In the infertile country of the arid regions of Australia, this is one hypothesis for the substantial diversity and coexistence of species relying on the same resources for survival.

In the following case study we examine a different hypothesis. A combination of the basic competition model together with the predator-prey model provides a mathematical argument for the stability and coexistence of competing species within an ecosystem. The case study supports the earlier comment in Section 5.1 that systems operating in isolation may be unstable, but when incorporated into more complex systems they typically display stability.

Note here the use of the word *stability*. A stable system in this sense is a system in which species coexist, whereas if one species suffers extinction, the system is referred to as unstable. This use appears in practice, but does not satisfy the strict definition we have met previously in the text. The following is adapted from Cramer and May (1972).

Observations have confirmed that while simple systems (for example, a single trophic level) are likely to be unstable in isolation, more complex systems comprising a collection of interacting trophic levels are often extremely stable.

Parrish and Saila (1970) attempted to validate this idea with some computer experiments. They proposed that while a two-species competition model might be unstable, the introduction of a predator could allow the system to stabilise, thus permitting all three species to coexist.

For the two species of prey, N_1 and N_2, and a predator N_3, the model they constructed was a simple competition model between N_1 and N_2 (both populations following a logistic growth pattern) with a predator N_3 included, whose birth rate was dependent on both N_1 and N_2, namely,

$$N_1' = N_1(\epsilon_1 - \alpha_{11}N_1 - \alpha_{12}N_2 - \alpha_{13}N_3),$$
$$N_2' = N_2(\epsilon_2 - \alpha_{21}N_1 - \alpha_{22}N_2 - \alpha_{23}N_3), \qquad (8.1)$$
$$N_3' = N_3(-\epsilon_3 + \alpha_{31}N_1 + \alpha_{32}N_2).$$

In the absence of the predator ($N_3 = 0$), we have a simple competition model that predicts the extinction of all but one species in most cases: that is, $\alpha_{11}\alpha_{22} - \alpha_{12}\alpha_{21} \leq 0$. Now starting with parameter values that ensure instability for the competition model, the predator is included. The new system can be stable, even when we do not have equal predation (that is, $\alpha_{13} = \alpha_{23}$ and $\alpha_{31} = \alpha_{32}$).

Figure 8.1 illustrates the case of the unstable competition model without predation. In this case, N_1 dies out and $N_2 \to 3.5 \times 10^5$. This is the usual case of competitive exclusion.

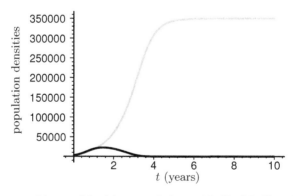

Figure 8.1: Unstable competition model without predation, with N_1 (black) and N_2(grey). Parameter values: $\alpha_{11} = 9 \times 10^{-5}$, $\alpha_{12} = 3 \times 10^{-5}$, $\alpha_{13} = 0$, $\alpha_{21} = 3 \times 10^{-5}$, $\alpha_{22} = 0.6 \times 10^{-5}$, $\alpha_{23} = 0$, $\epsilon_1 = 3$ and $\epsilon_2 = 2.1$. The initial populations are $N_1(0) = N_2(0) = 3 \times 10^3$.

Figure 8.2 corresponds to the case where a predator is included and predation of each of the prey species is the same, that is, $\alpha_{13} = \alpha_{23}$). In this case all populations approach fixed positive population sizes with $N_1 \to 11.66 \times 10^3$, $N_2 \to 8.34 \times 10^3$ and $N_3 \to 11.3$. So we get coexistence of all three species.

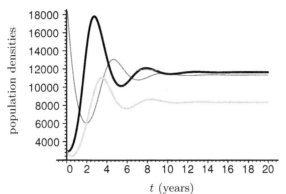

Figure 8.2: Unstable competition model with predation and N_1 (black), N_2 (grey) and N_3 (fine black). Parameter values are as in Figure 8.1 but with $\alpha_{13} = 0.15$, $\alpha_{23} = 0.15$, $\alpha_{31} = \alpha_{32} = 0.6 \times 10^{-4}$ and $\epsilon_3 = 1.2$. The initial populations are $N_1(0) = N_2(0) = 3 \times 10^3$ and $N_3(0) = 20$. In the displayed results, N_3 has been multiplied by 10^3.

Finally, Figure 8.3 illustrates the case of a stable system where all three populations coexist, but without the condition of equal predation. The results are that $N_1 \to 69.4 \times 10^3$, $N_2 \to 16.7 \times 10^3$ and $N_3 \to 11.1$.

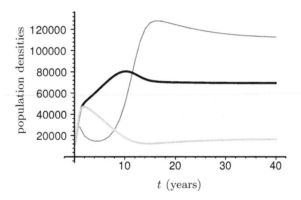

Figure 8.3: Unstable competition model with predation and N_1 (black), N_2 (grey) and N_3 (fine black). Parameter values: $\alpha_{11} = 0.9 \times 3.22 \times 10^{-5}$, $\alpha_{12} = 3.22 \times 10^{-5}$, $\alpha_{13} = 0.06$, $\alpha_{21} = 3.22 \times 10^{-5}$, $\alpha_{22} = 3.22 \times 10^{-5}$, $\alpha_{23} = 0.04$, $\alpha_{31} = 3 \times 10^{-5}$, $\alpha_{32} = 10^{-5}$, $\epsilon_1 = 3.22$, $\epsilon_2 = 3.22$ and $\epsilon_3 = 2.25$. The initial populations are $N_1(0) = N_2(0) = 10^4$ and $N_3(0) = 10$. In the displayed results, N_3 has been multiplied by 10^4.

The example serves to illustrate that while a single two-species population model may be unstable, the introduction of a further species (in this case a predator of both other species) may result in a stable system with the persistence and coexistence of all species. This provides an argument for the existence of a rich diversity of species, some of which may rely on the same resources, as is apparent in any natural ecosystem.

8.3   Extended predator-prey models

In this section we improve on the predator-prey model by incorporating further biological observations. The new model also exhibits oscillatory behaviour, which means that in the phase-plane we have a limit cycle or periodic solution.

Review of previous models

In Section 5.4 we developed a predator-prey model that predicted oscillations in the population densities, and we saw, in Section 6.6, how this led to closed curves, or a set of periodic solutions, in the phase-plane.

We then improved the model to include logistic growth and noted that the oscillations died out and the populations stabilised to some fixed equilibrium point. In the phase-plane this translated to an inwardly moving spiral approaching this equilibrium point. The differential equations for this model were

$$\frac{dX}{dt} = r_1 X \left(1 - \frac{X}{K}\right) - c_1 XY, \qquad \frac{dY}{dt} = c_2 XY - \alpha_2 Y. \tag{8.2}$$

Here $X(t)$ is the prey density and $Y(t)$ is the predator density. The constant K is the prey carrying capacity, r_1 is the prey per-capita intrinsic growth rate (the per-capita growth rate for small populations), α_2 is the predator per-capita death rate, and c_1 and c_2 are positive constants that can be interpreted in terms of the predator searching rate (see Section 5.4).

However, oscillating predator-prey pairs have been observed in nature, and this 'improved' model does not predict periodic solutions. This is a fundamental flaw, and we now suggest possible extensions that make the model more realistic.

Extending the model

The basic problem with the Lotka–Volterra predator-prey model is that the assumptions made regarding prey deaths and predator births and deaths are not sufficiently realistic. In practice, field ecologists studying predator populations measure two quantities: the *functional response* $F(X)$, and the *numerical response* $N(X,Y)$. The functional response is the rate at which a single predator kills prey, as a function of the prey population. The numerical response is the per-capita growth rate of the predators, which may be dependent on both prey and predator populations.

A general model for a predator-prey system, based upon functional and numerical response functions, and with X and Y the population densities, is

$$\frac{dX}{dt} = r_1 X \left(1 - \frac{X}{K}\right) - F(X)Y, \qquad \frac{dY}{dt} = N(X,Y)Y. \tag{8.3}$$

In system (8.2) the functional response is $F(X) = c_1 X$. This states that a single predator continues to increase its rate of killing prey as the prey density increases. In practice, it is observed that this function levels off to a constant amount as X increases, representing the maximum that a single predator will want to eat (in a given time). Another way of thinking about this is that if the prey are plentiful, the predator will not have to spend much time hunting for them, but will spend more time waiting until they need to hunt again. The following example uses the idea of a handling time to formulate a suitable functional response function.

Example 8.1: *Assuming that a single predator takes a time t_h to handle each prey, find a suitable functional response.*

Solution: *We modify the argument given in Section 5.4 for the interpretation of one of the parameters in terms of a searching rate c_1. Let T denote the time left available to the predator for searching (hunting prey). Consider a time interval Δt.*

The number of prey eaten is determined from the searching rate c_1 multiplied by the prey density $X(t)$:

$$\left\{ \begin{array}{c} \text{no. prey eaten} \\ \text{in time } \Delta t \\ \text{by one predator} \end{array} \right\} = c_1 T X(t). \tag{8.4}$$

The rate of prey (number per unit time) eaten by a single predator is then $c_1 T X(t)/\Delta t$. Now

$$\Delta t = T + t_h \times \left\{ \begin{array}{c} \text{no. prey eaten} \\ \text{in time } \Delta t \\ \text{by one predator} \end{array} \right\} = T(1 + t_h c_1 X(t)). \tag{8.5}$$

The functional response $F(X)$ is the rate of prey eaten by a single predator, which is given by dividing equation (8.4) by Δt. Using (8.5) we obtain

$$F(X) = \frac{c_1 X}{1 + t_h c_1 X}. \tag{8.6}$$

As $X \to 0$, we see that $F(X) \to 0$. For large prey densities, or as $X \to \infty$, then $F(X) \to t_h^{-1}$. This value is the maximum possible removal rate of prey: the functional response is *saturated* at this value. For large X, the prey are very easy for a predator to find, so this maximum rate is the reciprocal of the handling time t_h, for a single predator. This function also arises in the theory of reaction kinetics in chemistry, see Murray (1990), where it is known as the Michaelis–Menten function.

This particular form of the functional response $F(X)$ is called a Holling type II response function and is commonly used for invertebrate predators; see May (1981). Some examples of fitting the Holling type II model to field data are given in Hassell (1978, 1976) and Gotelli (1995), together with data for handling times and searching rates. For vertebrate predators, a Holling type III response is often used: this is obtained by setting the searching rate $c_1 = a'$ in (8.6) proportional to prey density. This could be thought of as reflecting the higher intelligence of vertebrates whose searching rates improve with higher prey density, as they learn how to track the prey.

For the numerical response $N(X, Y)$, we assume a logistic growth function for the predator population $r_2(1 - Y/K_2)$ but with the predator carrying capacity assumed to be proportional to the prey density, $K_2 = c_3 X$. Hence,

$$N(X, Y) = r_2 \left(1 - \frac{Y}{c_3 X} \right), \tag{8.7}$$

where r_2 and c_3 are positive constants. Note that as X increases, the deaths due to overcrowding decrease, but an increase in Y implies an overall increase in the deaths proportional to Y^2.

Our model becomes

$$\frac{dX}{dt} = r_1 X \left(1 - \frac{X}{K} \right) - \frac{c_1 X Y}{1 + t_h c_1 X}, \qquad \frac{dY}{dt} = r_2 Y \left(1 - \frac{Y}{c_3 X} \right). \tag{8.8}$$

This model is sometimes called the Holling–Tanner predator-prey model.

Scaling the equations

Before using `Maple` or `MATLAB` to examine the time-dependent graphs and phase-plane, we apply a technique often adopted in practice to simplify the analysis. We scale the equations. This process may offer several advantages.

After scaling, the parameters are in dimensionless form and provide a measure of the relative strengths of the interacting or competing effects within the model. This may provide important information towards improving the design of experimental processes, in turn creating a more applicable model. Also as a result of scaling, the mathematics may be simplified, or the number of parameters reduced. However, this is not necessarily true in all cases.

We can scale the above system of equations by defining new dimensionless variables x, y and τ, by

$$x = \frac{X}{K}, \qquad y = \frac{Y}{c_3 K}, \qquad \tau = r_2 t.$$

Then, using the chain rule, it can be shown that (see Exercise 8.3) the system becomes

$$\frac{dx}{d\tau} = \lambda_1 x (1 - x) - \frac{\lambda_2 x y}{(\lambda_3 + x)}, \qquad \frac{dy}{d\tau} = y \left(1 - \frac{y}{x} \right), \qquad (8.9)$$

with dimensionless parameters

$$\lambda_1 = \frac{r_1}{r_2}, \qquad \lambda_2 = \frac{c_3}{t_h r_2}, \qquad \lambda_3 = \frac{1}{t_h c_1 K}.$$

Interpretation of the parameters

The newly defined parameters can be interpreted in the following way. Since r_1 is the maximum specific growth rate of the prey with no predators and r_2 is the specific growth rate of predators with infinite prey density, then λ_1 is the ratio of the specific growth rates under ideal conditions.

The quantity λ_2 measures the predators' maximum prey removal rate t_h^{-1} relative to its ideal specific growth rate r_2. Parameter c_3 is already dimensionless as the product $c_3 X$ has the same units as Y, that is, numbers of animals.

Since t_h is the handling time per prey, the combination $t_h K$ represents a time scale (per unit area) for the total time to handle the maximum possible prey density (given by the carrying capacity K in the absence of predation). Since c_1 is the searching rate (area per unit time), c_1^{-1} is a time scale for searching a unit area. Hence the parameter λ_3 represents a ratio of time scales (searching time to handling time). For large λ_3, the handling time is insignificant, but for small values of λ_3, the predators ability to remove prey is more likely to rise to its saturation value of λ_2.

Numerical solution for scaled model

We use `Maple` to solve the system numerically with parameter values of $\lambda_1 = 5$, $\lambda_2 = 10$ and $\lambda_3 = 0.015$. Consider the phase-plane diagram as illustrated in Figure 8.4. Note that for all the initial conditions chosen from the illustrated domain, the solution is a single periodic solution.

For different parameter values, $\lambda_1 = 5$, $\lambda_2 = 1$ and $\lambda_3 = 0.6$, the equilibrium solution is a stable node similar to Figure 8.5. Trajectories with initial values close to the equilibrium are attracted to the equilibrium.

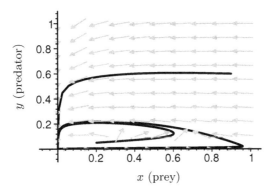

Figure 8.4: Phase-plane diagram for the scaled system (8.9) with $\lambda_1 = 5$, $\lambda_2 = 10$ and $\lambda_3 = 0.015$.

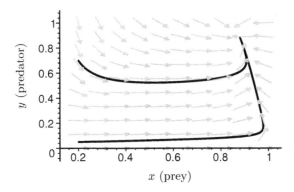

Figure 8.5: Phase-plane diagram for the scaled system (8.9) with $\lambda_1 = 5$, $\lambda_2 = 1$ and $\lambda_3 = 0.6$.

Discussion

Notice that this new system can give rise to a single periodic solution for certain parameters (Figure 8.4). In this case, the periodic cycle is always the same, regardless of the initial conditions chosen. This is very different from the infinite series of closed curves, each associated with a different set of initial values, which we obtained from our basic predator-prey model in Section 6.4.

The system also predicts populations stabilising to fixed densities for other parameter combinations (Figure 8.5). It would be useful to establish a diagram of the parameter values that lead to periodic solutions and those that do not. This is left as an exercise for the interested reader; however, such a diagram, establishing the parameter regimes that result in different dynamics is an important part of the system analysis process.

8.4 Case Study: Lemming mass suicides?

Dramatic fluctuations have been observed in the populations of several rodents. In the case of lemmings, several myths have evolved to account for these fluctuations, such as lemmings hurling themselves off cliffs in mass suicides. In the following case study, a simple model of rate equations is presented to describe these dramatic cycles. The model proposed is a further extension of the basic predator-prey model and makes use of the concept of modelling a single population in two parts, each part having different behavioural characteristics. This is similar to the notion we made use of when modelling the spread of a disease in a population by dividing the population into two subpopulations of infectives and susceptibles. Adapted from Dekker (1975).

Legend has it that lemmings periodically 'migrate' into the sea in an act of mass suicide. Early Scandinavians believed that lemmings fell from heaven in stormy weather! More recently it has been proposed by natural historians that epidemic disease, or possibly predation, intermittently wipe out large numbers of the lemming population. Each of the above attempts to explain the drastic fluctuations in the populations of lemmings, true also for certain other small rodents or voles, that have been observed.

It appears from collected data that the population cycles have a period of three to four years. Further, there appear to be no external forces causing these fluctuations, suggesting the existence of some internal control mechanism. To support this idea, the observations indicate changes in the population characteristics during a cycle, in terms of (amongst others) sex and age.

Following the work of Myers and Krebs (1974), populations can be divided into two main genotypes: one with a high reproduction rate and intolerant of high population density living conditions, and the other with a low reproduction rate and well adapted to crowded living conditions. The former populations, which deal with overcrowding by migrating (not necessarily jumping off a cliff into the ocean as was once believed for lemmings), are called the emigrants and their density will be denoted by $n_1(t)$, while the latter populations are known as the tolerants, and their density is denoted as $n_2(t)$. The mathematical model proposed to describe these interacting populations is given as

$$\frac{dn_1}{dt} = n_1 \left[a_1 - (b_1 - c_1)n_2 - c_1(n_1 + n_2) \right],$$

$$\frac{dn_2}{dt} = n_2 \left[-a_2 + b_2 n_1 \right],$$

(8.10)

where a_1, a_2, b_1, b_2 and c_1 are parameters that will be described below.

First, consider the growth of the emigrants in the absence of the tolerants. In that case the emigrants follow a logistic growth pattern and, since emigration depends on the total population density, the term that curbs the infinite exponential growth is $-n_1^2 c_1$. The individuals that migrate are those with a high reproductive potential, largely the young, lactating females.

On the other hand, the tolerants, in the absence of emigrants will die off exponentially. They have an average age far greater than that of the emigrant population, but a far lower reproductive potential. The tolerants interact with the emigrants in both a social and sexual manner, which leads to an improved birth rate for the tolerants.

Thus the parameter b_2 can be split into parts: b_2^0 governing the sexual creation of individuals and b_2^1 governing the possible creation of a tolerant from an emigrant through a social encounter. The latter process allows a flow of individuals from the emigrant to the tolerant population. Similarly, the parameter $(b_1 - c_1)$ describing the reduction in emigrant population consists of three parts: a social contribution where individuals flow to the tolerants, a sexual contribution where individuals are created, and a dispersive contribution where individuals leave the population (largely young females and usually when the total density of the populations is large).

A further consideration is that of seasonal fluctuations in the reproductive rates, a_1 and a_2, which reflect the observed maximum rate in late summer and minimum in late winter.

Assuming a simple sinusoidal rate, the time-dependent parameters are taken to be

$$a_1(t) = a_1^0 + a_1^1 \sin\left(\frac{\pi}{6}t\right), \qquad a_2(t) = a_2^0 - a_2^1 \sin\left(\frac{\pi}{6}t\right). \qquad (8.11)$$

The following figures (Figures 8.6 and 8.7) are constructed applying the above model and estimating the parameters from data collected by Krebs et al. (1973) and Myers and Krebs (1974) for the rodent voles Microtus pennsylvanicus (or meadow mouse).

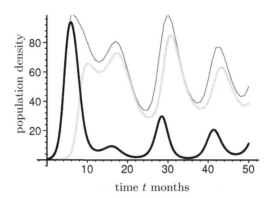

Figure 8.6: Population cycles for the meadow mouse. Parameters $a_1^0 = 1$, $a_1^1 = 0.35$, $b_1 = 1.75 \times 10^{-2}$, $c_1 = 1 \times 10^{-2}$, $a_2^0 = 0.140$, $a_2^1 = 0.075$ and $b_2 = 1.5 \times 10^{-2}$ with initial populations of $n_1 = 0.5$ and $n_2 = 0.5$. Emigrants n_1 (black), tolerants n_2 (grey) and total population (fine black).

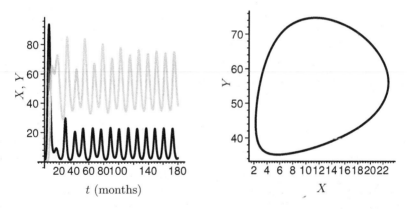

Figure 8.7: The time-dependent (first graph) and phase-plane (second graph) plots for the population cycles of the meadow mouse. Parameters as for Figure 8.6, but the model is run over a longer time period. (n_1 represented with black, and n_2 with grey in the first graph.)

For the figures, a monthly time step was taken and the parameter values are listed in the associated captions. From Figure 8.6 (following Dekker (1975)) it should be noted that the total population constitution of the two genotypes is very different in different phases of the first 48-month (4-year) cycle. Furthermore, for the total rodent population, the dramatic fluctuations are observed in the initial four year period. While the model is simple, easy to analyse, and provides a close resemblance to the observations in terms of the initial main cycles, it is not robust over long time periods.

If the time-dependent graph is considered over long periods (more than 20 years), then the population settles to a simple 1-year oscillation. This is illustrated in the time-dependent plot and phase-plane of Figure 8.7, and is not included or discussed in Dekker (1975).

However, if disease or predation (or some other mechanism) forced a reduction in the population of the emigrants in particular, then this model would predict dramatic fluctuations over the following decade. Thus the model requires further refinement, which might include different sex groups and an age distribution, as well as a better representation of the rodent dispersion with respect to distance and location, namely their spatial distribution. Furthermore, some statistics on the likelihood of interspecies and between-species interactions, such as predation, would provide an enhancement.

8.5   Case Study: Prickly pear meets its moth

In Chapter 5, we studied a predator-prey model describing a case where the introduction of an insect species to a region to eradicate a previously introduced pest was unsuccessful. Many such introductions have been and are currently being carried out, with many disasters and unforeseen ramifications for the environment, as we saw in Section 5.5 with the introduction of the Nile Perch to Lake Victoria. However, there have also been success stories, and the following case study examines one spectacular success.

The model proposed in this case study is a plant-herbivore system, which is similar to the predator-prey model. It includes extensions of the basic form, and the inclusion of a Michaelis–Menten type function. Adapted from a discussion in May (1981) and a description in Rolls (1969).

In 1864, Captain Phillip, on his way to Australia with the First Fleet, stopped off in Brazil to collect the cochineal insect and its host plant, the prickly pear (Opuntia inermis and Opuntia stricta). He was to introduce them to Australia to ensure a plentiful supply of red dye (from the cochineal insects) for his soldiers' coats. The insects did not fare well and several more introductions were tried but failed. However, the cacti, which were also planted as hedges for additional stock feed as well as garden plants, adapted well — far too well in fact. They ran wild. Extensive, dense stands of the cacti spread into the farmland of northern New South Wales and southern Queensland, averaging about 500 plants per acre (or 1,250 plants per hectare). They walled in homesteads with growth of impenetrable density, and destroyed the viability of thousands of square kilometres of farmland.

Attempts were made to eradicate the cacti by spraying them with arsenic pentoxide and sulphuric acid from horseback. While this destroyed the cacti it came into contact with, it was a drop in the ocean and, furthermore, it destroyed the mens' clothing, their boots and saddles and finally their horses, which lost their hair and developed sores which would not heal.

In 1925, a moth Cactoblastis cactorum was introduced from Argentina to combat the cactus growth. Its larvae bore through the plant, their breeding being restricted solely to the cacti. They were bred up in Brisbane, from where some 3,000 million eggs were distributed to farmers. Within only two years they had virtually wiped out the cacti.

The eggs of the moth are not laid at random, but in clumps of egg-sticks on the plants, with each egg-stick comprising about 80 eggs. Thus some plants were hit with large numbers of moths while others escaped completely. Since about 1.5 sticks of eggs produce sufficient larvae to destroy a plant, many plants received many more larvae than were required to destroy them.

A mathematical predator-prey type model is proposed in May (1981) to describe this plant-herbivore system. With V the plant mass (in plants per acre) and H the herbivore (moth) population size (in egg-sticks), the model is

$$
\begin{aligned}
\frac{dV}{dt} &= r_1 V \left(1 - \frac{V}{K}\right) - c_1 H \left(\frac{V}{V+D}\right), \\
\frac{dH}{dt} &= r_2 H \left(1 - \frac{JH}{V}\right).
\end{aligned}
\tag{8.12}
$$

Here K is the maximum biomass, r_1 is the intrinsic rate of increase in the cacti biomass, c_1 is the per-capita maximum rate of food intake by the moths, D is a grazing efficiency term at low plant density, J is the proportionality constant associated with the biomass required to sustain a moth at equilibrium, and r_2 is the per-capita rate of increase of the moth population.

Values for these parameters have been estimated from data collected by Dodd (1940) and Monro (1967), and the older unit of measurement, the acre, is retained to make use of the whole numbers from this collected data. (The conversion is 1 ha = 2.471 acres.) Since the root stock of the cacti can increase to 250 tons per acre in 2 years, r_1 is close to 2.7; however, this is only based on vegetative growth, and if we consider sexual reproduction r_1 is estimated to be 2, at maximum. The carrying capacity K is taken to be 5,000 plants per acre. Laboratory experiments were used to get an estimate of r_2 in which 2,750 eggs increased to 100,605 eggs in the following cycle, providing an estimate of 3.6. This compares with field data where 5,000 larvae could multiply to 10,000,000 in two years, suggesting a larger r_2 of 3.8 a year. Both figures may underestimate the rate and so r_2 was taken as 4.

From the work of Monro (1967), J was estimated as 2.23 in units of cactus plants per egg-stick. The parameters c_1 and D were more difficult to estimate. The former should be large, reflecting the damage to the plants by the feeding larvae, while the latter should be small to reflect the efficiency with which the female moths choose plants on which to lay eggs. Following the near total destruction of the cacti in a period of only two years, the model should predict this crash and then settle to a stable 11 plants per acre, as was observed . For this result, $c_1 = 6.2$ and $D = 4$.

Figure 8.8 plots the results of applying the proposed model with the parameter values as above.

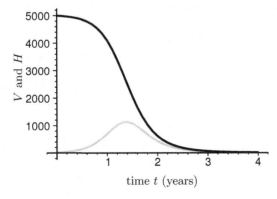

Figure 8.8: The crash of the prickly-pear cactus in just two years. The plants per acre V (black) and the moth population H in egg-sticks per acre (grey).

Clearly the destruction after two years is apparent, albeit with caution we should accept the model as it does not incorporate many of the features of the system under study.

Today, the cacti can still be found in certain small regions of southern Queensland, but they are well under control and no longer pose a threat to the environment. As a token of their gratitude to the moths, affected communities constructed memorial halls, still in current use, in honour of *Cactoblastis*.

8.6   Case Study: Geese defy mathematical convention

So far we have followed the convention that more available food or prey means more animals or predators: the result of the top-down approach of control theory. However, observations do not always support this theory, and the following case study provides an argument and model for a slightly different approach. The model presented below is based on the predator-prey model, in this case a plant-herbivore system, with a Michaelis–Menten type function to allow for an increase in predator (herbivore) numbers to a maximum, and then a reduction as the amount of prey (grass) is increased. The case study is adapted from a model considered by van de Koppel et al. (1996).

Classical exploitation theory has led us to believe that where more grazing is available, more herbivores will be found inflicting more pressure on the standing crop and thus keeping the crop height low; a top-down control in productive environments. This approach is a hotly debated topic, although still widely accepted. The article van de Koppel et al. (1996) contests the standard plant-herbivore models, arguing that herbivores do not 'control' the plant growth entirely.

Their models, supported by empirical evidence, include the possible coexistence of dense vegetation and low herbivore grazing pressure, all in the absence of any predator.

Field data were collected from a salt marsh on the Island of Schiermonnikoog in The Netherlands. It provided an ideal location as the island supports all types of vegetation (sparse, intermediate, and dense) as well as a collection of herbivores (rabbits, hares and two species of geese). Also, there are no predators for these herbivores, other than the occasional raptor or human.

The observation of the maximal grazing pressure by rabbits, hares and geese at the intermediate level of the crop can be explained by several hypotheses. For example, in the denser regions, lack of light may impact on the stem development of the plants, robbing the leaves of the plant of protein content and making them less attractive to the herbivores. Alternatively, because of the greater density, the herbivores in these regions may be under greater potential danger from the few predators on the island, and so may avoid them. Another theory is that the herbivores may have greater difficulty penetrating the thicker growth to forage. Whatever the reason, reduced foraging efficiency is observed in the dense regions of vegetation, and typically plant-herbivore models have not incorporated this aspect.

Let P and H be the plant and herbivore densities, respectively, with P measuring the plant mass per unit of area (g/m^2), and H the number of herbivores in the unit of area (number/m^2) calculated from the number of droppings found. The standard form of a plant-herbivore model is

$$\frac{dP}{dt} = G(P) - F(P)H, \qquad \frac{dH}{dt} = N(P)H,$$

where $G(P)$ is the plant growth function dependent on density, $F(P)$ is the per-capita consumption rate of a herbivore (the functional response), and $N(P)$ is the per-capita growth rate of herbivores (numerical response). Time t is measured in years. Typically $F(P)$ and $N(P)$ are monotonically increasing functions with increasing P. The proposed models now incorporate the reduced foraging efficiency.

Consider first the case of reduced digestion efficiency. This model assumes that the herbivore digestion is hampered by the change in stem structure of plants in regions of vegetation with greater density. To include this reduced foraging due to reduced digestion efficiency, the per-capita growth rate was taken to decline at high plant density. Let P_1^* and P_2^* be the two densities of vegetation for which herbivore growth is zero, so that below P_1^* there is too little vegetation to support a herbivore population, while above P_2^* there is too much vegetation for the herbivores to thrive and their density decreases.

The model proposed is

$$\frac{dP}{dt} = rP\left(1 - \frac{P}{K}\right) - c_{\max}\left(\frac{P}{a+P}\right)H,$$

$$\frac{dH}{dt} = \left(e_{\max}\left(\frac{P}{a+P}\right)e^{-bP} - d\right)H,$$

(8.13)

where clearly the growth rate for P is modelled as logistic. Parameter d is the herbivore mortality rate. The herbivore per-capita growth rate (numerical response),

$$N(P) = \left(e_{\max}\left(\frac{P}{a+P}\right)e^{-bP} - d\right),$$

and the saturating per-capita vegetation mortality rate (functional response),

$$F(P) = c_{\max}\left(\frac{P}{a+P}\right),$$

are illustrated in Figure 8.9 with parameter values $a = 10$, $b = 0.65$, $e_{\max} = 0.4$ and $d = 0.1$.

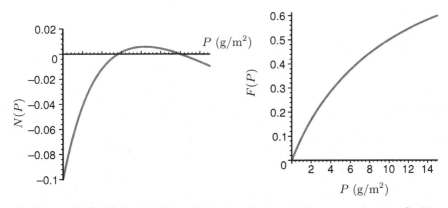

Figure 8.9: For model (8.13) the first figure illustrates the numerical response function (herbivore growth rate) and the second figure the functional response (herbivore consumption rate).

Note that the functional response is approximately $c_{\max}P/a$ when $a \gg P$, and is approximately $c_{\max}$ when $a \ll P$. Here it has been assumed that the consumption rate of the herbivore is an increasing function of plant density but the negative effect of the plant density, when it increases beyond some upper bound, is translated to growth in the herbivores via the factor e^{-bP}.

What is of interest is how this plant-herbivore system may behave with time. Does it settle to a situation of coexistence, or are there conditions under which the herbivore population dies out? To establish this, the phase-plane is examined.

Applying the above model, Figure 8.10 represents a typical phase-plane diagram with $K = 18$, $r = 1$, $c_{max} = 1$, $a = 10$, $b = 0.065$, $e_{max} = 0.4$ and $d = 0.1$.

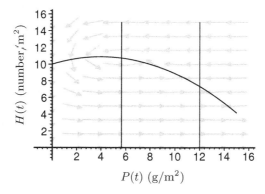

Figure 8.10: Phase-plane diagram with $K = 18$, including the H and P nullclines and the equilibrium points at their intersection.

The nullclines are included and it is clear from the figure (see enlargements in Figure 8.11) that there are three equilibrium points away from the origin. In this case, the points with $P = 0$ and $P \simeq 6$ are stable and the point with $P \simeq 12$ is unstable. Therefore, in only one case is there the possibility of long-term coexistence.

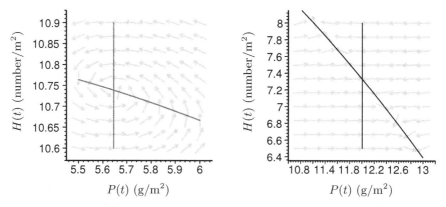

Figure 8.11: The phase-plane of Figure 8.10, magnified about two of the equilibrium points.

Now, considering K (the measure of the plant carrying capacity) as a variable parameter $0 < K < 25$ with all other parameters as before ($r = 1$, $c_{max} = 1$, $a = 10$, $b = 0.065$, $e_{max} = 0.4$ and $d = 0.1$), several phase-planes are examined to see how this situation may change with changing maximum density of the plant crop. These are illustrated in

Figure 8.12.

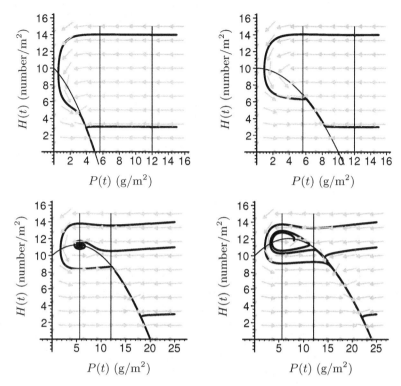

Figure 8.12: Phase-plane diagrams for varying K. From left to right in the first row, $K = 5$ and $K = 10$. In the second row, $K = 20$ and $K = 24$.

From Figure 8.12 it is clear that there is no herbivore population if $K < 5$ or $K > 25$. For $K = 10$, coexistence ensues from any initial densities. The more interesting situation occurs for $10 < K < 24$, where coexistence ensues from only a select number of initial conditions and the herbivores approach extinction otherwise, even with large initial densities.

A second scenario to consider is that of reduced consumption rate. In this case, the model can be further modified to include the increasing crop accompanied by a reduction in consumption rate. This is due to the possible situation where herbivores keep out of the regions of greater density for fear of predators. A further exponential is introduced into the term describing plant consumption as follows:

$$
\begin{aligned}
\frac{dP}{dt} &= rP\left(1 - \frac{P}{K}\right) - c_{\max}\left(\frac{P}{a+P}\right)e^{-bP}H, \\
\frac{dH}{dt} &= \left(e_{\max}\left(\frac{P}{a+P}\right)e^{-bP} - d\right)H.
\end{aligned}
\tag{8.14}
$$

This reduction in the consumption rate results in a feedback between plant growth of greater density and reduced grazing pressure, greater than in the earlier case of reduced digestion efficiency in system (8.13).

We can summarise the results from the models as follows. For K below P_1^* there is insufficient vegetation to sustain a herbivore population. For K between P_1^* and P_2^* the herbivores control the plant growth at a constant level (as is predicted in traditional plant-herbivore models where increases in plant mass result in increases of the herbivore population). For values of K above P_2^* there are two possible states in which the system can exist. In one there is intermediate plant density as well as herbivores, while in the other only dense vegetation exists. For certain initial conditions the system will tend to the coexistence stable state, while for others it will tend to the extinction of the herbivores. As K increases still further, the only possible stable state becomes that of herbivore extinction and dense vegetation: the vegetation 'escapes' herbivore 'control'.

One consequence of this perspective is that nutrient enrichment, such as fertiliser, is capable of triggering extra growth that may spark a relief from some herbivore control. Furthermore, there is the possibility of permanently destroying a plant-herbivore system; this may have its uses when certain herbivores are considered pests and not desired as part of the equilibrium.

8.7   Case Study: Possums threaten New Zealand cows

In Chapters 5 and 6, we examined the spread of an influenza in a school; however, we did not consider any strategies to combat the spread. In the following case study we examine a simple disease model that describes the spread of bovine tuberculosis in possums and the impact of different strategies to reduce the spread. Note that, unlike epidemics in human communities, we are not concerned about the survival of the New Zealand possum population, in fact, if it were wiped out, that would be a bonus! This model is adapted from two articles: Wake (1995) and Roberts (1992), and is a simplified version of that in Roberts (1992).

Sometime between 1830 and 1860 the brushtailed possum (Trichosurus vulpecula) was introduced into New Zealand from its native Australia. It is a cat-sized marsupial with large appealing eyes, a pink nose and a very warm fluffy brown coat. Back in the 1800s it was prized for this coat and thus it was introduced to New Zealand when the fur trade was booming. It thrived in its new environment without predators of any sort, and while the fur trade died out, the possum numbers grew to plague proportions. Its estimated population by 1995 was 70 million, with only a few possum-free regions remaining.

Not only does this introduced species have an impact on the native vegetation, where it creates a conservation problem of huge dimensions (by eating approximately 4,000 football fields of native forest each day!), but it also threatens bird species by robbing their nests of eggs, and as a final straw it harbours and spreads bovine tuberculosis (Mycobacterium bovis). This threatens the primary industry of New Zealand, livestock, posing an enormous and extremely costly problem for New Zealand agriculture.

In New Zealand possums are thought to be the main source for bovine tuberculosis (Tb) in cattle. While the actual mechanisms of transmission from possums to cattle are still being debated, one possibility is that they leave secretions containing the bovine tuberculosis bacteria on the pastures where cattle feed. This is a serious problem for the dairy industry where the disease may be passed through the milk and, without adequate pasteurisation, on to the human population. All animals are tested on a regular basis, and any meat or milk from diseased animals cannot be sold. Animals testing positive are slaughtered and whole dairies suspended until they are clear of infection. The economic impact is enormous.

To reduce the spread of the disease in cattle, it is necessary to examine the dynamics of its spread in the possum population. With such large numbers of animals, any feasible *plan of action is costly and thus requires extremely careful consideration. We will consider a very simple model for the disease dynamics, and then include some strategies to reduce its spread: culling, and vaccination. We are interested in examining the impact of these strategies on the population dynamics.*

Baseline model. *First we examine the possum populations with no control strategies as the baseline model for comparison. Consider the possum population to have density $N(t)$ per hectare. We divide this into two mutually exclusive groups: those with the disease, the contagious infectives $I(t)$, and those who are without disease, the susceptibles $S(t)$. Possums do not recover from having Tb. Then $N(t) = S(t) + I(t)$ at any time and subsequently $N' = S' + I'$.*

We assume a simple natural per-capita birth rate of b and a natural per-capita death rate of a. We will adopt α_d to describe the per-capita death rate attributable to the disease and β to describe the infection transmission coefficient assuming density-dependent transmission. These parameters are annual estimates and thus we adopt a yearly time scale for the model. Furthermore, since the population of possums in New Zealand is so large and successful, we assume that $b - a > 0$.

A system of differential equations describing the dynamics is

$$\frac{dS}{dt} = b(S + I) - aS - \beta SI,$$
$$\frac{dI}{dt} = \beta SI - (\alpha_d + a)I. \tag{8.15}$$

This system is a simplified version of that proposed in Roberts (1992) where the latent class has been omitted and density-dependent birth and death rates have been substituted with their constant density independent values. (Note that the symbols used here are different from those used by Roberts (1992), but are consistent with those used for disease models in this text.) It is easy to establish that there are at most two equilibrium points, the first at $(S, I) = (0, 0)$ and the other at

$$(S, I) = \left(\frac{\alpha_d + a}{\beta}, \frac{(b - a)(\alpha_d + a)}{\beta \left(\alpha_d + a - b \right)} \right).$$

We are interested in how the populations behave over time, and so we employ linearisation analysis and examine the system dynamics through the phase-plane. From above, the non-zero equilibrium point becomes very large when α_d is close to, or approaches, $(b - a)$. If this is a stable point, it is very bad news if we are hoping to eradicate the pest. We would like to alter this with the implementation of some control strategy, but first we examine the dynamics of system (8.15). Some of the details are given in Exercise 8.8.

For the $(0, 0)$ equilibrium point, we can calculate eigenvalues $\lambda_1 = (b - a)$ and $\lambda_2 = -(\alpha_d + a)$, and we can establish that this point is always an unstable saddle point. For the other equilibrium point, the algebra is not quite as simple; however, the real parts of the eigenvalues determine the stability and they can be established from the sign of the trace of the Jacobean matrix J, evaluated at this equilibrium point, giving

$$\text{trace}(J) = \frac{-b(b - a)}{\alpha_d - (b - a)}.$$

Clearly this value is negative when $\alpha_d > b - a$ and then the equilibrium is stable. Alternatively, when $\alpha_d < b - a$, the point is unstable, but in that case the positive populations I and S increase without bound as can be deduced from the gradient of the trajectories in the phase-plane. (It is worth noting here that the values of these parameters for New Zealand possums, as given by Roberts (1992), have $\alpha_d > b - a$.) Whether this point is a node or a *focus can also be established from the Jacobian matrix, and by examining the discriminant of the characteristic polynomial. It can be shown that the equilibrium point is a focus if certain conditions are satisfied by the parameters; otherwise it is a node.*

In summary, linear theory predicts that the equilibrium point at the origin is always unstable and a saddle point under our assumption that $b > a$. The non-zero equilibrium point is an unstable node (if $\alpha_d < b - a$) with the trajectories approaching infinity, or a stable node/focus (if $\alpha_d > b - a$) with the trajectories converging to this point. Adopting estimates for the parameter values in Roberts (1992) that relate to the possum population in New Zealand, we have taken $b = 0.305$, $a = 0.105$, $\alpha_d = 3$ and $\beta = 0.7$. This corresponds to a lifespan of $1/a = 9.5$ years for a healthy possum and $1/\alpha_d = 1/3$ years (4 months) for a Tb-infected possum. These values have been incorporated into the time-dependent graphs and phase-plane diagram of Figure 8.13.

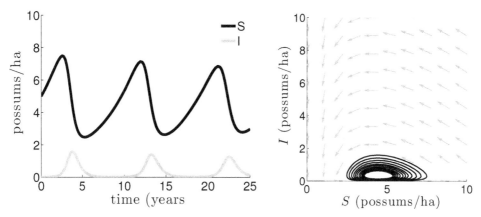

Figure 8.13: Baseline model — dynamics with no control strategy used. The time-dependent plot, S (black line) and I (grey line), and the phase-plane diagram for the NZ possum populations. Initial population densities are $S(0) = 5$ and $I(0) = 0.01$ possums per hectare with $\beta = 0.7$, $\alpha_d = 3$, $b = 0.307$, $a = 0.107$. In the phase-plane plot the time interval is $[0, 100]$ years.

Over a long time, the populations oscillate with decreasing amplitudes and eventually approach a stable population with densities $S = 4.44$ and $I = 0.32$ possums per hectare, thus predicting a total population of ≈ 5 animals per hectare. The time between outbreaks is approximately 15 years.

An important use of mathematical models is to examine how effective certain control strategies can be to help minimise (or in some cases maximise) the impact of a disease. Such strategies may include culling, where the rate of spread of the disease through the population may be slowed by reducing the rate of contact between individuals. Another strategy is to use some form of vaccination (delivered through food supplies). The use of mathematical modelling is to help determine which is the more effective strategy under different circumstances.

Culling strategy. *We now include in the model a constant per-capita rate of culling, κ, where κ represents the proportion of the total population culled per year. The system of equations (8.15) is now modified to*

$$\frac{dS}{dt} = b(S + I) - (a + \kappa)S - \beta SI,$$
$$\frac{dI}{dt} = \beta SI - (\alpha_d + a + \kappa)I. \tag{8.16}$$

This is equivalent to increasing the value of parameter a so that we can use the results from above to predict the dynamics. However, note that now we may have the situation where $b < a + \kappa$ so we need to include this in our interpretations. At this point it is informative to examine a graph of the nullclines to understand how the equilibrium points vary with parameter κ. This is illustrated in Figure 8.14.

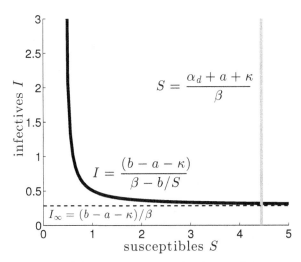

Figure 8.14: Graph of the nullclines ($S' = 0$-nullcline in black, $I' = 0$-nullcline in grey) for the system (8.16) to establish the impact of varying κ. The horizontal asymptote is at $I_\infty = (b - a - \kappa)/\beta$ (dashed line). The curves are drawn with same parameter values as used in Figure 8.13 and with $\kappa = 0$. We can use this diagram to think about what happens as κ increases from zero.

As κ increases from zero, so the hyperbole (with its asymptote at $(b - a - \kappa)/\beta$) shifts downwards and the I-nullcline at $(\alpha_d + a + \kappa)/\beta$ shifts to the right. This means that at equilibrium, the number of susceptibles increases while the number of infectives decreases compared to the baseline case $\kappa = 0$. However, what is more relevant for the dynamics is that the stability of both equilibrium points changes when $b \leq a + \kappa$ and we have Tb extinction ($I = 0$) at the stable equilibrium. For the particular parameter combination used above, we have Tb extinction for a cull rate $\kappa > b - a \simeq 20\%$ of the population per year.

Plotting the phase-plane with the parameters as before, but with $\kappa = 0.33$ included, we see that the dynamics match our theoretical predictions. The origin is now a stable equilibrium and the infected possum population rapidly approaches extinction as illustrated in Figure 8.15.

Vaccination strategy.. Another possibility for controlling Tb infection is the inclusion of a vaccination program. Vaccination is delivered through chemicals in food baits. We assume that a proportion ν of susceptible animals are vaccinated each year since vaccinating infected possums has no effect. We also assume that the vaccine will be permanent for that possum, so that once vaccinated it cannot be infected during its life. At the time Roberts (1992) was written, there was no vaccine available, but at the time of the writing of this book there is now one that protects possums from Tb for a couple of years. Here we seek to gain an understanding of the hypothetical benefit if a vaccine with perfect immunity were discovered, as was the aim of Roberts (1992).

We introduce a new dependent variable $Z(t)$ to represent the population of vaccinated animals, thus rendering three mutually exclusive groups: S the susceptibles, I the infectives

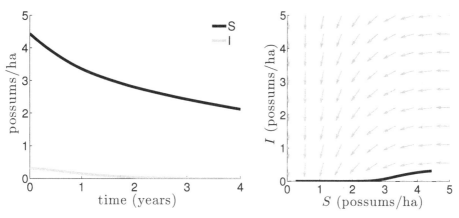

Figure 8.15: Culling control strategy. The time-dependent plot, $S(t)$ (black line) and $I(t)$ (grey line) and phase-plane diagram for the NZ possum populations with $\kappa = 0.33$. Other parameter values are the same as used in Figure 8.13.

and Z the vaccinated. The new model is

$$\frac{dS}{dt} = b(S + I + Z) - aS - \beta SI - \nu S,$$
$$\frac{dI}{dt} = \beta SI - (\alpha_d + a)I, \qquad\qquad (8.17)$$
$$\frac{dZ}{dt} = \nu S - aZ,$$

where ν is the per-capita vaccination rate of susceptibles.

We do not carry out any further algebraic analysis on this system here. However, we establish numerically some idea of how this vaccination program has affected the dynamics. Figure 8.16 illustrates the time-dependent graphs of the three populations, S, I and Z.

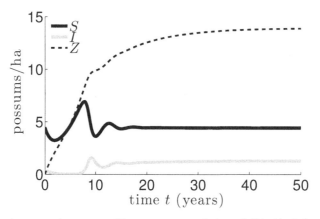

Figure 8.16: Vaccination control strategy. The possum population, S (black), I (grey) and Z (dashed), over time with the inclusion of a vaccination program with $\nu = 0.33$. Other parameters are the same as in Figure 8.13. The initial populations are ($S(0) = 10$, $I(0) = 5$ and $Z(0) = 0$ possums per hectare.

After a few initial damped oscillations the total population settles to the equilibrium value of approximately 19.64 possums per hectare in about 50 years. Thus there is a substantial *increase in the total possum population, as one might expect because vaccination also prevents possums from dying from Tb. This density is unrealistically high; however, it should be noted that the model used here does not have density dependence, so this density is higher than what would be observed in nature.*

Furthermore, the non-zero equilibrium point is given (approximately) as $(S, I, Z) = (4, 44, 1.31, 13.90)$ from `Maple` *or* `MATLAB`. *This equilibrium point is stable for the parameter combination of interest, and thus the infected population is larger than that where there is no implemented strategy. Not a very attractive scenario. This shift is due to the initial response to fewer deaths from disease. The scenario is not very attractive from a conservation perspective either, with the destruction of native vegetation increasing substantially. Although this model does not incorporate density dependence, it does indicate possible problems with the vaccination strategy.*

Comparison. *In Figure 8.17 we provide a comparison of the two control strategies on the possum infective population (I) over time, with the initial number of animals in each case given by the non-zero equilibrium of the baseline model (8.15). Here we compare the two strategies over a shorter time scale of four years, which might be of interest to groups concerned with the reduction of Tb as a short-term policy.*

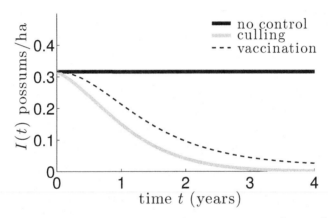

Figure 8.17: Comparison. The infective possum population over time under a culling strategy (gray), vaccinating a percentage of the population (dashed), and the infective population without any applied strategy (black). Parameter values are the same as in Figure 8.13 with vaccination rate $\nu = 0.33$ and culling rate $\kappa = 0.33$.

Culling still appears to be the most successful strategy in the short term, and we have seen in Figure 8.16 that vaccination will cause further Tb outbreaks in the longer term. This conclusion is consistent with the results of more complex models examined by Roberts (1996), which include a density-dependent contact rate and a latent infection compartment.

Further models. *Another possible strategy for reducing possum infections is that of sterilisation (reducing the number of possum births) using immuno-contraceptive chemicals in food. This was also considered by Roberts (1992) but found to be less effective than culling. This strategy is easily modelled by reducing the possum birth rate. We leave analysis of this strategy for the reader to investigate.*

The choices of parameter values used for possum Tb models have been much debated. Roberts (1996) has updated some of the parameters and extended the model of Roberts (1992). Barlow (2000) also has different opinions on the most appropriate values to use. Since the conclusions may change with different parameter values, accurate measurements from the field are of great value.

We note that there are still aspects of the possum and disease dynamics that are not included in these models, and that may have substantial impact. For example, the spatial distribution and movement of possums (see Fulford et al. (2002) for an associated model), and the rate at which cattle are infected or the characteristics of a particular vaccine agent, such as whether the vaccination protection wanes (i.e., reduces) with time. Furthermore, the direct and indirect costs associated with the strategies were not considered, such as the costs in hard dollar terms or the environmental impact of vaccination, which does not reduce the population size. However, we do gain some understanding of the impact on a population of a variety of disease eradication strategies. And the study highlights that although a control strategy might be beneficial for solving one problem (reducing Tb), it may exacerbate another (impact on the environment).

8.8   Exercises for Chapter 8

8.1. Species diversity. *With reference to the case study, competition, predation and diversity, of Section 8.2:*

(a) *Using* `Maple` *or* `MATLAB`, *generate the time-dependent graphs presented in the case study with the parameter values provided in the text.*

(b) *Show that for the simple competition model (when $N_3 = 0$), the condition $\alpha_{11}\alpha_{22} - \alpha_{12}\alpha_{21} \leq 0$ predicts extinction for all but one species in most cases.*

(c) *Find the equilibrium points for the system.*

8.2. Scaling. *Scaling can be used to simplify the mathematical analysis of a model by reducing the number of parameters and writing the equation in a dimensionless form. Consider the logistic equation*

$$\frac{dX}{dt} = rX\left(1 - \frac{X}{K}\right), \qquad X(0) = x_0.$$

Since X and K are in the same units, we choose a new dimensionless variable Y as a measure of X in terms of the carrying capacity K, such that $X/K = Y$.

(a) *Show that, with this change of variable, the above logistic equation becomes $dY/dt = r(1 - Y)Y$, with $Y(0) = y_0 = x_0/K$.*

(b) *Furthermore, the independent variable can be scaled in units of $1/r$, using $t = s/r$. Recall that the chain rule gives*

$$\frac{dY}{ds} = \frac{dY}{dt}\frac{dt}{ds}.$$

Show, using the chain rule, that with this scaling the logistic equation becomes

$$\frac{dY}{ds} = (1 - Y)Y, \qquad Y(0) = \lambda, \qquad \lambda = \frac{x_0}{K}.$$

Thus the model is reduced to a dimensionless form with only one parameter, λ.

8.3. Scaling the extended predator-prey model. *Consider the extended predator-prey model, the Holling–Tanner model developed in Section 8.3,*

$$\frac{dX}{dt} = r_1 X \left(1 - \frac{X}{K}\right) - \frac{c_1 XY}{1 + t_h c_1 X}, \qquad \frac{dY}{dt} = r_2 Y \left(1 - \frac{Y}{c_3 X}\right).$$

To simplify the analysis, the equations can be scaled as follows: Define dependent and independent variables x, y and τ by

$$x = \frac{X}{K}, \qquad y = \frac{Y}{c_3 K}, \qquad \tau = r_2 t.$$

The chain rule can be used to change both the dependent and independent variables in the equations since we have that

$$\frac{dx}{dt} = \frac{dX}{dt}\frac{1}{K}, \qquad \frac{dx}{d\tau} = \frac{dx}{dt}\frac{dt}{d\tau}.$$

In order to find the derivative $dx/d\tau$, use the chain rule as above to get expressions for X' in terms of x and $dx/d\tau$. Then substitute these expressions into the original equation for X'. In this way, all terms involving X, X' and Y can be replaced with terms in x, x' and y. Similarly, the equation for $dy/d\tau$ can be obtained. Show that

$$\frac{dx}{d\tau} = \lambda_1 x(1 - x) - \frac{\lambda_2 xy}{(\lambda_3 + x)}, \qquad \frac{dy}{d\tau} = y\left(1 - \frac{y}{x}\right),$$

with

$$\lambda_1 = \frac{r_1}{r_2}, \qquad \lambda_2 = \frac{c_3}{t_h r_2}, \qquad \lambda_3 = \frac{1}{t_h c_1 K}.$$

8.4. Extended predator-prey model. *Consider the scaled equations for the Holling–Tanner predator-prey model derived in Exercise 8.3,*

$$\frac{dx}{d\tau} = \lambda_1 x(1 - x) - \frac{\lambda_2 xy}{(\lambda_3 + x)}, \qquad \frac{dy}{d\tau} = y\left(1 - \frac{y}{x}\right),$$

where λ_1, λ_2 and λ_3 are dimensionless positive numbers, defined in the previous question.

(a) *Use* `Maple` *or* `MATLAB` *to draw the phase-plane for the scaled system of Exercise 8.3 above, with parameter values $\lambda_1 = 5$, $\lambda_2 = 10$ and $\lambda_3 = 0.015$. Show that, for a variety of initial conditions, the system settles to a single periodic solution over time as illustrated in Figure 8.4.*

(b) *By decreasing the value of $t_h c_1$, estimate roughly for what value of λ_3 this periodic solution collapses to a stable/unstable node/focus. What changes, if any, do you observe in the nature of the periodic solution as the parameter is decreased? What does this mean for the time-dependent graphs of the populations?*

(c) *Repeat this process for a variety of λ_2 values between 4 and 10. Then, using these results, plot the points (λ_3, λ_2), where the change in dynamics takes place, in the (λ_3, λ_2)-plane.*

Joining the points together provides the definition of a rough boundary in the parameter plane, which divides periodic behaviour from where all trajectories approach a stable point.

8.5. Prickly-pears. *Consider the plant-herbivore model presented as a case study on the spread of the prickly-pear in Australia, and its annihilation by an introduced moth Cactoblastis (see Section 8.5).*

(a) *What growth processes (exponential or logistic) are adopted for the plant biomass and the moths in this model? What are the numerical response and the functional response terms for this system?*

(b) The interaction term appearing in the rate equation for the plant biomass is given as

$$-c_1 H \left(\frac{V}{V + D} \right).$$

Explain clearly why you think a function of this nature has been used, in terms of its changing contribution when the number of plants increases, or decreases, significantly.

(c) In the case study, the claim is made that the system settles to a fixed and stable number of 11 plants per acre over a period of time. Using `Maple` or `MATLAB`, show that this is indeed the case for the suggested model with the parameter values given.

8.6. Lemmings. Consider the case study on Lemming mass suicides of Section 8.4.

(a) The model presented claims to support a four-year cycle. Using `Maple` or `MATLAB`, generate the time-dependent plots over both short and long time periods, as well as a phase-plane. Do your results support the claim of the authors? Provide clear reasons for your answers.

(b) Find the equilibrium values for the system in the case with constant parameter values (not time-dependent parameters).

(c) Linearise the system and classify these equilibrium points. Comment on whether your results support the claim made in (a) above.

8.7. Rabbits, hares and geese. The case study of Section 8.6, geese defy mathematical convention, presents improved plant-herbivore models appropriate to systems that do not exhibit the top-down control approach.

(a) Explain the suggested meaning of the parameters c_{max} used in the model.

(b) Sketch graphs of the numerical response and the functional response terms for the general case, in each of the two models. When do the two functional response terms differ significantly and when are they similar?

Explain how these differences and similarities reflect the different intentions of the two proposed models in the case study.

(c) Consider the reduced digestion efficiency model with parameter values as in the text, $K = 10$, $r = 1$, $c_{max} = 1$, $a = 10$, $b = 0.065$, $e_{max} = 0.04$ and $d = 0.1$.

The case study claims that the model can be used to predict when and how, by changing a parameter, the herbivore could move towards extinction. Thus, there is the possibility of eradicating a pest by encouraging plant growth beyond a threshold. It is also possible to control the vegetation by increasing herbivore numbers.

Use `Maple` or `MATLAB` to generate the phase-plane, together with direction trajectories and system nullclines. Show, considering the phase-plane, how increasing or decreasing the herbivore carrying capacity can force a change in the dynamics so that the two thresholds discussed above can be crossed. Make rough estimates of the values of K for which this change occurs.

(d) Using the second model proposed, the reduced consumption rate model, with the same parameter values as above, do the threshold values estimated in (c) change? Give reasons for your answer and find the new thresholds if they are different.

(e) Identify a value of K for which there is a periodic solution in each of the models proposed, and choose appropriate initial conditions to illustrate them in a phase-plane, using `Maple` or `MATLAB`.

8.8. **Bovine tuberculosis.** *A model for the bovine tuberculosis epidemic in NZ possums is presented in the case study, possums threaten New Zealand cows, of Section 8.7:*

$$\frac{dS}{dt} = b(S + I) - aS - \beta SI,$$

$$\frac{dI}{dt} = \beta SI - (\alpha_d + a)I.$$

(a) *Show that the only two equilibrium points are at $(S, I) = (0, 0)$ and*

$$(S, I) = \left(\frac{\alpha_d + a}{\beta}, \frac{(b - a)(\alpha_d + a)}{\beta (\alpha_d + a - b)} \right).$$

(b) *Construct the Jacobian matrix (J) at the non-zero equilibrium point from (a) and hence show that*

$$\text{trace}(J) = \frac{-b(b - a)}{\alpha_d + a - b},$$

and that the equilibrium point is a focus if

$$\frac{4(\alpha_d + a)(\alpha_d + a - b)^2}{b^2(b - a)} > 1,$$

and otherwise that it is a node.

(c) *Establish the nullclines for the system and sketch a graph to indicate how these nullclines, and thus the equilibrium points, shift when culling is included in the model.*

(d) *Alternatively, if a vaccination program is introduced how are the nullclines, and thus the equilibrium points, changed?*

(e) *Using* `Maple` *or* `MATLAB`, *draw, on the same system of axes, the infective population for the three models — the population with no control, that with a culling program, and that with a vaccination program. Use parameters values as in the case study.*

8.9. **Bovine tuberculosis with vaccination.** *Consider the equations for the possum bovine tuberculosis epidemic from Section 8.7, possums threaten New Zealand cows, with vaccination included:*

$$\frac{dS}{dt} = b(S + I + Z) - aS - \beta SI - \nu S,$$

$$\frac{dI}{dt} = \beta SI - (a_d + a)I,$$

$$\frac{dZ}{dt} = \nu S - aZ.$$

(a) *Draw a compartmental diagram for this model.*

(b) *Show that there are only two equilibrium points: one at the origin and find an expression for the non-zero equilibrium point.*

(c) *Construct the associated Jacobian matrix at the non-zero equilibrium point. (For known parameter values the trace and determinant can be calculated easily, but for general parameter values the algebra can get complicated and may yield little information.)*

Chapter 9

Formulating heat and mass transport models

This chapter is concerned with modelling processes that involve heat conduction. We concentrate on formulating differential equations for the heat and temperature of systems, and emphasise the difference between them. In the absence of heat conduction, we use the input-output principle, or the balance law introduced in Chapter 2, to determine the heat content of the object we are modelling. When conduction is involved, we also apply this input-output of heat energy, but to an arbitrarily thin section. The analytic solutions for these models are derived in the subsequent Chapters 10 and 11.

9.1 Introduction

We begin by introducing some problems that involve heating and cooling. It is of great importance to understand whether or not heat conduction is an important factor in any particular application, and to this end we consider a variety of problems, some of which require the consideration of heat conduction and others which do not.

Cooling of a cup of coffee

Suppose we have a hot cup of coffee sitting on a benchtop, too hot to drink. If the coffee is at a temperature of 60°C, how long will it take to cool down to 40°C? To begin to answer this, we introduce some simple physical concepts. In particular, we look at the distinction between temperature and heat, the two most important physical quantities in this problem.

First, consider *temperature*: This represents how hot the coffee is. If we can assume the temperature of the coffee is uniform throughout (we call this homogeneous), then the temperature will be a function of time alone. Temperature is measured in degrees Celsius °C, or in Kelvin K, where the temperature in °C is 273 less than the temperature in K, approximately. That is, °C + 273 = K.

Second, consider the quantity *heat*. The temperature of the coffee drops because heat energy is transferred to the surrounding air. This is due to the fact that the surrounding air is at a lower temperature, and heat is always transferred from a region of high temperature to a region of lower temperature. Since heat is a form of energy, it is measured in Joules, in units of the SI system.

Thus to answer the question about the cup of coffee, we need to formulate a suitable equation for the temperature. But to do this we need to determine what is happening to the heat. So we construct a compartmental diagram for the heat input and output for the cup of coffee (see Figure 9.1).

Equations that relate heat energy to temperature are derived in Section 9.2, using the concept of *specific heat*. We assume the temperature is homogeneous throughout the compartment, and thus a function of time alone. We also need an equation that describes

Figure 9.1: Input-output compartmental diagram for the heat content in a cup of coffee.

the rate of heat transfer to the surroundings. This too is introduced in Section 9.2, as is *Newton's law of cooling*.

A similar problem that we consider is that of determining the time taken for the body of water in a hot water tank to heat to a specified temperature. We again assume the temperature to be homogeneous, and thus a function of time. In this application we also need to account for the heat supplied to the system by the heating element.

Other heat conduction applications

Consider the problem of finding the rate at which heat can be transferred through a wall. This is important for determining the insulating properties of the wall, which are useful energy efficiency considerations.

The temperature on the outside of the wall will be different from that on the inside, which suggests we should also think of the temperature as a function of distance. There will be a temperature gradient in the wall, and heat will flow in the direction of hot to cold (from the inside to the outside of a wall in winter) as a result of heat conduction.

We also consider heat conduction through a hot water pipe insulating jacket. This is an interesting problem because we get counterintuitive results. We see that, under certain circumstances, insulating a hot water pipe is less efficient (since more heat escapes) than not insulating it at all! Also, we consider the conduction effect of heat fins that are used to conduct heat away from a source, such as in computers or on a motorbike.

9.2   Some basic physical laws

Before formulating equations for the problems mentioned above, we need to understand some basic physical concepts of heat transport, as well as the distinction and relationship between heat and temperature.

Heat and temperature

The result of applying heat, or heat energy, to an object is to raise its temperature. Similarly, as an object cools, its temperature drops as it loses heat. The greater the mass of the object, the more heat is required to change its temperature.

When formulating a mathematical model for a problem involving heating or cooling, we need to take into account the amount of heat flowing into or out of a system. While it makes little sense to talk about a 'flow' of temperature, it is a change in temperature that we observe. Thus we need to be able to relate heat and temperature, and this can be done using the specific heat of a substance.

We assume the change in heat is directly proportional to the change in temperature and also the mass of the object. Thus we write

$$
\left\{ \begin{array}{c} \textit{rate of} \\ \textit{change of} \\ \textit{heat content} \end{array} \right\} = c \times \{mass\} \times \left\{ \begin{array}{c} \textit{rate of} \\ \textit{change of} \\ \textit{temperature} \end{array} \right\}, \tag{9.1}
$$

where c is a positive constant of proportionality, known as the specific heat of the material. The definition of specific heat is the amount of heat required to raise the temperature of 1 kg of a substance at a given temperature, by 1°C. In symbols, if we define Q as the rate of change of heat with time (measured in Watts), m as the mass of the material being heated or cooled (measured in kilograms), and U as the temperature[1], then

$$
Q = cm\frac{dU}{dt}.
$$

We have assumed c to be independent of the mass of the object and the temperature, but we know that metals absorb heat more easily than water. Thus a greater amount of heat is needed to raise the temperature of 1 kg of water by one degree than is needed to raise the temperature of 1 kg of metal by one degree. This implies that metals have a smaller specific heat than water. In Table 9.1, the specific heats of some common materials are given.

Table 9.1: Table of specific heats for some common substances. Units are standard SI units (i.e., $\mathrm{J\,kg^{-1}\,°C^{-1}}$).

Substance	c	Substance	c
Aluminium	896	Asbestos	841
Copper	383	Brick	840
Stainless steel	461	Glass	800
Wood	2,385	Butter	2,300
Concrete	878	Lamb	3,430
Water (at 20°C)	4,187	Potatoes	3,520

The specific heat, c, is not actually constant over a large temperature range; however, if the temperature range is not too large use of (9.1), with constant specific heat c, provides reasonable predictions.

Newton's law of cooling

A mechanism for heat to be lost from an object is that of exchanging heat energy with its surroundings. This takes place from the surface of the object, and thus we would expect the rate of heat loss to increase with the exposed surface area of the object, and be directly proportional to this area. Further, if the difference in temperature between the surface of the object and the surroundings increases, then we would expect heat to be lost faster.

We therefore make the assumption that the rate of heat flow is directly proportional to the temperature difference between the surface and its immediate surroundings. Further, the existence of slowly moving air across the cooling object (a slight breeze) is assumed. Under these conditions we have Newton's law of cooling, which works equally well with

---

[1]U (not T) is used for temperature so that it is not confused with time t.

heating problems. Newton's law of cooling states that

$$\left\{ \begin{array}{c} \text{rate heat} \\ \text{exchanged with} \\ \text{surroundings} \end{array} \right\} = \pm hS\Delta U, \tag{9.2}$$

where ΔU is the temperature difference, S is the surface area from which heat is lost/gained (in units of m^2), and h is a positive constant of proportionality. The constant h is called the *convective heat transfer coefficient*, or the *Newton cooling coefficient*. The units for h are Watts/m^2/°C.

The correct sign for the temperature difference is determined in any specific problem by the direction of heat transfer and whether the surface is gaining or losing heat. If S is the surface area of an object and if the temperature U of the object is greater than the temperature u_s of the surroundings, then we would write

$$\left\{ \begin{array}{c} \text{rate heat} \\ \text{exchanged with} \\ \text{surroundings} \end{array} \right\} = hS\left(U - u_s\right) = hS\Delta U.$$

When the temperature of the surroundings is greater than the temperature of the object $(u_s > U)$, then heat is exchanged in the opposite direction and so

$$\left\{ \begin{array}{c} \text{rate heat} \\ \text{exchanged with} \\ \text{surroundings} \end{array} \right\} = hS\left(u_s - U\right) = -hS\Delta U.$$

Returning to the cup of coffee

Applying the above expressions, we can now link the rate of change of heat in (9.1) directly to the rate of change of temperature, as in the following example.

Example 9.1: *Formulate a differential equation for the temperature of a cup of coffee over time.*

Solution: *To describe the conservation of heat energy we have, from the heat energy input-output compartment diagram of Figure 9.1 in Section 9.1, that*

$$\left\{ \begin{array}{c} \text{rate of} \\ \text{change of} \\ \text{heat content} \end{array} \right\} = \left\{ \begin{array}{c} \text{rate heat} \\ \text{lost to} \\ \text{surroundings} \end{array} \right\}.$$

Using (9.2) and (9.1) we thus obtain the differential equation

$$cm\frac{dU}{dt} = -hS(U - u_s). \tag{9.3}$$

Note that a negative sign is chosen for the term on the RHS. The cup of coffee is cooling and $(dU/dt) < 0$; and then since $\Delta U = U(t) - u_s$ is always positive we need to include a negative sign for consistency.

Once again we have an example of an equation describing exponential growth or decay, depending on whether the object is warm or cold compared with its environment. The heat loss from the hot cup of coffee left to stand follows an exponential decay process and, as we see in Section 10.1, it stabilises at the temperature of the surroundings u_s as expected.

We have assumed h to be a constant; however, this is a limitation of the model as the rate of heat lost from a surface will be affected by the air flow near it. This is sometimes called a *wind-chill* effect. Table 9.2 illustrates this by showing how the coefficient h increases with the velocity of the air-flow passing over a metal plate, and indicates that the effect is nontrivial.

Table 9.2: The value of the convective heat transfer coefficient h from a plate of length $0.5\,\text{m}$ over which an airflow, given in metres per second (m/s), passes.

	h
Plate in still air	4.5
Air-flow at $2\,\text{m/s}$ over plate	12
Air-flow at $35\,\text{m/s}$ over plate	75

Summary of equations

To express a change of heat in terms of a change of temperature, we use the specific heat equation

$$\left\{ \begin{array}{c} \text{rate of} \\ \text{change of} \\ \text{heat content} \end{array} \right\} = cm \frac{dU}{dt},$$

where U is the temperature, m the mass of an object and c its specific heat. Newton's law of cooling is given by

$$\left\{ \begin{array}{c} \text{rate heat} \\ \text{exchanged with} \\ \text{surroundings} \end{array} \right\} = \pm h S \Delta U,$$

where S is the surface area, ΔU the temperature difference and h a positive constant, namely the convective heat transfer coefficient or the Newton cooling coefficient.

Summary of skills developed here:

- *Distinguish between heat and temperature.*
- *Relate heat to temperature using the specific heat of a substance.*
- *Formulate a differential equation for the temperature of an object that is either heating or cooling.*

9.3 Model for a hot water heater

We derive a differential equation for the temperature of water being heated by an electric heating element. The problem applies principles similar to those in the example above, except that in this case heat is added to the system via the heating element. We write a heat balance equation expressing the rates of change of heat inputs and outputs that leads to a differential equation for the temperature as a function of time.

Problem description

Some domestic examples of water heaters include hot water systems, urns and kettles. Our problem specifically refers to an electrically heated hot water system, which is the standard for an average home. It usually contains 250 litres of water and is cylindrical with dimensions of height $1.444\,\text{m}$ and diameter $0.564\,\text{m}$. Initially we assume the water to be at a temperature of $15°\text{C}$. The heating element when switched on supplies heat at a constant rate of $3.6\,\text{kW}$. (The other commonly used elements supply $4.8\,\text{kW}$.) In Figure 9.2 a schematic diagram illustrates the heat input and output for the system. We wish to determine how long it would take to heat the water to $60°\text{C}$.

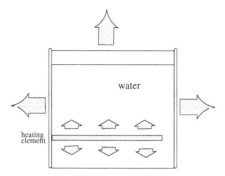

Figure 9.2: Schematic diagram of a hot water heater.

In order to study this problem we introduce some notation. We let $U(t)$ denote the temperature of the water at time t. We denote by u_0 the initial temperature of the water, u_f as the final temperature, m as the mass of water in the heater and q as the rate of heat energy supplied. S in m^2 will denote the surface area from which heat can escape, which in the case of the hot water heater is the surface of the tank.

Other symbols will be introduced as required. Note that by assigning symbols to these quantities, we increase the usefulness of our model, since we can later substitute various values for these symbols. We can thus use the model to explore the effect and sensitivity of change in these different physical quantities.

Model assumptions

We examine what simplifying assumptions can be made about the problem and build the model using these.

- First, let us assume that the water in the tank is well stirred so that the temperature remains homogeneous throughout. Without this assumption the problem would be far more complex, since temperature would then be a function of both time and position, and we would also have to take into account the heat transfer (by conduction and convection) between different points in the tank. By assuming that the water is well stirred, we only have to consider the temperature of the water as a function of time.

- We also assume that heat is lost from the surfaces of the tank according to Newton's law of cooling.

- Finally, we assume that thermal constants, such as the specific heat and the Newton cooling coefficient, remain constant in our applications.

General compartmental model

We start formulating the model by writing a word equation to describe the balance of heat energy in terms of rates of heat input and output. We must account for the heat produced by the heating element, and the heat lost from the surface of the water heater to the surroundings.

Example 9.2: *Give a suitable word equation describing the rate of change of heat.*

Solution: *We start by drawing an input-output diagram for the heat content of the system by applying the balance law. This is shown in Figure 9.3.*

A heat balance for the system is described by the word equation

$$\left\{ \begin{matrix} rate\ of \\ change \\ of\ heat \end{matrix} \right\} = \left\{ \begin{matrix} rate\ heat \\ produced\ by \\ heating\ element \end{matrix} \right\} - \left\{ \begin{matrix} rate\ heat \\ lost\ to \\ surroundings \end{matrix} \right\}. \tag{9.4}$$

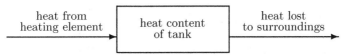

Figure 9.3: Input-output compartmental diagram for the heat content in the domestic hot water system.

Formulating the differential equation

As the temperature of the system rises, the net heat produced goes into raising the temperature of the system. To obtain a mathematical expression for the heat used to change the temperature, we use the fundamental equation relating heat to temperature through the specific heat from the first section of this chapter:

$$\left\{ \begin{matrix} \text{rate of} \\ \text{change of} \\ \text{heat content} \end{matrix} \right\} = cm\frac{dU}{dt},$$

where c is the specific heat of the water, m is the mass of the water and $U(t)$ denotes its temperature at time t.

Example 9.3: *Find expressions for the other terms in the word equation (9.4) and hence formulate a differential equation for the temperature.*

Solution: *It is assumed that the heating element produces heat at a constant rate per unit time, which we denote by q. Thus,*

$$\left\{ \begin{matrix} \text{rate heat} \\ \text{produced by} \\ \text{heating element} \end{matrix} \right\} = q. \tag{9.5}$$

For the rate of heat lost to the surroundings we use Newton's law of cooling, which states

$$\left\{ \begin{matrix} \text{rate heat} \\ \text{lost to} \\ \text{surroundings} \end{matrix} \right\} = hS \left\{ \begin{matrix} \text{temperature} \\ \text{difference} \end{matrix} \right\} = hS\left(U(t) - u_s\right), \tag{9.6}$$

where S is the surface area of the heater and h is the Newton cooling coefficient. The temperature difference is the difference between the current water temperature $U(t)$ and the (constant) temperature of the surroundings u_s.

Now that we have suitable expressions for each of the quantities in equation (9.4), we can obtain a differential equation for the temperature. Substituting equations (9.5) and (9.6) into equation (9.4) we obtain the differential equation

$$cm\frac{dU}{dt} = q - hS\left(U(t) - u_s\right), \tag{9.7}$$

which describes the temperature variation of the water with time.

It is generally simpler to formulate differential equations using rates of change, but in the following example we also give an alternative formulation using amounts of change in a finite time interval.

Example 9.4: *Repeat the derivation of the differential equation by considering the change in heat over a small but finite time interval, Δt, then take the limit as $\Delta t \to 0$.*

Solution: *For a small, but finite, time interval t to $t + \Delta t$ we can write the change in heat energy as*

$$\left\{ \begin{array}{c} \text{change in} \\ \text{heat energy} \\ \text{in time } \Delta t \end{array} \right\} = \left\{ \begin{array}{c} \text{amount of heat} \\ \text{produced by} \\ \text{water heater} \end{array} \right\} - \left\{ \begin{array}{c} \text{amount of heat} \\ \text{lost to} \\ \text{surroundings} \end{array} \right\}.$$

The amount of heat energy produced by the heater is the constant rate of heat produced multiplied by the length of time, $Q\Delta t$. Similarly, for the amount of heat lost to the surroundings, we have a rate of heat lost given by Newton's law of cooling, but this is not constant and changes over time. However, the time interval is short, so the amount of heat lost can be calculated using some temperature somewhere inside the interval $U(t^*)$. This is effectively using the mean value theorem from calculus. So the amount of heat lost to the surroundings is $hS(U(t^*) - u_s)\Delta t$, where $t < t' < t + \Delta t$. The change in heat energy is $mc(U(t + \Delta t) - U(t))$. Hence the word equation becomes

$$mc\left(U(t + \Delta t) - U(t)\right) = Q\Delta t - hS\left(U(t^*) - u_s\right)\Delta t.$$

Dividing by Δt gives

$$mc\frac{U(t + \Delta t) - U(t)}{\Delta t} = Q - hS\left(U(t^*) - u_s\right).$$

Now, take the limit as $\Delta t \to 0$ and use the definition of a derivative. Also note that in the limit as $\Delta t \to 0$, $U(t^*) \to U(t)$ since $t < t^* < t + \Delta t$, and we obtain

$$mc\frac{dU}{dt} = Q - hS(U - u_s),$$

where we abbreviate $U(t)$ to U since the time dependence of U is implied by the dU/dt term.

Numerical solution

Given an initial condition and suitable values for all the parameters, we can solve this differential equation numerically using `Maple` or `MATLAB`. (We also see how to obtain an analytic solution and make use of it in Section 10.2.)

For a typical set of parameters we consider the standard 250 litre tank, of height $1.444\,\text{m}$ and diameter $0.564\,\text{m}$. We take $m = 250\,\text{kg}$, $c = 4{,}200\,\text{W}\,\text{kg}^{-1}\,\text{s}^{-1}$, $h = 12\,\text{W}\,\text{m}^{-2}\,{}^\circ\text{C}^{-1}$, $S = 3.06\,\text{m}^2$, $q = 3{,}600\,\text{W}$ and $u_0 = u_s = 15{}^\circ\text{C}$ (all in SI units). The plot of temperature against time is given in Figure 9.4. The `Maple` code is given in Listing 9.1 and the `MATLAB` code is given in Listing 9.2. Figure 9.4 has units of time expressed in hours. In `Maple` or `MATLAB`, this is done by simply plotting $t/60^2$, which is easier than converting the units of all the constants into hours. Note that the model no longer applies after the temperature reaches boiling point ($100{}^\circ\text{C}$) since it fails to account for the additional latent heat needed to boil the water.

Listing 9.1: Maple code: c_he_waterheater.mpl

```
restart: with(plots): with(DEtools):
m:=250: c:=4200: h:=12: S:=3.06: q:=3600: us:=15:

de1:= c*m*diff(U(t),t) = q - h*S*(U(t) - us);
inits := U(0)=15;
plot1:=DEplot( de1, [U] , t =0..10*60^2,[[inits]]);
display(plot1);
```

Listing 9.2: MATLAB code: c_he_waterheater.m

```
function c_he_waterheater
global m c h S q us;
m=250; c=4200; h=12; S=3.06; q=3600; us=15;
```

```
u0 = 15;        %initial temperature
tend = 10*60^2; %end time in seconds
[tsol, Usol] = ode45(@rhs, [0 tend], u0);
plot(tsol/60^2, Usol); %plot for time in hours

function Udot = rhs(t, U)
global m c h S q us;
Udot = q/c/m - h*S/c/m*(U - us);
```

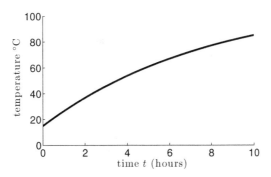

Figure 9.4: Plot of temperature against time using $m = 250\,\text{kg}$, $c = 4{,}200\,\text{W}\,\text{kg}^{-1}\,\text{s}^{-1}$, $h = 12\,\text{W}\,\text{m}^{-2}\,^{\circ}\text{C}^{-1}$, $S = 3.06\,\text{m}^2$, $q = 3{,}600\,\text{W}$ and $u_0 = u_s = 15^{\circ}\text{C}$.

In Figure 9.5 we repeat the solution with the heat input reduced considerably to only 500 W, and see that the temperature eventually approaches a constant value of approximately 30°C. This constant value is an equilibrium solution to the differential equation. We investigate this further in Section 10.2.

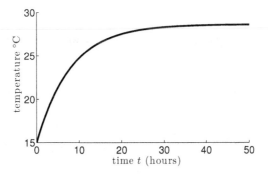

Figure 9.5: Plot of temperature against time (in hours) using $m = 250$, $c = 4{,}200$, $h = 12$, $S = 3.06$, $q = 500$ and $u_s = 15$ (all in SI units). The temperature approaches an equilibrium value.

9.4   Heat conduction and Fourier's law

Before starting to solve any specific problems, we first lay some groundwork in the basic physics of heat conduction. We define *heat flux* to describe the flow of heat and introduce *Fourier's law*, which relates the heat flux to the temperature.

Heat and temperature

In a given heating problem the temperature starts from some initial state and then, as a body is cooled or heated, the temperature changes. If the temperature is not the same at every point throughout the body, then heat will also be conducted through the body from regions of higher temperature to regions of lower temperature.

Eventually, the temperature will tend to some equilibrium state where there may still be a flow of heat through the body by conduction, but the temperature at any point within the body will not change with time. When the temperature is in this equilibrium state, none of the heat is used to increase the temperature, nor is any of the heat released to decrease the temperature. We can thus say, for *thermal equilibrium*, that

$$\left\{ \begin{array}{c} rate\ of \\ change\ of \\ heat\ content \end{array} \right\} = 0. \tag{9.8}$$

Heat conduction and heat flux

Heat conduction can occur in a solid, liquid or gas, and involves the transfer of heat energy by the vibration of molecules. In this book we are concerned with heat *conduction*; however, there are other heat transport mechanisms such as *convection*, the transfer of heat in a liquid or gas by the fluid flow, and *radiation*, which involves heat transfer by electromagnetic waves.

Observation demonstrates that the rate at which heat is conducted through a body is directly proportional to the cross-sectional area through which the heat flows. It is thus reasonable to talk about a rate of heat flow per unit area through a cross-section. This quantity is called the heat flux. We formally define *heat flux* $J(x)$ by

$$J(x) = \left\{ \begin{array}{c} rate\ of \\ flow\ of\ heat \\ per\ unit\ time \\ per\ unit\ area \end{array} \right\} \tag{9.9}$$

Heat flux is measured in SI units as Watts per square metre, where Watts are Joules per second.

Fourier's law of heat conduction

Common observation indicates that some substances conduct heat better than others. Heat flows more easily through certain metals than through substances such as brick and stone.

Fourier did some controlled experiments where he measured the heat conducted through thin plates of different materials. In his experiments he held the temperature constant, but at different values on either side of the plate. He noticed that the heat flux increased with the temperature difference, and decreased as the thickness of the plates increased. He thus came to the conclusion that the heat flux (rate of heat flow per unit area per unit time) is proportional to the temperature gradient, which is the gradient of the temperature expressed as a function of distance through the plate.

We can express this result mathematically. If $J(x)$ denotes the heat flux at x, and $U(x)$ the temperature at x, then *Fourier's law* states

$$J(x) = -k\frac{dU(x)}{dx}. \qquad (9.10)$$

The positive constant k is called the *conductivity*, which will be different for different materials.

The minus sign is necessary to ensure a positive rate for the conduction of heat. This means that when the temperature is decreasing (i.e., when the temperature gradient is negative), the heat flows in the positive direction. When the temperature is increasing (i.e., when the temperature gradient is positive), the heat flows in the negative direction. This reflects the fact that heat flows from regions of higher temperature to regions of cooler temperatures, and not the other way around.

Table 9.3 gives values of the conductivity k for some different materials. Note that generally metals have higher conductivities than other materials, which means that heat flows more easily through them.

Indeed, it is sometimes desirable to use materials with low conductivities such as rock wool or polystyrene to provide good insulation, while at other times high conductivity is desired as in heating elements.

Table 9.3: Table of heat conductivities for some common materials (standard SI units).

Substance	k	Substance	k
Copper	386	Brick	0.38–0.52
Aluminium	204	Asbestos	0.113
Iron	73	Concrete	0.128
Stainless steel	14	Glass	0.81
Water (at 0°C)	0.57	Wood	0.15
Lamb (at 5°C)	0.42	Rock wool	0.04
Butter (at 5°C)	0.20	Polystyrene	0.157

Over large temperature ranges the conductivity of a material k is not strictly constant. However, the assumption of constant conductivity over temperature ranges that do not vary dramatically still enables us to make sufficiently accurate predictions.

Summary of key ideas

The important equations of heat transport are summarised as follows: For thermal equilibrium,

$$\left\{ \begin{array}{c} \text{rate of} \\ \text{change of} \\ \text{heat} \end{array} \right\} = 0.$$

Fourier's law of heat conduction, where J is the heat flux and k the conductivity, is given by

$$J(x) = -k\frac{dU(x)}{dx}.$$

Summary of skills developed here:

- *Understand the physical meaning of thermal equilibrium and relate it to a zero time-derivative for temperature.*
- *Describe the flow of heat in terms of heat flux.*
- *Understand the assumptions for, and application of, Fourier's law.*

9.5   Heat conduction through a wall

We now formulate a differential equation for the simplest heat conduction problem. This involves the conduction of heat through a slab of some material in just one direction. We develop a differential equation for the equilibrium temperature reached as time increases.

Problem description

Imagine we have a plate with width L of some material. One side is hot and the other relatively cold, so that heat flows through the material in the x-direction towards the cooler side. A typical application is the wall of a building where the inside and outside are at different temperatures, as in Figure 9.6.

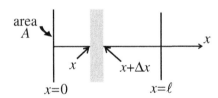

Figure 9.6: Coordinate system for the problem. We focus on the rate of heat going into and coming out of an imaginary section between the surfaces x and $x + \Delta x$.

General compartmental model

Our aim is to formulate a differential equation for the equilibrium temperature inside the material at any point x. We do this by first writing a word equation describing the rate of heat input to and output from a small section of the material, assuming no heat is used to heat up the material. We then let the thickness of the section tend to zero.

We consider a thin section, or slice, of the material from x to $x+\Delta x$ as shown in Figure 9.6 as our compartment. This is a modified version of the approach we have taken before. The surface area of the face of this slice is A. We must account for all the inputs and outputs of heat for this section, that contribute to the rate of change of heat. We do this in the following example.

Example 9.5: *Obtain an input-output word equation for the arbitrary section of width Δx.*

Solution:

The only heat input is due to conduction into the section at x, and the only heat output is due to the heat conducted out of the section at $x + \Delta x$. The input-output diagram in Figure 9.7 illustrates this.

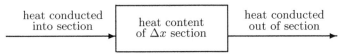

Figure 9.7: Input-output compartmental diagram describing the heat inputs and outputs to the arbitrary section of thickness Δx.

Hence, we have the word-equation

$$\left\{\begin{array}{c} \text{rate of} \\ \text{change of} \\ \text{heat in section} \end{array}\right\} = \left\{\begin{array}{c} \text{rate heat} \\ \text{conducted} \\ \text{in at } x \end{array}\right\} - \left\{\begin{array}{c} \text{rate heat} \\ \text{conducted} \\ \text{out at } x + \Delta x \end{array}\right\}. \tag{9.11}$$

Formulating the differential equation

First we introduce some notation. We let $U(x)$ denote the equilibrium temperature at the point x and let $J(x)$ be the flux of heat at x. Recall that this is the rate of flow of heat, per unit area per unit time, through the material at the point x.

The rate of heat entering the section is obtained by multiplying the heat flux $J(x)$ by the cross-sectional area (surface area of face of the section) A. Thus,

$$\left\{\begin{array}{c} \text{rate heat} \\ \text{conducted} \\ \text{in at } x \end{array}\right\} = J(x)A. \tag{9.12}$$

The rate of heat leaving the section is denoted by $J(x + \Delta x)$. Hence the rate of heat leaving the section, obtained by multiplying the flux by the area A, is

$$\left\{\begin{array}{c} \text{rate heat} \\ \text{conducted} \\ \text{out at } x + \Delta x \end{array}\right\} = J(x + \Delta x)A. \tag{9.13}$$

For equilibrium temperatures, the LHS of the heat balance word equation (9.11) is zero. Substituting (9.12) and (9.13) back into (9.11), with the LHS set to zero, gives

$$J(x)A - J(x + \Delta x)A = 0. \tag{9.14}$$

We are going to let $\Delta x \to 0$, but first we divide (9.14) by Δx. This leads to an expression involving the derivative with respect to x.

Example 9.6: *Obtain a differential equation for the temperature by taking the limit as $\Delta x \to 0$ in (9.14) and then applying Fourier's law, $J = -k\, dU/dx$.*

Solution: *We divide (9.14) by Δx and rearrange to obtain*

$$\frac{J(x + \Delta x) - J(x)}{\Delta x} = 0. \tag{9.15}$$

From the definition of the derivative[2] of J with respect to x, equation (9.15) becomes

$$\frac{dJ}{dx} = 0. \tag{9.16}$$

Since the heat flux through the material is due to conduction only, we can apply Fourier's law to express the differential equation in terms of temperature. Substituting this into equation (9.16) gives

$$\frac{d}{dx}\left(-k\frac{dU}{dx}\right) = 0. \tag{9.17}$$

Since we assume the conductivity k to be a constant, equation (9.17) reduces to

$$\frac{d^2U}{dx^2} = 0. \tag{9.18}$$

This is a second-order differential equation for the equilibrium temperature $U(x)$. It is a trivial example of a differential equation but serves to illustrate how such equations arise. More complicated differential equations can occur when we include volumetric heat sources, as is the case when heat is produced by an electric current or chemical reaction.

Numerical solution

Using `Maple`, we can obtain a numerical solution where we specify both the temperature and heat flux at $x = 0$ (see Figure 9.8). The `Maple` code to produce this is given in Listing 9.3. Equivalent `MATLAB` code is given in Listing 9.4, where the second-order differential equation has been written as a system of two first-order differential equations using $J = -kdU/dx$.

Listing 9.3: Maple code: c_he_wall.mpl

```
restart: with(plots): with(DEtools):
k := 1.0:
de1 := diff(U(x),x$2) = 0;
inits := U(0)=10, D(U)(0)=-1/k;
plot1 := DEplot(de1, U, x=0..1, [[inits]] ):
display(plot1);
```

Listing 9.4: MATLAB code: c_he_wall.m

```
function c_he_wall
global k
k = 1;
u0 = 10; %temp at x=0
J0 = 1; %flux at x=0
y0 = [u0; J0]; %set initial condition vector
xend = 1;
[xsol, ysol] = ode45(@rhs, [0 xend], y0);
Usol = ysol(:,1);
plot(xsol, Usol);

function ydot = rhs(x, y)
global k
U = y(1); J = y(2);
```

---

[2]From first principles, the derivative of $J(x)$ with respect to x is

$$\frac{dJ}{dx} = \lim_{\Delta x \to 0} \frac{J(x + \Delta x) - J(x)}{\Delta x}.$$

```
Udot = -k*J;
Jdot = 0;
ydot = [Udot; Jdot];
```

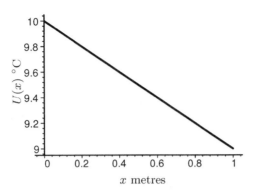

Figure 9.8: Numerical solution with $k = 1\,\mathrm{W\,m^{-1}\,{}^{\circ}C^{-1}}$, $U(0) = 10^{\circ}C$ and $J(0) = 1\,\mathrm{W/m^2}$.

However, more usually we would know either temperature or heat flux at both $x = 0$ and $x = l$. This changes the problem into a *boundary value problem* instead of an IVP. We consider such problems in greater detail later and discuss methods for solving them in Chapter 11.

When the temperature is not at equilibrium, the temperatures are dependent on time as well as position. Furthermore, we must account for the amount of heat 'used up' in raising the temperature. This is examined in detail in Chapter 12 where we consider a problem similar to the present one, but derive a partial differential equation with two independent variables, x and t.

Modelling approach summary

The procedure adopted to derive models for heat conduction depends on a set of physical principles and is slightly different from what we have done before. A summary of the basic procedures for formulating differential equations describing heat transport problems is:

- Consider all inputs and outputs of heat from a thin section of width Δx.

- Develop a word equation for the rate of change of heat.

- Divide by Δx and then take the limit as $\Delta x \to 0$ to derive a differential equation.

- Using Fourier's law, substitute to express heat flux J in terms of temperature U.

Summary of skills developed here:

- *Be able to reproduce the derivations of the equilibrium temperature equation.*
- *Reproduce the derivation of the second-order differential equation.*
- *Modify the word equation if there is an internal heat source, such as an electric current, providing heat at a given rate.*

9.6   Radial heat conduction

Radial heat flow occurs in problems involving cylinders and spheres where the temperatures are the same at any given distance from the centre of the cylinder or sphere. We now formulate a differential equation for the equilibrium temperature in a cylinder. The procedure is similar to that used in the previous two sections but we must account for the fact that the area through which heat is conducted now changes with the radial distance from the centre of the cylinder or sphere.

Fourier's law for radial heat conduction

Radial heat flow occurs in cylinders and spheres. In radial heat flow we need to account for the heat spreading out as it is transferred into or out of a cylindrical or spherical shell

If r denotes the radial distance of a point from the centre of a cylinder or sphere, then we denote the heat flux at r by $J(r)$. The overall rate of flow of heat is

$$\left\{\begin{array}{c}\text{rate of}\\\text{flow of}\\\text{heat}\end{array}\right\} = J(r)A(r),$$

where $A(r)$ denotes the area through which the heat flows. For a cylindrical surface of radius r and length l, $A(r) = 2\pi r l$, while for a spherical surface of radius r, $A(r) = 4\pi r^2$.

For radial heat conduction, Fourier's law takes the form

$$J(r) = -k\frac{dU(r)}{dr}, \tag{9.19}$$

where $J(r)$ is the heat flux at r, $U(r)$ is the temperature at r and k is the conductivity. Thus Fourier's law states that the heat flux is directly proportional to the temperature gradient.

Model assumptions and approach

Let us consider a cylindrical object that is hottest on the inside and coolest on the outer surface (see Figure 9.9). Define the outer radius of the cylinder at b and the inner radius at a; then as a approaches zero the cylinder becomes solid, while for a close to b the cylinder models a hollow pipe. Take the length of the cylinder as ℓ, and let r denote the radial distance from its centreline.

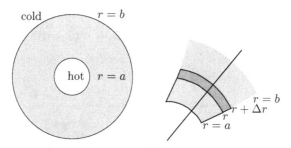

Figure 9.9: Radial heat flow in a cylinder. The first diagram illustrates the cross-section of a cylinder. Heat flows from the centre of the cylinder to the outside in all directions. The second diagram is of an arbitrary cylindrical shell, inside the cylindrical region, from r to $r + \Delta r$.

We now make some assumptions on which to build the mathematical model.

- We assume that heat flows in the radial direction only; the temperature inside the cylinder will then depend on r and the time t only.

- If we also assume thermal equilibrium, then the equilibrium temperature will be a function of the radial distance r alone.

General compartmental model

Let us consider an imaginary small annular shell inside the cylinder, from r to $r + \Delta r$. This is shown in Figure 9.9 and is the compartment. Then, applying the balance law, we consider any heat inputs into and outputs from this shell. Assuming no heat is generated inside the shell, heat input is through conduction only, as is the heat output, and thus we have the compartmental diagram in Figure 9.10.

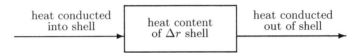

Figure 9.10: Input-output compartmental diagram describing all the heat inputs and outputs for the arbitrary cylindrical shell.

From this we can formulate a word equation.

Example 9.7: *Obtain a suitable word equation for a thin annular shell inside which no heat is generated.*

Solution:

Heat enters the shell at r and leaves at $r + \Delta r$ through the process of conduction. For this problem there is no heat generated internally in the shell. A word equation describing this balance of heat entering and leaving the cylindrical shell is thus

$$\left\{ \begin{array}{c} \text{rate of} \\ \text{change of} \\ \text{heat in shell} \end{array} \right\} = \left\{ \begin{array}{c} \text{rate heat} \\ \text{conducted} \\ \text{in at } r \end{array} \right\} - \left\{ \begin{array}{c} \text{rate heat} \\ \text{conducted} \\ \text{out at } r + \Delta r \end{array} \right\}. \tag{9.20}$$

Formulating the differential equation

We let $U(r)$ denote the equilibrium temperature at a distance r from the centreline. We also let $J(r)$ be the heat flux at a distance r from the origin.

Example 9.8: *Obtain a differential equation for the equilibrium temperature inside the annular shell.*

Solution: *The heat entering the cylindrical shell flows through an area $A(r) = 2\pi r \ell$. Similarly, the heat leaving the annular section flows through an area $A(r + \Delta r) = 2\pi(r + \Delta r)\ell$. In terms of the heat flux $J(r)$, we have*

$$\left\{ \begin{array}{c} \text{rate heat} \\ \text{conducted} \\ \text{in at } r \end{array} \right\} = J(r)A(r),$$

$$\left\{ \begin{array}{c} \text{rate heat} \\ \text{conducted} \\ \text{out at } r + \Delta r \end{array} \right\} = J(r + \Delta r)A(r + \Delta r). \tag{9.21}$$

For equilibrium temperatures, the LHS of the word equation (9.20) is zero. Substituting (9.21) back into (9.20) gives

$$J(r)A(r) - J(r + \Delta r)A(r + \Delta r) = 0.$$

We can write this as

$$-\left[J(r + \Delta r)A(r + \Delta r) - J(r)A(r)\right] = 0. \tag{9.22}$$

Dividing by Δr gives

$$-\left[\frac{J(r + \Delta r)A(r + \Delta r) - J(r)A(r)}{\Delta r}\right] = 0 \tag{9.23}$$

and then taking the limit as $\Delta r \to 0$ we have, from the definition of a derivative[3],

$$-\frac{d}{dr}\left[J(r)A(r)\right] = 0. \tag{9.24}$$

Thus, regardless of the radius, this equation implies that the combination of the heat flux multiplied by the area is the same at any point inside the cylinder.

Substituting for the area $A(r) = 2\pi r\ell$ and using Fourier's law ($J = -k\,dU/dr$), we obtain the second-order differential equation

$$-\frac{d}{dr}\left(-2k\pi\ell r\frac{dU}{dr}\right) = 0. \tag{9.25}$$

Since $2k\pi\ell$ is a constant, for the equilibrium temperature $U(r)$ this simplifies to

$$\frac{d}{dr}\left(r\frac{dU}{dr}\right) = 0. \tag{9.26}$$

Summary of skills developed here:

- Derive a differential equation for the equilibrium temperatures in spherical heat flow.
- Derive the equilibrium temperatures for radial (cylindrical or spherical) heat flow with an internal generation of heat.

9.7   Heat fins

We extend the approach adopted above to develop a mathematical model for a heat fin, which is a cooling device. Such a model describes the heat distribution along the fin, which will change with distance from the heat source.

---

[3]Let $f(r) = J(r)A(r)$. Note that the derivative of f with respect to r is

$$\frac{df}{dr} = \lim_{\Delta r \to 0}\frac{f(r + \Delta r) - f(r)}{\Delta r} = \lim_{\Delta r \to 0}\frac{J(r + \Delta r)A(r + \Delta r) - J(r)A(r)}{\Delta r}.$$

Background

Heat fins are used to enhance the dissipation of heat from machinery in which damage may be caused if heat is allowed to build up. Thus they are found on many types of machinery such as in motorcycle engines, refrigerators and computers. These fins work by increasing the surface area through which heat can be lost, and often have fans attached to them to force air across the fins, thus reducing the temperature more efficiently. There are many different forms that these fins can take, and some types are illustrated in Figure 9.11.

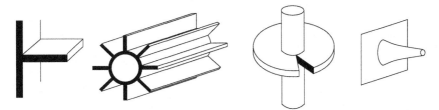

Figure 9.11: Some different types of heat fins.

A typical problem is to calculate the temperature distribution along a heat fin. From this, the rate of heat dissipation can be calculated, and the cooling efficiency for the heat fin determined. But first we need to formulate a differential equation for the changing temperature along the heat fin.

Model assumptions and approach

As our mathematical model we consider a rectangular heat fin of length ℓ, width w and thickness b (see Figure 9.12).

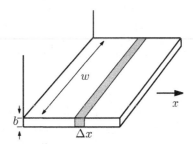

Figure 9.12: Schematic diagram for mathematical model.

We make some assumptions and build our mathematical model on these:

- We assume the thickness of the heat fin b to be small compared with its length ℓ.

- We assume that heat is lost from the surfaces of the heat fin to the surroundings (at temperature u_s) according to Newton's law of cooling.

- Because most of the heat loss occurs from the faces of the fin, both top and bottom, in our initial model we neglect the heat lost from the sides. Improving the model by including heat lost from the sides is not difficult and is explored in the exercises.

- A further assumption is to neglect the temperature variation over the cross-section of the fin. This assumption is reasonable, provided the heat fin is thin (that is, $b \ll \ell$).

In doing this, we can effectively assume the equilibrium temperature to be a function of x alone.

General compartmental model

Let us consider a small section of the heat fin from x to $x + \Delta x$, as shown in Figure 9.12. We first consider all the heat inputs into and outputs from this section, and later we let Δx tend to zero to establish the differential equation.

Example 9.9: *Determine an input-output word equation for an arbitrary section of the heat fin.*

Solution: *The thin section of the heat fin is shown in Figure 9.12 and we use the balance law to establish the inputs and outputs in Figure 9.13. Heat is conducted into the section at x and out from the section at $x + \Delta x$. We should also account for the heat lost from the top and bottom surfaces to the surroundings and for this we use Newton's law of cooling. Thus there is a single means by which heat enters the section and two mechanisms by which it leaves. (Recall that we have neglected any heat loss from the sides.)*

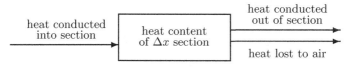

Figure 9.13: Input-output diagram for an arbitrary thin section of the heat fin.

The basic heat balance equation for the section is then

$$\left\{ \begin{array}{c} rate\ of \\ change\ of \\ heat\ in\ section \end{array} \right\} = \left\{ \begin{array}{c} rate\ heat \\ conducted \\ in\ at\ x \end{array} \right\} - \left\{ \begin{array}{c} rate\ heat \\ conducted \\ out\ at\ x + \Delta x \end{array} \right\} - \left\{ \begin{array}{c} rate\ heat \\ lost\ to \\ surroundings \end{array} \right\}. \qquad (9.27)$$

For the equilibrium temperature, the rate of change of heat is zero and so

$$\left\{ \begin{array}{c} rate\ heat \\ conducted \\ in\ at\ x \end{array} \right\} - \left\{ \begin{array}{c} rate\ heat \\ conducted \\ out\ at\ x + \Delta x \end{array} \right\} - \left\{ \begin{array}{c} rate\ of\ heat \\ lost\ to \\ surroundings \end{array} \right\} = 0. \qquad (9.28)$$

Formulating the differential equation

The next step is to assign appropriate symbols to each of the terms in the word equation (9.28). We express the heat conduction quantities in terms of the heat flux J and later use Fourier's law of heat conduction to express these in terms of temperature. Once again we develop the model with symbols so that we can examine the effect of changing some of these physical quantities easily.

Example 9.10: *Express each of the quantities in equation (9.28) in terms of the heat flux J or the temperature U.*

Solution: *The area through which heat is being conducted is bw. In terms of the heat flux $J(x)$,*

$$\left\{ \begin{array}{c} rate\ heat \\ conducted \\ in\ at\ x \end{array} \right\} = J(x)bw, \qquad \left\{ \begin{array}{c} rate\ heat \\ conducted \\ out\ at\ x + \Delta x \end{array} \right\} = J(x + \Delta x)bw. \qquad (9.29)$$

We use Newton's law of cooling for the heat loss from the surface of the section in the time interval. Recall that Newton's law of cooling states that the rate of heat loss is proportional to the surface area and the temperature difference between the object and its surroundings. The surface area of the section through which heat is lost to the surroundings is $2w\,\Delta x$. This corresponds to the top and bottom faces only, as we are neglecting heat lost from the sides.

To calculate the approximate heat loss by Newton's law of cooling, we could use the temperature at x to give $hS(U(x) - u_s)$. (Recall that S is the surface area, u_s is the temperature of the surroundings and h is the Newton cooling coefficient.) But this would give a slight overestimate of the temperature since $U(x)$ decreases with x. Similarly, if we used the temperature at $x + \Delta x$, we would obtain a slight underestimate. The true value would use the temperature somewhere in between, say at x^*, where $x < x^* < x + \Delta x$. Thus we write

$$\left\{ \begin{array}{c} \text{rate heat} \\ \text{lost to} \\ \text{surroundings} \end{array} \right\} = 2hw\,\Delta x\,[U(x^*) - u_s]\,. \tag{9.30}$$

(Although we do not know the exact value of x^*, it does not matter as eventually we let Δx tend to zero and then $x* \to x$.)

We now substitute the individual terms (9.29) and (9.30) into the heat balance equation (9.28) to obtain

$$J(x)bw - J(x + \Delta x)bw - 2hw\Delta x\,[U(x + \lambda\Delta x) - u_s] = 0, \tag{9.31}$$

where $x^* = x + \lambda\Delta x$ and $0 \le \lambda \le 1$.

Taking the limit as $\Delta x \to 0$ to get the derivative, and then applying Fourier's law to give the equation in terms of temperature, we establish the differential equation required.

Example 9.11: Let $\Delta x \to 0$ in equation (9.31) and obtain a differential equation for the temperature.

Solution: Dividing (9.31) by $bw\Delta x$ we are left with

$$-\left(\frac{J(x + \Delta x) - J(x)}{\Delta x} \right) - \frac{2h}{b}\,(U(x + \lambda\Delta x) - u_s) = 0. \tag{9.32}$$

Then, taking the limit as $\Delta x \to 0$ and applying the definition of the derivative[4],

$$-\frac{dJ}{dx} - \frac{2h}{b}\,(U(x) - u_s) = 0. \tag{9.33}$$

From (9.33), substituting for J and applying Fourier's law $(J(x) = -k\,dU/dx)$, we find

$$k\frac{d^2U}{dx^2} - \frac{2h}{b}\,(U(x) - u_s) = 0. \tag{9.34}$$

This equation can be written in the compact form

$$\frac{d^2U}{dx^2} = \beta(U - u_s) \qquad \text{where} \quad \beta = \frac{2h}{kb}, \tag{9.35}$$

establishing a second-order differential equation describing the heat flux in terms of temperature.

---

[4]By first principles, the derivative of $J(x)$ with respect to x is

$$\frac{dJ}{dx} = \lim_{\Delta x \to 0} \frac{J(x + \Delta x) - J(x)}{\Delta x}.$$

Extensions and further problems

Heat fins come in many shapes and sizes. The derivation of the differential equation for heat fins with different cross-sections, such as circular, is almost identical to the above. In fact, we end up with the same equation (9.35) but with a different value for β.

Obtaining a differential equation for when the cross-section varies with distance from the origin is more difficult but still follows the same procedure. Together with the previous section that deals with radial heat conduction, you have the required theory to deal with such problems.

Other possible extensions include the incorporation of a heat production term. Another case is that where the temperature of the surroundings may be higher than the temperature of the fin, so that the fin gains heat rather than loses it.

Summary of skills developed here:

- *Derive the differential equation for a solid cylindrical heat fin (using ideas from this section and Section 9.6).*
- *Formulate differential equations for heat fins of rectangular, or other, cross-sections in the case of an equilibrium temperature.*
- *Formulate a differential equation for one-dimensional heat flow with a non-constant internal heat source.*

9.8   Diffusion

Particles that are highly concentrated in a volume of fluid will tend to disperse by random movement so that they become less concentrated. This process is called diffusion and is one mechanism for mass transport. Particles appear to flow from high concentration to low concentration as in Figure 9.14, in this case towards a concentration that is uniform across the available space. This process is analogous to heat flow from high temperatures to low temperatures, and the mathematics describing mass transport of this type is very similar to the mathematics describing heat transport.

Figure 9.14: The random movement of small particles. On the left the particles are concentrated into a smaller area. Random movement of particles causes the particles to spread out so they become less concentrated.

Concentration and mass flux

For mass transport the key variable is the concentration, which can be defined in a number of ways. The most common definition is the mass of the particles divided by the volume containing the particles, giving it SI units of kg/litres (or mg/litre). However, it can be also defined as the number of particles in a volume divided by the volume, or as the volume of all particles given as a fraction of the total volume containing the particles. In chemistry the particles are called the solute, while the mixture of particles and fluid is called the solution.

To formulate equations describing diffusion, we use $C(x)$ for the equilibrium concentration and $C(x,t)$ for the time-dependent concentration at some point x and time t. For the flow of particles, we define the *mass flux*, $J(x)$, as the rate of flow of particles through a cross-section at location x, per unit time per unit cross-sectional area. Comparing mass transport through diffusion with heat transport by conduction, mass flux is analogous to heat flux and concentration is analogous to temperature.

Similar to Fourier's law of heat conduction, in mass transport there is Fick's law of diffusion with

$$J(x) = -D\frac{dC}{dx}$$

for equilibrium concentrations, and

$$J(x,t) = -D\frac{\partial C}{\partial x}$$

for time-dependent concentrations. The constant of proportionality, D, is called the *diffusion coefficient*, or the diffusivity. The diffusion coefficient has SI units of $m^2\,s^{-1}$ and takes different values depending on the size of particles and the type of fluid in which the particles are diffusing (see Table 9.4). It can also depend on the temperature, with particles diffusing more easily at higher temperatures. However, in general, diffusion as a mechanism for mass transport is a slow process, as is evident from the values provided in Table 9.4.

Table 9.4: Diffusion coefficient, D, for some molecules in SI units of m^2/s.

Particles/fluid	D
Oxygen in air at 25°C	1.76×10^{-5}
Oxygen in water at 25°C	2.10×10^{-9}
Air in water at 25°C	2.82×10^{-5}
Carbon dioxide in water	1.92×10^{-9}

Governing equations

To derive a differential equation for the equilibrium concentration we use the same process as that to derive equations for equilibrium temperature (see Section 9.5). Consider a pipe filled with fluid that is not moving, but which contains particles. To establish the diffusion process of these particles within the fluid, we set up a mass balance in region x to $x + \Delta x$, a slice across the cylinder with cross-sectional area A.

The rate of change for the mass of particles inside the region, or slice, is determined by the net amount flowing into and out of the region. Expressed as a word equation, this says that

$$\left\{\begin{array}{c} \text{rate of change} \\ \text{of mass} \\ \text{within volume} \end{array}\right\} = \left\{\begin{array}{c} \text{rate of mass} \\ \text{flowing in at } x \end{array}\right\} - \left\{\begin{array}{c} \text{rate of mass} \\ \text{flowing out at } x + \Delta x \end{array}\right\}. \tag{9.36}$$

At equilibrium the rate of change of mass with time will be zero, so the LHS is zero. The rate at which mass flows in at x is given by the mass flux (mass per unit time per unit area) multiplied by the cross-sectional area, and similarly for the amount flowing out. Hence

$$J(x)A - J(x + \Delta x)A = 0.$$

Dividing by $-A\Delta x$ gives

$$\left[\frac{J(x + \Delta x) - J(x)}{\Delta x} \right] = 0,$$

and then taking the limit as $\Delta x \to 0$ leads to

$$\frac{dJ}{dx} = 0.$$

Substituting Fick's law, $J = -D(dC/dx)$, gives an equation for the concentration at equilibrium:

$$\frac{d^2 C}{dx^2} = 0.$$

Above we have considered the general diffusion process in a linear geometry and we now relate this to different geometries. For diffusion in cylindrical geometries, where diffusion is inwards (or outwards) along the radius r of the cylinder, the equilibrium diffusion equation for any location along the cylinder can be obtained as

$$\frac{d}{dr} \left(r \frac{dC}{dr} \right) = 0, \tag{9.37}$$

similar to that obtained in Section 9.7 for heat conduction in a layer of insulation surrounding a pipe. This follows since the area through which diffusion occurs changes with distance from the centre, r. For spherical geometries with radius r, where diffusion is towards the centre from the surface (or away from the centre towards the surface), the diffusion equation becomes

$$\frac{d}{dr} \left(r^2 \frac{dC}{dr} \right) = 0, \tag{9.38}$$

since the area through which diffusion occurs changes with the square of the distance from the centre. This provides a means of establishing concentration at distance r from the centre of the sphere. In the event that there is a further source of particles, or sink (absorption of particles), these equations change. Assuming a volumetric mass source at distance r from the centre, producing mass at a rate of $M(r)\,\text{kg/m}^3/\text{sec}$, the diffusion equation for spheres is given by

$$D\frac{1}{r^2} \frac{d}{dr} \left(r \frac{dC}{dr} \right) + M(r) = 0, \tag{9.39}$$

where D is the diffusion coefficient. Note that when there is no volumetric source term, the constant D cancels, and further, the term $A(r)$ becomes negative if mass is lost instead of gained, such as in living organisms where cells consume oxygen.

9.9 Exercises for Chapter 9

9.1. Simple heat calculations. *The following questions require the use of appropriate equations from Chapter 9. (You are <u>not</u> required to solve any differential equations).*

(a) *If the heat flux through a $10\,\mathrm{m}^2$ wall is 30 (in SI units), how much heat is lost in 1 hour?*

(b) *How much heat does it take to heat a solid copper sphere of radius $1\,\mathrm{cm}$ from $5°\mathrm{C}$ to $25°\mathrm{C}$?*

(c) *If the outside of a $10\,\mathrm{m}^2$ wall of a house loses heat to the surrounding at a rate of $900\,\mathrm{W}$, and the temperature of the surroundings is $5°\mathrm{C}$, what is the outside temperature of the wall? You may assume Newton's law of cooling with heat transfer coefficient $h = 10$ (SI units).*

9.2. Including an internal heat source. *Heat flows inside a slab of material from left to right. Inside the material, an electric current generates heat at a constant rate Q_0 Watts m^{-3}.*

(a) *Write a suitable word equation for the rate of change of heat in a section x to $x + \Delta x$ of the slab, where x is in the direction of heat flow.*

(b) *Hence deduce that the equilibrium temperature satisfies the differential equation of the form*

$$\frac{d^2 U}{dx^2} + \beta = 0,$$

where β is a constant you must determine.

9.3. Formulating model extensions. *Consider the formulation of the ordinary differential equation in Section 9.5 for the equilibrium temperature.*

(a) *Suppose, instead, heat flows in the opposite direction (negative x-direction). Does the differential equation (9.18) remain the same? Justify your answer.*

(b) *Suppose, instead, the conductivity is not a constant but depends on temperature. Does the differential equation (9.18) remain the same?*

9.4. Model formulation with circular cross-section. *A cylindrical heat fin has circular cross-section of radius a. The heat fin loses heat to the surrounding air, which is at temperature u_a. Assuming heat flow in the linear x-direction, start from a suitable word equation to show that the equilibrium temperature satisfies the differential equation of the form*

$$\frac{d^2 U}{dx^2} - \alpha U = -\beta,$$

where α and β are constants you must determine.

9.5. Nuclear radiation. *In a shielding wall for a nuclear reactor, gamma-ray radiation causes the wall to heat up internally at a given rate $q(x)$ per unit time per unit volume, within the wall. Formulate the differential equation for the equilibrium temperature.*

9.6. Including an internal heat source. *Heat flows radially in a solid cylinder of length ℓ and radius a. Heat is also generated internally by an electric current at a constant rate Q_0 Watts per unit volume. The temperature on the inside is maintained at temperature u_a and heat is lost from the outer surface to the surroundings at temperature u_s, according to Newton's law of cooling.*

(a) *Write a word equation for the rate of change in heat for a cylindrical shell r to $r + \Delta r$.*

(b) *What is the volume of the shell in (a)? Hence give an expression for the rate of heat generated by the electric current.*

(c) *Obtain a differential equation for the equilibrium temperature. (Hint: Not all of the information in the question needs to be used.)*

9.7. Equilibrium temperature inside a sphere. *Heat flows radially, by conduction, through a spherical shell of radius a. Formulate the differential equation for the equilibrium temperature inside the sphere.*

9.8. Cylindrical heat fin. *In the thin cylindrical heat fin (shown in Figure 9.15) of thickness ℓ, inner radius a and outer radius b, heat flows radially. Also, heat is lost from the surface of the fin to the atmosphere, at temperature u_s, according to Newton's law of cooling.*

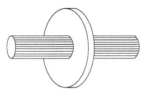

Figure 9.15: A cylindrical heat fin (see Exercise 9.8).

Deduce the differential equation for the equilibrium temperature assuming the temperature variation over the cross-section of the heat fin is negligible.

9.9. Diffusion with source in a sphere. *Derive the differential equation for diffusion in a spherical geometry with a volumetric mass source that produces mass at a rate $M(r) \, \mathrm{kg \, m^{-3} \, s^{-1}}$.*

9.10. Diffusion with source in a cylinder. *Derive the differential equation for diffusion in a cylindrical geometry with a volumetric mass source which produces mass at a rate $M(r) \, \mathrm{kg \, m^{-3} \, s^{-1}}$.*

Chapter 10

Solving time-dependent heat problems

In this chapter, we revisit the time-dependent heating and cooling problems introduced in Chapter 9. These resulted in ODEs that we now solve using standard analytic methods. Given such analytic solutions, we can answer some general questions about the problems. We also consider some extensions of time-dependent heat transport, without conduction, in two case studies: a model describing house temperatures and a model for predicting the spontaneous combustion of fish and chip crumble.

10.1 The cooling coffee problem revisited

In this section, we revisit the problem of determining the time it takes for a hot cup of coffee to cool. The exponential growth/decay differential equation that arises can be solved analytically using different techniques, and this solution provides a more complete understanding of the cooling process than the particular numerical solution generated in Chapter 9.

Review of the model

In Section 9.1, we first introduced the problem of determining the temperature of a cup of coffee. In Section 9.2, we formulated the general differential equation for the temperature $U(t)$ as a function of time, which was

$$\frac{dU}{dt} = \frac{hS}{cm}(U - u_s), \tag{10.1}$$

where c is the specific heat of the coffee, m is the mass of the coffee, h is the Newton cooling coefficient, S is the surface area through which heat is lost and u_s is the constant temperature of the surroundings.

Analytic solution

We can write the differential equation as

$$\frac{dU}{dt} = -\lambda(U - u_s), \tag{10.2}$$

where λ is a constant of proportionality and equivalent to hS/cm. The solution to this differential equation can be found using the separable technique, as in the following example.

Example 10.1: *Solve the differential equation (10.2) subject to the initial condition $U(0) = u_0$.*

Solution: *Separating the variables,*

$$\frac{1}{U - u_s} \frac{dU}{dt} = -\lambda.$$

Integrating both sides with respect to t, assuming $U > u_s$, gives

$$\ell n(U - u_s) = -\lambda t + K$$
$$U = e^{-\lambda t + K} + u_s.$$

Applying the initial condition $U(0) = u_0$ gives $e^K = u_0 - u_s$ and so

$$U(t) = (u_0 - u_s)e^{-\lambda t} + u_s. \tag{10.3}$$

Applying the coffee cooling model

We return to our original problem (see Section 9.1) to find how long it takes a cup of coffee to cool to 40°C, and find the answer using the analytic solution we have just derived. A hot cup of coffee is at a temperature of 60°C and after 10 minutes the coffee has cooled to 50°C. The room temperature is 20°C. The following example shows how to find this time.

Example 10.2: *Find how long it will take for the cup of coffee to cool to 40°C.*

Solution: *We can find λ from the given information that $U = 50$ at time $t = 10$. Substituting this into the solution (10.3) we obtain*

$$50 = (60 - 20)e^{-10\lambda} + 20, \quad \text{and} \quad \lambda \approx 0.0288.$$

Now, if U is 20°C then

$$40 = (60 - 20)e^{-0.029t} + 20 \quad \text{and} \quad t \approx 24 \text{ minutes.}$$

Thus the coffee will reach the temperature of 40°C about 24 minutes after it was put into the room.

An alternative cooling law

The differential equation (10.2) applied Newton's law of cooling, which assumes that air at room temperature is moving over and around the object. This is not necessarily the case in a room, or on a still day. A better model for cooling in still air is one in which the rate of the temperature increase/decrease is directly proportional to the (5/4)th power of the difference in temperature between the object and its surroundings. This is the law of *natural cooling* and is given by

$$\left\{ \begin{matrix} \text{rate of} \\ \text{change of} \\ \text{heat} \end{matrix} \right\} = \pm h_1 S(U - u_s)^{5/4}, \tag{10.4}$$

where the $\pm$ sign depends on whether the object is cooling or heating, h_1 is a positive constant of proportionality and S is the surface area. Note that $h_1 \neq h$, the Newton cooling coefficient in (10.1).

For the problem of cooling a cup of coffee, the differential equation for the temperature is

$$\frac{dU}{dt} = -\lambda(U - u_s)^{5/4}, \qquad \lambda = \frac{h_1 S}{cm}. \tag{10.5}$$

(Recall that m is the mass of the coffee and c its specific heat.) This differential equation is also separable and can be solved using the same technique as before. (See Exercises Exercise 10.1.) Using its solution we can show that the model predicts it will take approximately 26 minutes for the coffee to cool from 60°C to 40°C in a room at temperature 20°C, given that it takes 10.3 minutes to cool to 50°C. This estimate for the time is slightly longer than that obtained using Newton's law of cooling. A comparison between these two cooling models is illustrated in Figure 10.1. The difference is not large for this problem.

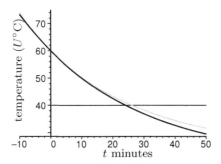

Figure 10.1: A comparison between Newton cooling (black line) and natural cooling (grey line). For comparison purposes, it has been assumed that the coffee takes 10.3 minutes to cool to 50°C for natural cooling.

The physical meaning of this is that the natural cooling law gives a slower rate of cooling than Newton's law. While the factor $(U - u_s)^{5/4}$ is larger than $(U - u_s)$, the associated values for λ (0.0115 and 0.0288, respectively) effectively render the rate for natural cooling less than that for Newton cooling. This makes sense because we expect cooling in an environment with a slight breeze (Newton's law of cooling) to be more effective than cooling in perfectly still air (natural law of cooling).

Summary of equations

Newton's law of cooling (cooling in a breeze) is

$$\frac{dU}{dt} = \pm\frac{hS}{cm}(U - u_s).$$

The natural law of cooling (applicable when there is no breeze) is

$$\frac{dU}{dt} = \pm\frac{h_1 S}{cm}(U - u_s)^{5/4}.$$

> **Summary of skills developed here:**
>
> - *Solving a differential equation with a cooling law (e.g., Newton's law of cooling).*
> - *Solving for a constant of proportionality and a constant of integration using a set of known results and an initial condition.*
> - *Using a solution for the temperature to calculate the time of cooling (or time of heating).*

10.2   The hot water heater problem revisited

We now return to the water heater problem introduced in Section 9.3. Again, our aim is to determine and interpret the analytic solution to indicate how such a solution provides a better general understanding of the system, compared with any particular numerical solution.

Review of the model

Suppose we have a domestic hot water heater. We wish to determine a formula for how long it takes to heat the water to, say 60°C, and what physical quantities this depends on. One interesting question that we can pose is:

Does doubling the power output of the heating element halve the heating time?

We formulated the differential equation for the water temperature $U(t)$ in Section 9.3 as

$$cm\frac{dU}{dt} = q - hS\left(U - u_s\right), \tag{10.6}$$

which describes the temperature variation of the water with time. Here q is the rate of heat supplied, m the mass of water, S the surface area through which heat is lost, c the specific heat of the water, h the Newton cooling coefficient, u_s the temperature of the surroundings and u_0 the initial temperature of the water. The final temperature will be denoted by u_f, and here $u_f = 60°C$.

Analytic solution

To solve the differential equation, we first write it in the form

$$\frac{dU}{dt} = \beta - \alpha U, \qquad U(0) = u_0, \tag{10.7}$$

where the constants α and β are defined as

$$\alpha = \frac{hS}{cm}, \quad \beta = \frac{q + hSu_s}{cm}.$$

In this form, it is easier to see that the differential equation is a first-order linear equation.

Example 10.3: *Solve the differential equation (10.7) using the integrating factor technique.*

Solution: *Since the differential equation is first-order linear, we can find an integrating factor,*

$$R(t) = e^{\int_0^t \alpha \, dt} = e^{\alpha t}.$$

Hence, we can write the differential equation as

$$\frac{d}{dt}\left(Ue^{\alpha t}\right) = \beta e^{\alpha t}.$$

Integrating both sides with respect to t,

$$e^{\alpha t}U = \frac{\beta}{\alpha}e^{\alpha t} + K,$$

$$\text{and} \quad U = \frac{\beta}{\alpha} + Ke^{-\alpha t},$$

where K is the constant of integration. Now applying the initial condition $U(0) = u_0$ gives $u_0 = \beta/\alpha + K$ and so $K = u_0 - \beta/\alpha$. Hence the solution is

$$U = u_0 e^{-\alpha t} + \frac{\beta}{\alpha}\left(1 - e^{-\alpha t}\right).$$

Substituting for α and β this gives

$$U(t) = \left(u_0 - u_s - \frac{q}{hS}\right)e^{-(hS/cm)t} + u_s + \frac{q}{hS}. \tag{10.8}$$

Note that this equation is also separable and could have been solved using that technique. (See Exercise 10.3.)

Applying the hot water heater model

We apply the above theory to a typical hot water cylinder used in an average Australian home (of four individuals) to provide hot water. Then, by considering that energy used to heat the system has a cost, which most of us would like to minimise, we discuss how better to operate the system.

Suppose that we have a standard home hot water tank that holds 250 litres of water, and is cylindrical with a height of 1.444 m and a diameter of 0.564 m. The water is heated by a heating element immersed in the water that supplies heat at a constant rate of 3,600 Watts ($q = 3.6\,\text{kW} = 3,600\,\text{J sec}^{-1}$). We assume the mass of the water to be 250 kg (as a litre of water has the mass of approximately 1 kg) with the surface area of the tank approximately $3.06\,\text{m}^2$. We know the specific heat of water is $c = 4,200\,\text{J kg}^{-1}\,{}^\circ\text{C}^{-1}$ and the heat transfer coefficient is $h = 12\,\text{W m}^{-2}\,{}^\circ\text{C}^{-1}$. Given that the temperature of the surroundings is $u_s = 15^\circ\text{C}$, and that initially the water in the tank is the same temperature as its environment ($u_s = u_0$), we are interested in how long it would take for the water to reach a temperature of 60°C.

Example 10.4: Find the time it would take to heat the tank of water to 60°C. First find a general formula for the time and then substitute the parameter values.

Solution: Using solution equation (10.8) we need to solve for t. Again, we let

$$\alpha = \frac{hS}{cm}, \quad \beta = \frac{q + hSu_s}{cm},$$

and then the solution is

$$U = \left(u_0 - \frac{\beta}{\alpha}\right) e^{-\alpha t} + \frac{\beta}{\alpha}.$$

Solving for $e^{-\alpha t}$ gives

$$e^{-\alpha t} = \frac{U - \beta/\alpha}{u_0 - \beta/\alpha}.$$

Hence we obtain

$$t = \frac{1}{\alpha} \ell n \left| \frac{\beta/\alpha - u_0}{\beta/\alpha - U} \right|.$$

With the above parameter values substituted into this equation, the time required to heat the water from $u_0 = u_s = 15°C$ to $60°C$ is ≈ 4 hours and 50 minutes. This result can be compared with the numerical solution in Figure 9.4 in Section 9.3.

It is possible to improve on the efficiency of the tank by insulating it well, and the standard used in Australia is approximately 3 cm of compacted fibreglass. Then, assuming some cost of power per Watt, we can compare the efficiency of the above tank with that for a tank that is insulated. (See Exercise 10.5.)

What is of major interest is how to run the most cost effective system. Clearly, the cost of heating the tank is proportional to the time it takes to heat the water, or the length of time for which the heater is on. Assuming all the constants as above, would it be cheaper to use a thermostat that switches the heater on each time the temperature drops below 50°C, or to switch the heater off for 8 hours a day relying on the thermostat at other times? Would this result change if the temperature of the environment were to fall to 10°C? (See Exercise 10.4.)

We are also in a position to answer the question posed at the beginning of this section. This is done in the following example.

Example 10.5: Does doubling the power output halve the heating time?

Solution: We start with equation (10.6),

$$cm\frac{dU}{dt} = q - hS\left(U(t) - u_s\right)$$

and its solution as an expression for t is

$$t = \frac{cm}{hS} \ell n \left| \frac{\frac{q}{hS} + u_s - u_0}{\frac{q}{hS} + u_s - U} \right|. \tag{10.9}$$

It is clear that if q is doubled, its effect on t is not to halve it. The effect of q is not quite so simple. While it will certainly reduce the natural logarithm function with the effect of decreasing the time, in general it will not halve it. Figure 10.2 illustrates the behaviour of t as q increases or decreases.

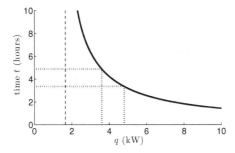

Figure 10.2: A plot of q in Watts as a function of time t in hours for a water heater to heat from 15°C to 60°C where the outside temperature is 15°C. The plot is of the formula (10.9). Other parameters are the same as for Example 10.5. The thick dashed vertical line is an asymptote; below $q \simeq 1.6\,\text{kW}$ the heating element cannot supply enough energy to heat the mass of water to 60°C. The dotted lines indicate the time in hours to heat the water for a 3.6 kW element and for a 4.8 kW element.

The equilibrium temperature

Physically the differential equation (10.6) represents a balance between heat produced by the heating element and heat lost from the surface. This suggests the existence of an equilibrium temperature, or thermal equilibrium. To find the equilibrium temperature we set $dU/dt = 0$ and then solve for U. Thus,

$$q - hS\left(U - u_s\right) = 0.$$

It follows that the equilibrium temperature U is

$$U = u_s + \frac{q}{hS}. \tag{10.10}$$

Note that, because q, h and S are all positive constants, the equilibrium temperature is always greater than the temperature of the surroundings. This makes sense intuitively.

Now consider the solution to differential equation (10.6), which is given by (10.8). If we take the limit of this solution as t becomes large, we see that

$$U \to u_s + \frac{q}{hS} \qquad \text{as } t \to \infty.$$

We see that this limiting temperature, called the steady-state temperature, is just the equilibrium temperature of equation (10.10).

The physical significance of this is that, as the steady-state temperature is approached, heat is no longer used up to change the temperature. Instead, a balance is attained where the heat input is equal to the heat escaping from the surface. Thus, in mathematical terms from the differential equation describing the heat balance, equation (10.6), the derivative term is zero, which is the condition for an equilibrium temperature.

Summary of skills developed here:

- *Formulate a differential equation for a hot water heater with insulation of varying degrees of effectiveness.*
- *Formulate a model to describe a cooling device, such as a refrigerator.*
- *Be able to ascertain how each parameter in the model affects the heating/cooling times.*

10.3   Case Study: It's hot and stuffy in the attic

We have considered some simple applications of heat transport, but how do these relate
to larger systems? In the case study that follows, we adapt these ideas to consider the
changing temperatures in a building. This is essential for good insulation characteristics.
The following was based on (and substantially modified from) Sansgiry and Edwards (1996).

*In different parts of the world, dwellings and public buildings take on many different
and imaginative forms. Mud, corrugated iron, fibro (asbestos) sheeting and glass are just
some of a huge variety of materials used. One aspect considered in their construction is
the climate. Certain materials used, or the aspect and design chosen, may effectively use
the environment to enhance the heating/cooling mechanisms to suit a particular climate.
Others, lacking this forethought, may be stuffy or cold.*

*To understand heat transfer in complicated structures, we need a model for the heat
transfer between two regions at different temperatures. We assume*

$$\left\{ \begin{array}{c} \text{rate} \\ \text{heat} \\ \text{transferred} \end{array} \right\} = \frac{S \times \Delta U}{R},$$

*where S is the surface area between the two regions, ΔU is the temperature difference
between them and R is the resistance to heat transport. For walls of houses, R represents
the total heat resistance due to the resistance of the material making up the wall, which
might include insulation materials. The value of R for a given wall can be determined
experimentally, or theoretically (as we see later in Chapter 11).*

*To understand how the materials used for heating/cooling processes work, we consider a
model for the temperature of a simple brick house with a tiled roof, shown in the Figure
10.3. The building has two compartments: the interior at temperature $U_1(t)$ and the roof
cavity (between the ceiling and the roof) at temperature $U_2(t)$. The temperature of the
surrounding air is assumed to be $u_s = 35°C$ corresponding to a house in a hot climate.*

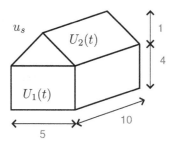

Figure 10.3: Schematic diagram for a model of a house with two compartments: an interior and a roof
cavity. Here, $V_1 = 200\,\text{m}^3$ and $V_2 = 25\,\text{m}^3$.

*A system of differential equations, dependent on heat conduction, can be derived by
considering the heat inputs and outputs in each of the two compartments. The resulting*

governing equations are

$$M_1 \frac{dU_1}{dt} = \frac{S_{10}}{R_{10}}(u_s - U_1) + \frac{S_{12}}{R_{12}}(U_2 - U_1),$$

$$M_2 \frac{dU_2}{dt} = \frac{S_{20}}{R_{20}}(u_s - U_2) + \frac{S_{12}}{R_{12}}(U_1 - U_2),$$

(10.11)

where S is surface area, R total heat resistance due to the building material, and M_1 and M_2 are parameters representing the overall thermal mass (air plus solid materials) of the two regions. For example, $M_1 = \rho c V_1 + \rho_b c_b V_b$, where ρ is the density and c the specific heat of air, V_1 the volume of region 1, ρ_b is the brick wall density, c_b the brick wall specific heat and V_b is the volume of the brick wall. (For the results generated here, these values are calculated as $M_1 = 0.6498018 \times 10^8$ and $M_2 = 0.5184672 \times 10^7$, both in SI units.) Also S_{10}, S_{12} and S_{20} denote the surface areas of the various boundaries of the regions (or compartments) and R_{10}, R_{12} and R_{20} are the corresponding thermal resistances. Typical values for these heat resistances are $R_{10} = 0.3$ for walls without insulation, $R_{12} = 0.1$ for ceilings (without insulation) and $R_{20} = 0.2$ for the roof. With wall and ceiling insulation, these R values increase, with typically $R_{10} = 2$ and $R_{12} = 3$. Clearly, which building and insulation materials are chosen has a substantial impact on the resulting house temperatures.

Using `Maple` or `MATLAB` to find a numerical solution to the system provides the results illustrated in Figure 10.4.

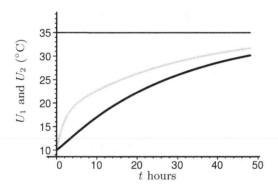

Figure 10.4: Plot of temperatures against time for the model without radiation from the roof tiles. The black line is U_1, the interior temperature, the grey line is U_2, the roof cavity temperature, and the horizontal line is the temperature of the outside environment.

A valid criticism of the above model is that over long periods of time the roof cavity temperature tends towards the outside temperature. In reality, in warm climates the temperature of the roof cavity is often much hotter than the outside temperature, particularly on very hot days, as in the case under consideration. One reason for this is that the roof tiles heat up to a temperature substantially greater than that of the surroundings, due to their exposure to the sun, and then radiate this heat into the roof cavity. To incorporate this aspect into the model, we include the transfer of heat by radiation.

Let us assume the temperature of the roof tiles to be $80°C$. A model for the heat transfer by radiation is given by the Stefan–Boltzmann law,

$$\left\{ \begin{array}{c} \text{rate} \\ \text{heat} \\ \text{transferred} \end{array} \right\} = \epsilon \sigma S \left(T_r^4 - T_2^4\right),$$

where $\sigma = 5.6697 \times 10^{-8}$ is the Stefan–Boltzmann constant, the parameter ϵ is the emittance (typically $\epsilon \simeq 0.8$), T_r is the absolute temperature of the roof tiles (measured in degrees Kelvin) and T_2 is the absolute temperature of the roof cavity (also in degrees Kelvin). S is the surface area of the roof, as before. The quantities $u_s + 273$, $U_1(t) + 273$, $U_2(t) + 273$ and $u_r + 273$ now represent the absolute temperatures in degrees Kelvin (K), where u_r is the temperature of the roof tiles.

Incorporating this extra heat source into our model results in a modification of the second equation; note that the temperature is now expressed in degrees Kelvin. Setting $T_r = (u_r + 273)\,\mathrm{K}$ and $T_2 = (U_2 + 273)\,\mathrm{K}$, the resulting governing equations are now

$$
\begin{aligned}
M_1 \frac{dU_1}{dt} &= \frac{S_{10}}{R_{10}}(u_s - U_1) + \frac{S_{12}}{R_{12}}(U_2 - U_1), \\
M_2 \frac{dU_2}{dt} &= \frac{S_{20}}{R_{20}}(u_s - U_2) + \frac{S_{12}}{R_{12}}(U_1 - U_2) \\
&\quad + \epsilon \sigma S_{12}\left((u_r + 273)^4 - (U_2 + 273)^4\right).
\end{aligned}
\tag{10.12}
$$

Temperature predictions for this new system are given in Figure 10.5. Note that the temperature in the roof cavity $U_2(t)$ is now greater than the outside temperature. This has also caused the temperature of the interior $U_1(t)$ to increase compared with the first model, as would be expected.

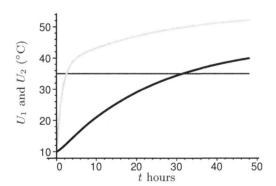

Figure 10.5: Inclusion of radiation from roof tiles. Plot of roof cavity temperature U_2 (grey) and interior temperature U_1 (black) against time in hours. The roof cavity temperature exceeds the temperature of the surroundings, $u_s = 35°\mathrm{C}$ (horizontal line).

Clearly, over a number of days, the outside temperature would change and also vary cyclically with a 24-hour period, and this time dependence could be incorporated into the model. However, it would still be appropriate to use this model to explore the effects of different insulation strategies. For example, would it be more cost efficient and effective to increase the insulation in the walls or in the ceiling? Or would different roofing materials improve, and possibly solve, the problem of a hot stuffy attic?

10.4   Spontaneous combustion

Spontaneous combustion may occur in a variety of situations when heat produced by a chemical reaction cannot escape from a system fast enough. Thus heat may build up and result in ignition. In this section, the basic mathematical theory describing this process is developed, with both an analytical and numerical analysis. The complete theory (including conduction) is complex, beyond the scope of this book; thus, what is presented here is the simplest case, that of a *stirred chemical reactor* where there is no conduction of heat.

Background

Spontaneous combustion occurs when the temperature inside a body increases at a rate faster than heat is able to escape from its surface. The higher the temperature, the more heat is produced by the chemical reaction (usually oxidation), so it is important when designing systems that involve heat production to ensure that heat can escape sufficiently easily.

   The important mechanisms are the rate of heat production and the rate of heat conduction away from this heat source. The reaction causing the heat production may be one where some organic material intersects with the atmosphere or oxygen, such as woollen fibres in bales. It is characterised by its *exothermicity*, that is, the amount of heat produced by the reaction, and by the *rate of the reaction*, which determines the rate at which the heat is produced. The process by which heat is able to escape is typically conduction, and thus the surrounding materials and their ability to conduct heat also have a substantial effect on the system. Before formulating the differential equations governing this process, we need to introduce some further physical principles.

Arrhenius law for heat production in a reaction

The rate of the reaction (associated with the rate of heat gain) can be described by an exponential function $Ae^{-E/(RT)}$, where E is the *activation energy*, with R the universal gas constant[1], given by $R = 8.314 \, \mathrm{J \, K^{-1} \, mol^{-1}}$, and T is the absolute temperature, measured in degrees Kelvin (K) from absolute zero. T will be used here to distinguish between absolute temperature, measured in Kelvin, K, and the temperature measured in °C, denoted by U, where $T = U + 273.15$. The parameter A is a further constant dependent on the type of the reaction. It is called the *pre-exponential factor* and has units of $\mathrm{s^{-1}}$, where s is the SI unit of time, seconds. The rate of gain of heat (per unit volume) is given by

$$\left\{ \begin{array}{c} \text{rate} \\ \text{heat generated} \\ \text{by reaction} \\ \text{per unit volume} \end{array} \right\} = \rho Q A e^{-E/(RT)} \tag{10.13}$$

where ρ is the density (mass per unit volume) and Q is the 'heat of reaction', measured in Joules per unit mass.

   Clearly, this function increases monotonically as the temperature rises, with

$$\lim_{T \to \infty} Ae^{-E/RT} = A.$$

From a graph of this reaction rate, we note that there is a domain of temperatures for which a small increase in T will result in a large increase in the reaction rate (see Figure 10.7 for

---

[1]The universal gas constant also occurs in the famous law $PV = nRT$ relating pressure, volume and temperature.

the scaled equation, to come). This is associated with a substantial increase in the heat production.

Model assumptions and approach

We now make some simplifying assumptions on which to build the mathematical model.

- We neglect heat conduction by assuming a uniform temperature throughout the reacting material.
- The rate of heat created is given by the Arrhenius law.
- Heat is lost from the surface of the reacting material according to Newton's law of cooling.

General compartmental model

We can apply the balance law to this process, where the compartment is the container of material (drums, wool-bales, etc.). The heat input is via some exothermic reaction and the output is through the mechanism of heat conduction away from the surface of this compartment. (We have assumed above that the temperature inside the compartment is uniform.)

The process is illustrated in the input-output diagram of Figure 10.6.

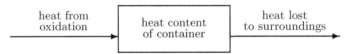

Figure 10.6: Input-output compartmental diagram for the heat production and dissipation process.

In words, the heat balance equation is

$$\left\{\begin{array}{c} rate\ of \\ change\ of \\ heat\ content \end{array}\right\} = \left\{\begin{array}{c} rate \\ heat\ generated \\ by\ reaction \end{array}\right\} - \left\{\begin{array}{c} rate \\ heat\ lost \\ to\ surroundings \end{array}\right\}. \tag{10.14}$$

Formulating the differential equation

The rate of heat loss is proportional to the surface area (S) of the body of material. It also depends on the type of material, for which there is an associated heat transfer coefficient h. Thus mathematically, the heat loss is proportional to the temperature difference between the material and its surroundings, and can be represented as

$$\left\{\begin{array}{c} rate\ of \\ heat \\ loss \end{array}\right\} = hS(T - T_a), \tag{10.15}$$

where h is the Newton cooling coefficient, S is the surface area from which heat is lost, and T_a is the temperature of the surroundings, namely the ambient temperature (measured in degrees Kelvin).

Combining the expressions for heat loss (10.15) and gain (10.13) into an equation describing the rate of temperature change (10.14), and using the specific heat to evaluate the rate of change of heat in terms of the rate of change of temperature, we obtain the differential equation

$$\rho V c \frac{dT}{dt} = \rho V Q A e^{-E/(RT)} - hS(T - T_a), \tag{10.16}$$

where ρ is the reacting substance density (moles per unit volume), Q is a measure of heat gain associated with the reaction (Joules per mole) and c is the specific heat of the reacting substance. V and S are the volume and surface area of the material, respectively, and T_a is the ambient temperature (measured in degrees Kelvin from absolute zero). This model is known as the *Seminov* model of spontaneous combustion; see Jones (1993).

Numerical solution

In equation (10.16) there are numerous parameters that make the analysis cumbersome. We therefore simplify by scaling the temperature. We define a dimensionless temperature θ (since E/R has the dimension of temperature)

$$\theta = \frac{T}{E/R}. \tag{10.17}$$

Then, with this change of variable, the system can be rewritten as

$$\sigma \frac{d\theta}{dt} = \lambda e^{-1/\theta} - (\theta - \theta_a), \tag{10.18}$$

with

$$\theta_a = \frac{T_a}{E/R}, \qquad \lambda = \frac{\rho A V Q}{hS(E/R)}, \qquad \sigma = \frac{\rho V c}{hS}. \tag{10.19}$$

The constant θ_a denotes the dimensionless ambient temperature. The parameter λ is known as the reaction efficiency parameter, since high values of λ correspond to high levels of heat generation from the chemical reaction, compared with the heat loss through Newton cooling. Both λ and θ_a are dimensionless parameters. Parameter σ is a measure of the speed of the reaction. Equation (10.18) contains only three parameters[2] θ_a, λ, and σ.

Using MATLAB or Maple, we can solve the differential equation (10.18) numerically for different values of the reaction efficiency parameter λ. The results are illustrated in Figure 10.7. The MATLAB code used to obtain the graph is given in Listing 10.1. In each case the solution tends to a steady state; however, as we vary λ from $\lambda = 2.84$ through to $\lambda = 2.85$, the amount by which the steady state jumps increases dramatically.

Listing 10.1: MATLAB code: c_ht_sponcombust.m

```
function c_ht_sponcomp
global lambda sigma theta_a;

sigma = 1.0;
theta_a = 0.2;
tend = 300;
y0 = theta_a;
lambda = 2.84;
[tsol, ysol] = ode45(@rhs, [0 tend] , y0);
plot(tsol, ysol, '-.');
hold on;
lambda = 2.85;
[tsol, ysol] = ode45(@rhs, [0 tend] , y0);
plot(tsol, ysol, ':');
lambda = 2.86;
[tsol, ysol] = ode45(@rhs, [0 tend] , y0);
plot(tsol, ysol, '-');
```

---

[2]We could reduce this to only two parameters by writing $t = \sigma\tau$ and using the chain rule, $\sigma d\theta/dt$ becomes $d\theta/d\tau$. The equation would then be completely dimensionless.

```
axis([0, tend, 0, 2])

function thetadot = rhs(t, theta)
global lambda sigma theta_a;
thetadot = lambda/sigma*exp(-1/theta) - 1/sigma*(theta - theta_a);
```

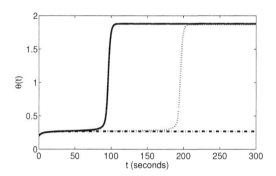

Figure 10.7: Numerical results for dimensionless temperature $\theta(t)$ against t, on solving equation (10.18). The parameter λ is gradually increased: $\lambda = 2.84$ (dot-dashed line), $\lambda = 2.85$ (solid line) and $\lambda = 2.86$ (dotted line).

With reference to equations (10.19), increasing λ can be considered as increasing the volume-to-surface area ratio. The system finds it more difficult to release heat to the surroundings, and so the reaction increases in intensity. The sudden jump in equilibrium temperature, apparent in the right-hand diagram of Figure 10.7, corresponds to the spontaneous ignition of the material.

Analytic solution

Ideally, we would like to determine exactly which values of λ correspond to the sudden jump in equilibrium temperature. To find this value, we set $d\theta/dt = 0$, for equilibrium, and solve the resulting equation graphically. Thus for equilibrium,

$$\lambda e^{-1/\theta} = \theta - \theta_a, \qquad \lambda = \frac{\rho V A Q}{hS(E/R)} = \frac{\rho R V A Q}{hSE}. \tag{10.20}$$

Figure 10.8 illustrates the two curves for heat loss and heat gain given by

$$y = e^{-1/\theta}, \qquad y = \frac{1}{\lambda}(\theta - \theta_a),$$

on the same system of axes. Clearly there can be at most three points of intersection where the heat loss equals the heat gain, that is, $(d\theta/dt) = 0$ or $(dT/dt) = 0$. From (10.18) if the heat loss is greater than the heat gain then $d\theta/dt < 0$, and if the heat loss is less than the heat gain then $d\theta/dt > 0$.

With reference to the solid straight line for heat loss in the figure (Figure 10.8), if the temperature is low so that $\theta < \theta_1$, then the heat gain is greater than the heat loss and the temperature increases with $\theta \to \theta_1$. If the temperature is such that $\theta_1 < \theta < \theta_2$, then the heat loss is greater than the heat gain and the temperature will decrease so that $\theta \to \theta_1$. Thus we consider the intersection at θ_1 as a stable equilibrium for the system. Arguing in a similar manner, we see that at θ_2 there is an unstable equilibrium, and at θ_3 another stable point.

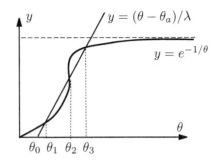

Figure 10.8: Sketch of the graphical solution of equation (10.18). There are, at most, three equilibrium solutions (θ_1, θ_2 and θ_3), although for some values of λ there may be only one solution.

If the initial condition is such that the system approaches the equilibrium at θ_3, we have the conditions for spontaneous ignition. However, if the initial conditions are different, ignition can be avoided. If we start with an initial temperature such that $\theta(0) = \theta_a$, then the temperature tends to the closest equilibrium. However, if λ is sufficiently large (i.e., the slope of the line, $1/\lambda$, is sufficiently small), for only one equilibrium point (corresponding to θ_3) to exist, then ignition will occur.

As we increase λ, which decreases the slope of the line $1/\lambda(\theta - \theta_a)$, we reach critical values where the number of equilibrium values changes: from one through two to three, or from three through two to one. The values of λ for which this change occurs are called bifurcation points.

Example 10.6: *Given $\theta_a = 0.2$ find the critical value of the parameter λ where ignition occurs.*

Solution: *The critical value occurs where the slope of the line in Figure 10.8 is tangential to the curve (which may occur in two places). Hence we solve the simultaneous equations*

$$e^{-1/\theta_c} = \frac{1}{\lambda_c}(\theta_c - 0.2), \qquad \frac{1}{\theta_c^2}e^{-1/\theta_c} = 1/\lambda_c,$$

where the first equation describes the point(s) of intersection, and the second equates the first derivatives of each, corresponding to the line and curve having the same slope. With a little algebra, we can deduce

$$\lambda_c = \theta_c^2 e^{1/\theta_c}, \qquad \theta_c^2 - \theta_c + 0.2 = 0.$$

Solving the quadratic equation gives two roots

$$\theta_{c_1} = \frac{1}{2}\left(1 - \sqrt{1 - 4 \times 0.2}\right) \simeq 0.277,$$

$$\theta_{c_2} = \frac{1}{2}\left(1 + \sqrt{1 - 4 \times 0.2}\right) \simeq 0.724.$$

After substituting these back into the expression, for λ_c we get critical values of λ corresponding to a bifurcation of the equilibrium temperatures,

$$\lambda_c \simeq 2.847, \qquad \lambda_c \simeq 2.085.$$

As we have seen above, it is the larger value $\lambda_c = 2.847$ that corresponds to spontaneous ignition. This value is consistent with the results of Figure 10.7.

Alternatively, we can hold λ fixed and think of varying θ_a. This corresponds to the slope of the line in Figure 10.8 remaining constant, but the intersection point it makes with the θ-axis moving to the right as θ_a increases. Again, we move from one equilibrium temperature (θ_1) to three equilibrium temperatures (θ_1, θ_2, θ_3) and back to just one equilibrium temperature

(θ_3), for the larger values of θ_a. It is the second bifurcation here that corresponds to thermal ignition, since, owing to the stability of the equilibrium points, the temperature to which the system evolves jumps suddenly from θ_1 to θ_3. The value of θ_a where this occurs is the critical ambient temperature, θ_{ac}. (See Exercise 10.10 for a numerical example of determining the critical ambient temperature.)

Further interpretation

Clearly the critical ambient temperature is crucial to any strategy for avoiding spontaneous combustion. The analysis above indicates that both the storage conditions and temperature of the material, as well as the ambient temperature, need to be carefully monitored in processes where spontaneous ignition may occur.

Spontaneous combustion can be avoided completely if the ambient temperature is sufficiently low so that there is only one stable equilibrium point on the lower branch of the heat gain curve. This occurs for low temperatures, below the ambient temperature θ_{ac}. However, such temperatures are not always practicable in the 'real world' and the regular monitoring of all conditions may be required.

Limitations and extensions

There are some important factors that our model has not taken into account.

- As the reaction proceeds, fuel is used up. Here we have neglected the consumption of fuel.

- We have assumed the reactants are mixed (a stirred chemical reactor). In practice, perfect mixing will not occur, and the temperature is different at different points throughout the reacting material, which allows heat conduction to occur. This factor is especially important in solid materials.

Both these assumptions can be relaxed, but this leads to more complicated models. If we relax the first assumption, the model will comprise two simultaneous ODEs, one for the temperature and one for the concentration. Equilibrium values can be found by solving the two appropriate algebraic equations simultaneously, and bifurcations can be established as a parameter is varied. This is discussed briefly in Fulford and Broadbridge (2000).

If we relax the second assumption (no conduction), then our model results in a PDE for the temperature (see Section 12.1 for how partial derivatives arise). For further details of how to deal with this, see Fulford and Broadbridge (2000); Jones (1993).

In Section 10.5, we discuss a case study involving the spontaneous ignition of fish and chip crumble. Another practical example involving spontaneous ignition of moist milk powder is given in Rivers et al. (1996). In Lignola and Maio (1990), inclusion of reactant consumption in a stirred reactor shows that some interesting oscillatory behaviour can occur in these processes.

> ### Summary of skills developed here:
> - *Use* `Maple` *or* `MATLAB` *to find a numerical solution for the temperature in a stirred chemical reactor.*
> - *For a given dimensionless ambient temperature, find the critical value of the reaction efficiency.*
> - *For a given reaction efficiency parameter, find the critical ambient temperature such that spontaneous ignition will occur.*

10.5   Scenario: Fish and chips explode

There have been many recorded cases of spontaneous combustion, for example, the sudden combustion of a barn filled with hay, the spontaneous ignition of wool bales in transit or storage, and a case of spontaneous ignition in a fish and chips shop in Napier (New Zealand), which is the subject of the scenario that follows. Such events are of great interest and importance to insurance companies, which are called upon to pay for the damage. The information for the following case derives from such a report, written for a New Zealand insurance company. This discussion is adapted from a report by Smedley and Wake (1987).

Fish and chips shops are not usually associated with explosives; however, in 1986 in Napier, New Zealand, one such shop was severely damaged when a drum of waste from the cooking scraps ignited spontaneously.

Fish and chips are cooked in large sieves submerged in vats of animal fat that are heated to about 180°C. The hot crumble, or waste, which detaches during cooking, is allowed to cool and drain above the vat before being tipped into drums. The concern was that over a period of time the pile of crumble would (and the case in New Zealand proved that it could) spontaneously ignite through some heat-producing mechanism within the waste.

The authors approached the problem from the point of view of constructing a mathematical model applicable to general problems of this nature, but which would be sufficiently simple for non-mathematicians to apply *in situ*. That is, they needed to provide adequate conditions to test for on site in order to prevent such 'explosive' events.

Spontaneous combustion is possible if the temperature inside the pile of crumble increases at a faster rate than heat is able to escape from the surface, where the crumble meets the air and drum wall. That is, if the exothermicity and the reaction rate are sufficiently great, such that the heat cannot dissipate, but instead builds up to a point of ignition. These processes need to be examined closely.

The exothermic reaction, which is the reaction producing the heat, is that of the organic material consisting of the fish and chip residue reacting with the atmosphere or oxygen. Since spontaneous combustion has occurred, this indicates that the exothermicity of this reaction is sufficient to cause ignition. In the case of oxidation for fish and chip residue, the activation energy was measured to be $E/R = 30,000$. (In comparison, for wool $E/R = 9,036$, or for cotton $E/R = 18,000$.)

As mentioned above, heat is lost through the interface between the crumble and the air as well as through the crumble-drum wall interface. Thus a change in this area will affect this heat loss, and there is the possibility of changing the dimensions of the drums in order to exercise a degree of control with respect to the heat loss. For the fish and chips shop in question, the drums used were cylinders with a diameter and height of 38 cm.

In order to avoid spontaneous combustion, which is the aim, consider the relationship between these heat loss and heat gain processes, illustrated in Figure 10.8 of Section 10.4. The diagram illustrates the existence of a critical temperature (T_c), above which spontaneous combustion will always occur, and also a critical initial temperature (T_{0c}), dependent on the ambient temperature, below which combustion can be avoided. In order to predict the dynamics of such a system, particularly to avoid the event of spontaneous combustion, both the storage conditions and temperatures, as well as the ambient temperatures, need to be monitored closely in the waste disposal process.

We return to the formulation of strategies to adopt *in situ*, which will avoid a repeated disaster in the fish and chips shop and keep the insurance company happy. The authors used a modified version of the model in Section 10.4. It was based on a heat balance, but was more complex, including details of the combustion process and an approximation of the Arrhenius term (see Jones (1993)). Using the predictions of this more sophisticated model, they were able to provide a simple look-up table of values from which the proprietor could read the critical ambient temperature and the associated critical initial temperature. Such figures could be derived for a variety of convenient drum sizes.

In practice, the safety procedure is simple. A quick reference to the look-up table and some intermittent monitoring of the temperatures will provide the required conditions for an uneventful evening of business down at the fish and chips shop.

10.6   Exercises for Chapter 10

10.1. *Coffee cooling.* *Imagine you have a cup of coffee cooling according to the natural law of cooling, so that the temperature $U(t)$ satisfies*

$$\frac{dU}{dt} = -\lambda(U - u_s)^{5/4},$$

where λ is a positive constant and $u_s = 20°C$ is the temperature of the surroundings.

(a) *Assuming $U(0) = u_0$, separate the variables to obtain*

$$\lambda t = 4\left[(U - u_s)^{-1/4} - (u_0 - u_s)^{-1/4}\right].$$

(b) *If the cup of coffee cools from an initial temperature $u_0 = 60°C$ to $50°C$ in 10 minutes, obtain λ and hence find the time it takes for the cup of coffee to cool to $40°C$.*

(c) *Using `Maple` or `MATLAB`, compare the solution of this natural cooling model with the solution for Newton's law of cooling by plotting them on the same system of axes, as in Figure 10.1. (Assume that with Newton's law of cooling, the coffee cools from $u_0 = 60°C$ to $50°C$ in 10 minutes.)*

10.2. *Beer warming.* *A cold beer is at a temperature of $10°C$. After 10 minutes, the beer has warmed to a temperature of $15°C$. If the room temperature is $30°C$, how long will it take the beer to warm to $20°C$, assuming that Newton's law of cooling applies?*

10.3. *Separating variables.* *Solve the equation developed to describe the hot water heater (equation (10.7) from Section 10.2), using the separation of variables technique, and show the solution to be equivalent to that provided in the text.*

10.4. *The hot water heater.* *Consider the problem of a water heater, with parameters and model as in Section 10.2,*

$$\frac{dU}{dt} = \beta - \alpha U, \qquad U(0) = u_0, \tag{10.21}$$

where the constants α and β are defined as

$$\alpha = \frac{hS}{cm}, \quad \beta = \frac{q + hSu_s}{cm}.$$

(Parameters $q = 3{,}600\,\text{W}$, $S = 2.7\,\text{m}^2$ and $u_s = 15°C$ with constants $h = 12\,\text{W}\,\text{m}^{-2}\,°\text{C}^{-1}$ and $c = 4\,200\,\text{J}\,\text{kg}^{-1}\,°\text{C}^{-1}$, $m = 250\,\text{kg}$.)

(a) *Suppose the temperature of the water in the hot water tank is $60°C$, the heater is turned off so that no heat is supplied and the water is allowed to cool. What is the water temperature after 8 hours?*

(b) *With heat once again supplied ($q = 3{,}600\,\text{W}$), how long would it take to heat the water back up to $60°C$, after it has cooled for 8 hours in (a)?*

(c) *Suppose again the water is* 60°C, *that the heater is turned off so that no heat is supplied and the water is allowed to cool. How long will it take for the water to reach* 50°C?

(d) *With the heater turned on again, how long will it take to heat the tank of water from* 50°C *to* 60°C?

(e) *What is of paramount interest is how to run the most cost-efficient system. Clearly the cost of heating the tank is directly related to the time for which the heater is on.*

Assuming all constants as above, would it be cheaper to use a thermostat that switches the heater on each time the temperature drops below 50°C *and off each time the temperature reaches* 60°C, *or to switch the heater off for 8 hours each day relying on the thermostat at other times? Use the model and the results found above to justify this.*

(Although actual costs are not needed to calculate the most economical solution, typical values in Australia in 2005 are 8.22 c per kW hr.)

10.5. Insulating the hot water heater. *Assume the model and parameters of Exercise 10.4. Suppose with good insulation no heat is lost to the environment. Adapt the model to accommodate this change and then, with a surrounding temperature of* 15°C, *and the water in the tank at this temperature, establish how long it would take to heat the tank to* 60°C.

10.6. Thermometer temperatures. *A vet is taking the temperature of a sick horse. Initially, the temperature of the thermometer is* 27.8°C. *Three minutes after insertion, the reading is* 32.2°C *and three minutes later it is* 34.4°C. *The horse then has a violent convulsion that destroys the thermometer completely so that no final reading can be taken.*

You can assume that the rate of change of temperature of the thermometer is directly proportional to the difference between the temperature of the thermometer and that of the horse.

(a) *Let* $U(t)$ *be temperature of the thermometer at time* t, *with* $U(0) = u_0$ *its initial temperature and* u_h *the temperature of the horse (assumed constant). Model the system with a differential equation and show that its solution is*

$$U(t) = u_h + (u_0 - u_h)e^{-\lambda t},$$

where λ *is the constant of proportionality.*

(b) *Find the temperature of the horse,* u_h, *and also the value of the constant* λ.

10.7. Modelling house temperatures. *Develop a model for the temperature inside a building comprising two rooms with a flat ceiling (no roof cavity), which is described below.*

One room has a heater that supplies heat at a constant rate q_0. *You may assume that heat is lost to the surroundings, from both rooms, according to Newton's law of cooling, and that the heat transfer from one room to the other is proportional to the temperature difference between the two rooms. You may also assume the air inside the building is well mixed so that heat conduction can be ignored. Any radiation can also be ignored.*

Give either a compartmental diagram or suitable word equations to help explain your formulation and define any symbols introduced. You are not required to solve any equations.

10.8. Stirred chemical reactor. *For a stirred chemical reactor, the heat balance equation can be given in terms of the scaled temperature* (θ) *as*

$$\sigma \frac{d\theta}{dt} = \lambda e^{-1/\theta} - (\theta - \theta_a),$$

where θ_a, λ *and* σ *are positive constants.*

(a) *Refer to Figure 10.8, which illustrates the three equilibrium solutions. By interpreting each of the two curves as a heat gain or heat loss term in the differential equation, determine the stability of each of the equilibrium solutions* $\theta^{(1)}$, $\theta^{(2)}$, $\theta^{(3)}$.

(b) Discuss what happens to $\theta(t)$ as $t \to \infty$ when θ_0 is small and when θ_0 is large.

10.9. Spontaneous ignition and critical values. *Consider the equation for the scaled absolute temperature in an exothermic oxidation reaction*

$$\sigma \frac{d\theta}{dt} = \lambda e^{-1/\theta} - (\theta - \theta_a).$$

(a) For $\theta_a = 0.3$, $\lambda = 2$, by plotting the curves, determine all the equilibrium solutions.

(b) Repeat for $\theta_a = 0.1$ and $\lambda = 5$.

10.10. Critical ambient temperature. *Consider the equation for the scaled absolute temperature in an exothermic oxidation reaction, with $\lambda = 2$,*

$$\sigma \frac{d\theta}{dt} = 2e^{-1/\theta} - (\theta - \theta_a).$$

(a) Deduce that the critical ambient temperature θ_{ac} where bifurcations occur satisfies the equations

$$\theta_a = \theta - \theta^2, \qquad \theta_2 = 2e^{-1/\theta},$$

where θ is the critical value for which bifurcations occur.

(b) Find the two critical values of θ and hence determine the approximate critical ambient temperature θ_a above which spontaneous ignition occurs.

Chapter 11

Solving heat conduction and diffusion problems

In the previous chapter we formulated differential equations for some heat conduction problems. We now see how to solve these differential equations and apply suitable boundary conditions to obtain the temperature distribution.

11.1   Boundary value problems

In the previous chapter we discussed the formulation of differential equations. To solve these, we first need to solve the equations, and then evaluate the arbitrary constants using some known conditions. The problems we look at involve specifying conditions on the boundaries of a region. This is, technically, slightly different from applying initial conditions, although in practice the process is similar. We concentrate here on how to apply boundary conditions to the general solution of a differential equation. Such problems are known as boundary value problems.

Typical problems

The first problem we consider is the overall rate of heat loss through the wall of a building. We solve the appropriate differential equation to find the equilibrium temperature at any point inside the wall and then use Fourier's law to calculate the heat flux through the wall. The second problem analyses the rate of heat lost through an insulating jacket surrounding a water pipe. This problem is similar to the previous one, but involves radial heat flow rather than heat flow in one direction. In our final problem, we examine the cooling of a computer chip by mounted heat fins. A problem similar to this is the cooling of a motorcycle engine, also using heat fins.

Each of these problems involves first solving a differential equation for the equilibrium temperature. After finding the general solution, we then apply suitable boundary conditions to find the arbitrary constants, a process that requires the solution of simultaneous equations.

Boundary conditions

When we solve a second-order differential equation, we obtain a solution with two arbitrary constants. In general, the number of arbitrary constants is the same as the order of the differential equation. (General principles used in solving second-order equations can be found in Appendix A.2 and Appendix A.5.)

For example, consider the differential equation

$$\frac{d^2U}{dx^2} - 4U = 0.$$

Since the differential equation is linear and its coefficients are constant, we look for a solution of the form $U(x) = e^{\lambda x}$. Substitution of this into the differential equation gives

$$\lambda^2 e^{\lambda x} - 4e^{\lambda x} = 0,$$

and hence we obtain $\lambda^2 - 4 = 0$. Thus we find two possible values for λ: $\lambda = 2$ or $\lambda = -2$. This yields two solutions: e^{2x} and e^{-2x}.

The general solution of the differential equation is obtained by taking a linear combination of these two solutions

$$U(x) = C_1 e^{2x} + C_2 e^{-2x}. \tag{11.1}$$

This solution has two arbitrary constants C_1 and C_2 whose values are determined by applying boundary conditions.

Boundary conditions are conditions applied on either boundary of a region. Suppose the region is $0 \le x \le 1$. One pair of boundary conditions might be, for example,

$$U(0) = 0, \qquad U(1) = 3.$$

Another example is

$$U(0) = 1, \qquad U'(1) = 0.$$

This second example involves the derivative of U as one of the boundary conditions.

In solving differential equations where the independent variable is time, we have so far encountered initial conditions. The difference between initial and boundary conditions is the value of the independent variable they are applied to. Initial conditions are always applied at the same value (usually $t = 0$). Boundary conditions, however, are applied at two different values (e.g., $x = 0$ and $x = 1$).

Applying boundary conditions

The process of applying boundary conditions is exactly the same as applying initial conditions. Applying the conditions leads to, in general, two simultaneous equations for the two arbitrary constants.

As an example, consider the following differential equation with boundary conditions:

$$\frac{d^2 U}{dx^2} - 4U = 0, \qquad U(0) = 0, \quad U(1) = 3.$$

The solution to this differential equation is given by (11.1). Substituting $x = 0$ and then $x = 1$ into this solution, and then applying the boundary conditions, produces the pair of simultaneous equations

$$C_1 + C_2 = 0, \qquad C_1 e^2 + C_2 e^{-2} = 3.$$

Solving simultaneously for C_1 and C_2 gives $C_1 = 3/(e^2 - e^{-2})$ and $C_2 = -3/(e^2 - e^{-2})$. We now substitute these values for C_1 and C_2 back into the general solution (11.1). Hence the solution U that satisfies both boundary conditions is

$$U(x) = 3\frac{e^{2x} - e^{-2x}}{e^2 - e^{-2}}$$

for x between 0 and 1.

This can also be expressed in terms of the hyperbolic function $\sinh(2x) = (e^{2x} - e^{-2x})/2$ as

$$U(x) = 3\frac{\sinh(2x)}{\sinh(2)}.$$

(For details on hyperbolic functions, see Appendix B.4.)

In the following sections we consider some practical problems. These involve solving differential equations derived in the previous chapter and then applying suitable boundary conditions.

Summary of skills developed here:
- *Solve simultaneous equations to find the constants of integration.*
- *Differentiate between, and apply, initial conditions and boundary conditions.*
- *Understand the principles involved in solving simple second-order differential equations, with boundary conditions.*

11.2   Heat loss through a wall

We solve a differential equation and apply boundary conditions to find the equilibrium temperature inside a brick wall. The boundary condition on the outside of the wall follows from Newton's law of cooling, while on the inside the wall is kept at a constant temperature.

Problem description

The design of energy-efficient buildings has become more important as energy costs have risen, and there is a greater awareness of the responsible use of resources. As was seen in the earlier case study (Section 10.3), which examined the temperatures in a building, a crucial concern in such design is that of calculating the rate of heat loss from a building wall. We take a closer look at this heat loss process.

Consider a building with walls of thickness $\ell = 20$ cm. The inside of the house is maintained at a comfortable temperature $u_i = 20°$C. The outside of the wall is exposed to the surrounding air at a temperature $5°$C. Here we assume the inside wall temperature is fixed at the room temperature, $u_i = 20°$C. However, we assume the outside wall temperature is higher than $5°$C, and loses heat to the surroundings, at temperature $u_o = 5°$C (see Figure 11.1). (Subscripts i and o denote variables on the inside and outside of the wall, respectively.)

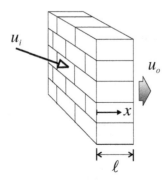

Figure 11.1: Heat flow through the wall of a house, $0 \leq x \leq \ell$.

This is not completely realistic. The inner wall also will not be exactly the same temperature as the room temperature. However, for the purpose of illustrating two different types of boundary conditions, we will continue with this assumption for now and later modify it to account for the inner wall also being at a different (lower) temperature than the surrounding air.

We are interested in determining the rate of heat flow through the wall. One possible question we might ask is:

> *If the thickness of the wall is doubled, what happens to the rate of flow of heat through the wall?*

Model assumptions and approach

We let the x-axis measure distance from the inside of the wall, outwards. Closer to the inside the temperature will be closer to 20°C. Hence we may think of the temperature as a function of two variables x and t. For the equilibrium temperature, however, the temperature will be a function of x only.

We assume heat flows directly from the inside to the outside through the wall. Thus we neglect any complicated heat flow at the edges, top and bottom. For most of the wall, the heat flow will be in the x-direction. We also assume the wall to have no windows. Since the inside and outside temperatures are constant once thermal equilibrium is established, the temperature inside the wall will be constant. We assume the outside temperature of the wall satisfies Newton's law of cooling. The temperature of the outer surface of the wall can be determined from the solution for the temperature in the wall.

The differential equation

In the previous chapter we formulated a differential equation for the equilibrium temperature of some material through which heat was conducted in the x-direction. The differential equation (from Section 9.5, equation (9.18)) is

$$\frac{d^2U}{dx^2} = 0. \tag{11.2}$$

This is a second-order differential equation so we expect two arbitrary constants in the solution.

The equation arose from a heat balance in thermal equilibrium. Note that the differential equation does not depend on the conductivity of the material. The conductivity will be incorporated through the boundary conditions.

Analytic solution

It is a simple matter to solve this differential equation. We can integrate both sides twice. By integrating with respect to x once we obtain

$$\frac{dU}{dx} = C_1, \tag{11.3}$$

where C_1 is an arbitrary constant and then integrating again gives

$$U(x) = C_1 x + C_2. \tag{11.4}$$

Here C_1 and C_2 are the two arbitrary constants. In order to solve for these constants, we need two boundary conditions, one at each boundary: $x = 0$ and $x = \ell$.

The boundary conditions

In modelling this problem we assumed the temperature of the inside of the wall is always at a given temperature u_i. Thus the boundary condition at $x = 0$ can be written as

$$U(0) = u_i. \tag{11.5}$$

We also assumed that the outside of the wall loses heat to the surroundings according to Newton's law of cooling, which states that the rate of heat loss is proportional to the surface area of the wall face and the temperature difference between the wall and the surroundings. We can equate this to the rate of heat conducted through the wall at $x = \ell$, given in terms of the heat flux as $J(\ell)A$, where A is the area through which the heat passes. Then, dividing by the surface area,

$$J(\ell) = h(U(\ell) - u_o), \tag{11.6}$$

where h is the convective heat transfer coefficient and u_o is the temperature of the surroundings outside the wall. Since $U(\ell) > u_o$, this correctly gives a positive heat flux.

We now apply both boundary conditions (11.5) and (11.6) to determine the arbitrary constants C_1 and C_2. This gives us two equations for the two unknowns C_1 and C_2.

Example 11.1: *Apply the boundary conditions (11.5) and (11.6) to determine the arbitrary constants C_1 and C_2 of equation (11.4). Hence obtain an expression for the temperature profile in the wall.*

Solution: *The boundary condition $U(0) = u_i$ states that $U = u_i$ at $x = 0$. To apply this we substitute $x = 0$ into the general solution (11.4) and obtain the equation*

$$C_1 \times 0 + C_2 = u_i. \tag{11.7}$$

We can then solve directly for C_2 giving

$$C_2 = u_i. \tag{11.8}$$

To apply the Newton cooling boundary condition (11.6), we first calculate $J(x)$ using Fourier's law $J = -k(dU/dx)$, where k is the conductivity. Using equation (11.4), we obtain

$$J(x) = -k\frac{dU(x)}{dx} = -kC_1.$$

At $x = \ell$, equation (11.6) becomes

$$-kC_1 = h\left(C_1\ell + C_2 - u_o\right). \tag{11.9}$$

Substituting (11.8) into (11.9) and dividing through by h we obtain

$$-\frac{k}{h}C_1 = C_1\ell + u_i - u_o.$$

Rearranging gives

$$-\left(\ell + \frac{k}{h}\right)C_1 = u_i - u_o,$$

which implies that

$$C_1 = \frac{-(u_i - u_o)}{\ell + \frac{k}{h}}. \tag{11.10}$$

Substituting the above expressions for the arbitrary constants (11.8) and (11.10) back into the general solution (11.4) gives the solution for the equilibrium temperature,

$$U(x) = u_i - \left(\frac{u_i - u_o}{\ell + \frac{k}{h}}\right)x. \tag{11.11}$$

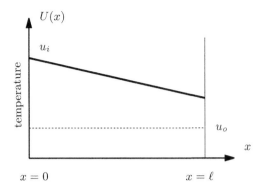

Figure 11.2: Equilibrium temperature in a wall where the inside of the wall is at temperature u_i and on the outside wall heat is lost to the surroundings, which are at temperature u_o.

This is a simple linear function that is illustrated in Figure 11.2. The temperature $U(x)$ decreases with x. But note that the temperature of the outside surface of the wall, $U(\ell) = u_o/(1 + k/(h\ell))$, is higher than that of the surroundings, u_o.

As a quick check on the solution, note that as $h \to 0$, $U(\ell) \to u_i$. Physically, $h \to 0$ corresponds to not allowing any heat to escape so that the temperature on the outside would be the same as the temperature on the inside. Alternatively, consider the limit as $h \to \infty$. In this case, $U(\ell) \to u_o$. This corresponds physically to the situation where the temperature of the outside wall is the same as the temperature of the outside surroundings.

Rate of heat loss from a wall

The rate of heat loss from the wall can be calculated using Fourier's law. This gives the heat flux (the rate of flow of heat per unit time per unit area). From Fourier's law, the heat flux at $x = \ell$, $J(\ell)$, can be evaluated from

$$J(x) = -k\frac{dU(x)}{dx}.$$

Example 11.2: *Find an expression for the heat flux through the wall.*

Solution: *Using the solution for the temperature (11.11) we obtain*

$$J(\ell) = -k \times -\frac{u_i - u_o}{\ell + \frac{k}{h}} = \frac{u_i - u_o}{\frac{\ell}{k} + \frac{1}{h}}. \tag{11.12}$$

As a quick check we see that this gives a positive value since $u_i > u_o$.

Let us carry out some calculations with typical numerical values. Suppose the inside temperature is $u_i = 20°C$, whereas the outside temperature is $5°C$. Consider a brick wall of thickness $\ell = 15\,\text{cm} = 0.15\,\text{m}$. The conductivity of brick is $k = 0.5$ SI units, and the convective heat transfer coefficient $h = 10$ SI units. With these values we calculate $\ell/k = 0.33$ and $1/h = 0.1$, so the the heat flux through the wall is

$$J = \frac{15}{0.33 + 0.1} = 37.5\,\text{W}\,\text{m}^{-2}. \tag{11.13}$$

Thus we can obtain an estimate for the total rate of heat loss from the walls of a brick house by multiplying this value by the total area of the walls.

Note that J is directly proportional to the temperature difference but not directly proportional to the conductivity k or to the thickness of the wall ℓ. Thus doubling the wall thickness will not halve the heat loss from the wall, and nor would doubling the conductivity (except in cases where $1/h$ is much smaller than ℓ/k). However, for decreased k and h there is a reduction in the heat flux J.

We can do a similar calculation as that above to determine heat loss through windows, using the same temperature difference. For a 5-mm thick pane of glass, $\ell = 5 \times 10^{-3}$m. The conductivity of glass is $k = 0.8$ SI units and we take $h = 10$ SI units. The heat loss through a single glass pane is then,

$$J = \frac{15}{0.0062 + 0.1} \approx 141 \, \mathrm{Wm}^{-2}.$$

This indicates why heat loss through windows is generally far more significant than that through walls.

Heat resistance R-values

There is a neat interpretation to equation (11.12), which we can write as

$$J = \frac{u_i - u_o}{R}, \qquad \text{where} \quad R = \frac{\ell}{k} + \frac{1}{h}. \tag{11.14}$$

The parameter R is interpreted as a combined thermal resistance. We can think of R as made up of two separate resistances:

- $R_1 = \ell/k$, the resistance to heat flow through the material
- $R_2 = 1/h$, the surface resistance to heat flow to or from the air

Note that the resistance to heat flow through the material, ℓ/k, increases with the width of the material ℓ and decreases with the conductivity k, as we might expect.

If we include a surface resistance for both the inside and outside surfaces of a wall, using resistances it is simple to calculate

$$J = \frac{\Delta U}{1/h_i + \ell/k + 1/h_o},$$

where $1/h_i$ is surface resistance of the inside of the wall and $1/h_0$ is the surface resistance of the outside of the wall. Repeating the earlier calculation of heat flux through a brick wall and assuming $h_i = h_o = 10$ SI units,

$$J = \frac{15}{0.1 + 0.33 + 0.1} = 28.3 \, \mathrm{W\,m}^{-2}.$$

Companies that provide insulation batts often talk about the R-value of their batts. For example, $R2.0$ batts have a thermal resistance of $\ell/k = 2.0 \, \mathrm{W\,{}^\circ C}^{-1}$. Using this notion of thermal resistances, we can calculate the rate of heat loss for more complicated structures, such as an insulated wall of a house. Here, the total resistance is

$$R = \frac{1}{h_i} + \frac{\ell_1}{k_1} + \frac{\ell}{k} + \frac{1}{h_o},$$

where k_1 and ℓ_1 are the conductivity and width of the insulation batts, respectively, and k and ℓ are those values corresponding to brick. We have assumed the same surface resistance on both sides of the wall. With $R2.0$ batts ($\ell_1/k_1 = 2.0$), a brick wall ($\ell/k = 0.33$) and $h_i = h_o = 10$ SI units, this gives a total heat flux of

$$J = \frac{u_i - u_o}{R} = \frac{15}{0.1 + 2.0 + 0.33 + 0.1} = 5.9\,\mathrm{W\,m^{-2}}.$$

This is much lower than the previously calculated flux through a single brick wall of $28.3\,\mathrm{W\,m^{-2}}$, showing how effective wall insulation can be.

For windows, double or triple glazing is used to reduce the loss of heat and is remarkably effective. Over 90% of the heat loss can be prevented as is shown in the following case study (Section 11.3).

Summary of skills developed here:

- *From a word description, write in mathematical form the boundary conditions corresponding to:*
  - *prescribed temperature on a boundary;*
  - *prescribed heat flux on a boundary; and*
  - *Newton heat exchange with surroundings on a boundary.*
- *Be able to apply the above boundary conditions to find the temperature for heat flow in one direction.*
- *Obtain the heat flux from the temperature.*
- *Be able to interpret the heat flux in terms of thermal resistances.*

11.3   Case Study: Double glazing: What's it worth?

The following case study is adapted from Mesterton-Gibbons (1989) and information brochures on double glazing from manufacturing companies.

Is double glazing what it's cracked up to be? We saw in Section 11.2 that there is a simple relation between the heat flux and the temperature difference for heat flow through a slab of any material. It is

$$J = \frac{u_1 - u_2}{R},$$

where R is the heat resistance (commonly called the R-value). We consider here a case study that makes use of this result in an examination of the heat transfer process in windows. In particular, we are interested in the extent to which double glazing reduces this heat loss and improves efficiency.

Advertising promotes double glazing as a very effective way of preventing heat loss from your home; however, how effective is it? The windows are expensive to buy and install so we need to know to what extent the benefits outweigh the cost.

A wall of any building transfers heat in one direction if the temperatures on either side are different. In the case of glass window panes, there is a heat transfer outwards in winter and inwards in summer, neither of which may be desirable. Double glazing companies manufacture windows typically consisting of two 3 mm panes of glass separated by a 12 mm gap. Using the theory of heat flux we can get a very good idea of just how much heat loss is prevented by installing these double glazed windows.

Let u_1 be the temperature inside a room and u_2 the temperature outside. We are concerned with winter heat loss, so $u_1 > u_2$. Let ℓ_g be the thickness of the glass in each pane and ℓ_a the width of the air cavity between the sheets of glass. We take u_a and u_b to be the temperatures on either side of the air cavity, as illustrated in Figure 11.3.

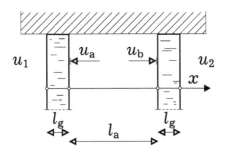

Figure 11.3: Schematic diagram of double glazing.

Let J (the heat flux) be the amount of heat transfered through a unit area (per unit time) of a plane perpendicular to the direction of the x-axis, at x. We are not interested here in the initial heat transfer to the glass or to the cavity, but rather in the behaviour of the system after it has settled down to a state of thermal equilibrium, where J is a constant.

The heat flux J is given by

$$J = \frac{u_1 - u_2}{R}, \tag{11.15}$$

where R is the total heat resistance and $R = 2R_g + R_a$. This can be derived from the relation

$$J = \frac{u_1 - u_a}{R_g} = \frac{u_a - u_b}{R_a} = \frac{u_b - u_2}{R_g},$$

which expresses the equality of heat flux through each part of the system resulting from thermal equilibrium, or constant J. R above is the sum of the three individual heat resistances, due to the two panes of glass R_g and the air cavity R_a. We later relate the heat resistances R_g and R_a to the glass and air conductivities and thicknesses.

We need to compare this flux with that in the case of a single paned window, that is, one that is not double glazed ($\ell_a = 0$), but where the pane now has thickness $2\ell_g$. The heat flux for the *single-paned* window, J_s, is

$$J_s = \frac{(u_1 - u_2)}{2R_g}.$$

(We have taken the single pane to have a thickness of $2\ell_g$ so that the only difference between J and J_s is due to the air gap.) We define a relative heat loss Δ by

$$\Delta = \frac{J_s - J}{J_s} = \frac{R_a}{2R_g + R_a}. \tag{11.16}$$

If we assume the temperature of the window is exactly the temperature of the air, then $R_g = \ell_g/k_g$ where ℓ_g is the width of a single pane of glass and k_g is its conductivity. Similarly, $R_a = \ell_a/k_a$ where ℓ_a is the distance between the panes and k_a is the conductivity of the air. This essentially follows the calculations of Mesterton-Gibbons (1989). Although this assumption is not very accurate, we shall see where it leads and then extend the model with this assumption relaxed.

Using approximate values of the conductivities,

$$k_g = 0.8 \, \mathrm{J \, m^{-1} \, s^{-1} \, {}^\circ C^{-1}}, \qquad k_a = 0.05 \, \mathrm{J \, m^{-1} \, s^{-1} \, {}^\circ C^{-1}},$$

we have $k_g/k_a \simeq 16$. Also, we set $\ell_a/\ell_g = r$. Substituting these values into (11.16) we obtain the relative heat loss Δ_0 as

$$\Delta_0 = \frac{16r}{16r + 2} = \frac{r}{r + 1/8}.$$

The graph of this function can be seen in Figure 11.4. It is a monotonically increasing function with $\Delta = 0$ when $r = 0$ and $\lim_{r \to \infty}(r/(r + 1/8)) = 1$.

Thus, for $r = 1$, the reduction in heat loss is 89%. For $r = 4$ the reduction has increased substantially to 97% but, for $r = 8$, the reduction is not much better at 98%. To conclude then, if $\ell_a \approx 4\ell_g$, the heat loss can be reduced by 97% and not much is gained by increasing the gap (ℓ_a) between the glass panes. This is indeed a substantial reduction in heat loss, and these specifications are exactly those manufactured as standard by the companies considered.

However, the above calculations assumed the surface temperature of the glass to be the same as the air temperature. This is a unrealistic assumption for a typical pane of glass and can result in an overestimate of the heat flux. In an improved model, we do not take the temperature of the glass surface to be the same as that of the air, but instead replace this with a Newton cooling condition. This results in heat resistances being increased by an amount $1/h$ for each glass-air contact, where h is the Newton cooling coefficient. If we now include the Newton cooling terms for only the external glass-air contacts (but neglect the ones for the internal air cavity), our expression for R_g becomes $R_g = \ell_g/k_g + 1/h$ and, with $k_g/k_a = 16$, $\ell_a/\ell_g = r$, the relative heat loss Δ_1 is

$$\Delta_1 = \frac{16r}{2\sigma + 16r + 2}, \qquad \text{where} \quad \sigma = \frac{k_g}{h\ell_g}.$$

If we also allow for Newton cooling from the glass to the air cavity and from the air cavity to the glass, then R_a becomes $R_a = \ell_a/k_a + 2/h$ and we obtain, from (11.16), Δ_2

$$\Delta_2 = \frac{2\sigma + 16r}{4\sigma + 16r + 2}, \qquad \text{where} \quad \sigma = \frac{k_g}{h\ell_g}.$$

A comparison of the results from these three models, Δ_0, Δ_1 and Δ_2, is illustrated in Figure 11.4.

The reduction in heat loss here is not quite the 97% of the first model; however, there is still a substantial improvement, particularly as window panes are a major cause of heat loss from homes during winter. Furthermore, in summer, the transmission of heat inwards is reduced to a significant degree, and double glazing provides a substantial reduction in sound transmission as well.

It should be mentioned here that k_a applies to dry, still air, and so this model is not applicable where air may contain moisture (which increases k_a and also the heat loss). For this reason double glazing requires a fully sealed cavity without moisture. Wall cavities, floors and ceilings use other materials for insulation, as it is not feasible to seal the cavities involved. In these cases, where there is moving air, heat can be convected (transport by a moving medium) whereas in the above double glazing example, only the conduction of heat was considered (transport through a medium at rest).

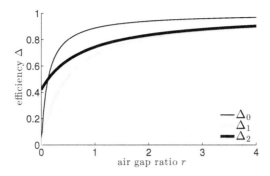

Figure 11.4: Relative heat loss efficiency calculations Δ_0 (thin black), Δ_1 (grey) and Δ_2 (thick black) plotted against r, the ratio of air gap to glass width. We have set $h = 10\,\text{W}\,\text{m}^{-2}\,{}^\circ\text{C}^{-1}$, $\ell_g = 0.003\,\text{m} = 3\,\text{mm}$ and $k_g/k_a = 16$.

Clearly, from the results of the above models, a substantial reduction in heat loss can be achieved through windows that are double glazed. Perhaps they are nearly as good as the manufacturers claim and worth the cost. Those without the economic resources, or in rental accommodation, have been known to use plastic-wrap (transparent sheet of plastic) attached against the inside of single window panes to create a crude type of double glazing in the colder months — with remarkably rewarding effects.

11.4 Insulating a water pipe

In this section we adapt the theory developed in Section 9.6, concerning the radial conduction of heat, to consider the insulation of a cylindrical pipe. After examining the physical processes, we arrive at an unusual result — that sometimes we can be worse off with insulation than with no insulation at all!

Problem description

In cold climates, a significant heat loss can occur from water pipes exposed to cold air. In extreme conditions, the heat loss may be such that the temperature of the water in the pipes drops below 0°C and freezes. As the water freezes it expands and can burst the pipes. To restrict this heat loss and prevent such problems, a jacket of insulating material is placed around the pipe, or the pipe is lagged.

It is therefore useful to calculate the rate of heat loss from a pipe of given length. One of the things we should investigate is:

> *How does this rate of heat loss change as the thickness of the insulating material increases?*

Model assumptions and approach

For a mathematical model we examine radial heat conduction inside the insulation jacket. We let the inner radius of the jacket be a and the outer radius be b. Hence the thickness is $(b - a)$ and the temperatures u_a and u_b are those on the inside and outside of the insulated pipe, respectively. We denote by r the radial distance from the centre of the pipe. The geometry is shown in Figure 11.5.

We make some simplifying assumptions and then build our model on these:

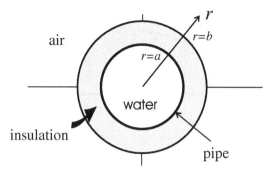

Figure 11.5: Geometry of the pipe insulation problem.

- Since metal is such a good conductor of heat, we assume the pipe metal to be at the same temperature as the insulation jacket, and to have good contact with it. We assume the inside of the insulation jacket is always at that temperature.

- On the outside of the insulation jacket the air temperature is u_b. We assume that heat is lost from this surface according to Newton's law of cooling.

The differential equation

In a previous chapter we formulated the differential equation for radial heat conduction in a cylinder or cylindrical shell. The equilibrium temperature satisfied the differential equation (Section 9.6, equation (9.26))

$$\frac{d}{dr}\left(r\frac{dU}{dr}\right) = 0. \tag{11.17}$$

This differential equation is second-order, but is not a constant coefficient differential equation. Using the product rule we can write this differential equation in the form

$$r\frac{d^2U}{dr^2} + \frac{dU}{dr} = 0.$$

This shows that the differential equation has coefficients that are not constant with respect to r. However, the original form is more convenient here for finding its solution.

Analytic solution

To obtain the solution to this differential equation, we integrate both sides twice with respect to r.

Example 11.3: Solve the differential equation (11.17).

Solution: Integrating once gives

$$r\frac{dU}{dr} = C_1,$$

where C_1 is an arbitrary constant of integration. Hence,

$$\frac{dU}{dr} = \frac{C_1}{r}.$$

Integrating again with respect to r ($r > 0$) gives

$$U(r) = C_1\ln(r) + C_2, \tag{11.18}$$

where C_2 is the second arbitrary constant of integration.

We have two arbitrary constants that are required for the general solution of a second-order differential equation. The equilibrium temperature depends on the logarithm of the radial distance from the centreline, rather than the linear distance as for the problem solved in Section 11.2.

The boundary conditions

We need two boundary conditions in order to solve for the two arbitrary constants. For this problem the boundary conditions are at the edges of the insulation at $r = a$ and $r = b$.

At $r = a$, we specified the temperature as u_a. In mathematical form this is written as

$$U(a) = u_a. \tag{11.19}$$

At $r = b$, the surface of the insulation loses heat to the surroundings. Newton's law of cooling states that the rate of loss of heat to the surroundings is proportional to the surface area and the temperature difference. Since the heat flux is the rate of heat loss divided by the surface area, we can write

$$J(b) = h(U(b) - u_b), \tag{11.20}$$

where u_b is the temperature of the surrounding air. And since the temperature of the insulation is greater than the temperature of the air, this gives a positive heat flux for $U(b) > u_b$.

Example 11.4: *Apply the boundary conditions (11.19) and (11.20) and hence find an expression for the equilibrium temperature $U(x)$.*

Solution: *Applying the first boundary condition (11.19) at $r = a$, we obtain the equation*

$$C_1 \ell n(a) + C_2 = u_a. \tag{11.21}$$

To apply Newton's law of cooling at the boundary (11.20), we first calculate the heat flux. Using the general solution (11.18) and Fourier's law,

$$J(r) = -k\frac{dU(r)}{dr} = -k\frac{C_1}{r} \tag{11.22}$$

and hence

$$J(b) = -k\frac{C_1}{b}. \tag{11.23}$$

The Newton cooling condition at $r = b$ now reduces to the equation

$$-k\frac{C_1}{b} = h(C_1 \ell n(b) + C_2 - u_b). \tag{11.24}$$

Equations (11.21) and (11.24) give two simultaneous equations for C_1 and C_2:

$$\ell n(a)C_1 + C_2 = u_a, \tag{11.25}$$

$$\left(\ell n(b) + \frac{k}{bh}\right)C_1 + C_2 = u_b. \tag{11.26}$$

Solving these equations we get

$$C_1 = \frac{-(u_a - u_b)}{\ell n(b/a) + \frac{k}{bh}} \quad \text{and} \quad C_2 = u_a + \frac{(u_a - u_b)}{\ell n(b/a) + \frac{k}{bh}} \ell n(a),$$

and substituting these values for C_1 and C_2 back into the general solution (11.18) we have

$$U(r) = u_a - \frac{u_a - u_b}{\ell n(b/a) + \frac{k}{bh}} \ell n(r/a). \tag{11.27}$$

Rate of heat loss from a pipe

We calculate the total rate of heat loss from the pipe. For this we use Fourier's law of heat conduction to find the heat flux at $r = b$.

Example 11.5: *Calculate the total rate of heat loss at $r = b$ of an insulated pipe of length ℓ.*

Solution: Let Q denote the total rate of heat loss from a pipe of length ℓ. We can calculate this quantity by multiplying the heat flux at $r = b$ by the surface area of the insulation jacket (at $r = b$). The total surface area of the insulation jacket is $2\pi b\ell$. Hence,

$$Q = 2\pi b\ell J(b).$$

Using Fourier's law, $J = -k\,dU/dr$, the rate of heat loss per unit length becomes

$$Q = -2k\pi b\ell \frac{dU(b)}{dr},$$

where we evaluate dU/dr at $r = b$.

Using the solution obtained for $U(r)$ from equation (11.27), we obtain

$$Q = -2k\pi b\ell \frac{d}{dr}\left(u_a - \frac{u_a - u_b}{\ell n(b/a) + \frac{k}{bh}}\,\ell n\left(\frac{r}{a}\right)\right)\Bigg|_{r=b}$$

$$= 2k\pi b\ell \frac{u_a - u_b}{\ell n(b/a) + \frac{k}{bh}}\,\frac{d}{dr}\,\ell n\left(\frac{r}{a}\right)\Bigg|_{r=b}.$$

Evaluating the derivative and then substituting for $r = b$ gives

$$Q = 2k\pi b\ell \frac{u_a - u_b}{\ell n(b/a) + \frac{k}{bh}} \times \frac{1}{b}. \tag{11.28}$$

Hence

$$Q = \frac{2k\pi\ell(u_a - u_b)}{\ell n(b/a) + \frac{k}{bh}}. \tag{11.29}$$

As a quick check, note that $u_a > u_b$ and $b/a > 1$ so we get a positive rate of heat loss per unit length, as would be expected.

Results and interpretation

In order to predict how the rate of heat loss varies as the insulation thickness is increased, we consider the above model with some appropriate parameter values.

Take $\ell = 1\,\text{m}$, $a = 15\,\text{mm}$, $u_a = 60°\text{C}$, $u_b = 15°\text{C}$ and $h = 5\,\text{W}\,\text{m}^{-2}\,°\text{C}^{-1}$. We vary the outer radius b. The computed values of Q are given in Table 11.1 where we look at two different values of conductivity $k = 0.05$ and $k = 0.17$ SI units. The first value $k = 0.05$ corresponds to polyurethane and the value $k = 0.17$ corresponds to asbestos felt.

Note from Table 11.1 that the rate of heat loss for conductivity $k = 0.05$ decreases as the thickness of the insulation increases, as we might expect. However, for the larger conductivity of $k = 0.17$, we see that the rate of heat loss increases as the thickness of the insulation is increased. In this case it appears that we are better off not insulating the pipe at all!

Figure 11.6 gives a graph of the rate of heat loss from the insulated water pipe as a function of the outer radius of the insulation b. Note that there is a maximum rate of heat loss at the value $b = k/h$.

Table 11.1: Table of values of heat loss from an insulated water pipe for different thicknesses (in mm) of insulation and for two different conductivities (in SI units).

Insulation thickness $(b - a)$	Rate of heat loss $k = 0.05$	Rate of heat loss $k = 0.17$
0	21.2	21.2
1	20.5	22.0
2	19.8	22.6
5	18.0	24.2
10	15.5	25.7

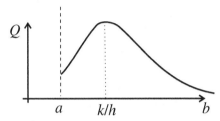

Figure 11.6: Diagram of the rate of heat loss from an insulated water pipe versus the outer radius of the insulation. There is a maximum at $b = k/h$. It illustrates the general behaviour for any values of k, h and b.

This suggests that there is a critical value for the outer radius b. For the case cited above, where the conductivity is $k = 0.05$, the critical value of the outer radius is 10 mm, but for $k = 0.17$ the critical radius is 34 mm. In this latter case, one would need an outer radius greater than 34 mm before the insulation would have the effect of reducing the heat loss from the pipe. If this critical value is less than the inner radius, then placing insulation around the pipe has the desired effect of reducing the heat loss from it. If the critical value is greater than the inner radius, however, then it may be better not to insulate the pipe at all.

The physical explanation for this is that there is a balance between two competing effects. Increasing the thickness of the insulation means that there is more insulation through which the heat need flow. However, increasing the thickness of the insulation also increases the surface area through which heat is lost.

Solid cylinders and spheres

A different problem is that of a solid cylinder (or sphere). For this type of problem we usually do not have a boundary condition given at $r = 0$. Note that the general solution for a cylinder (11.18) cannot be evaluated at $r = 0$ due to the $\ln(r)$ term, which tends to $-\infty$. And for a solid sphere, there is a term r^{-1} that becomes infinite at $r = 0$.

For these problems, a 'hidden', or implicit, boundary condition is applied. Physically we require the equilibrium temperature inside the cylinder to be finite. The only way this can be true is if the arbitrary constant coefficient in the $\ln(r)$ term is identically zero. This condition sets the value of one of the arbitrary constants and hence acts like a boundary condition, but in an implicit way.

Summary of skills developed here:

- *Solve for the temperature in a cylindrical shell, with any combination of pre-scribed temperature, heat flux or Newton cooling boundary conditions.*
- *Solve for temperature in a spherical shell with any combination of prescribed temperature, heat flux or Newton cooling boundary conditions.*
- *Extend to a cylinder or sphere with internal heat production.*
- *Solve for a solid cylinder or sphere with an implicit boundary condition, using that the temperature is finite at the centre (r = 0) to set one of the arbitrary constants to zero.*

11.5   Cooling a computer chip

We now consider a single heat fin from a set of heat fins attached to a computer chip. We solve the appropriate differential equation to find an analytic solution, and then examine this to understand what it tells us about the general behaviour of the temperature in the chip.

Problem description

In a computer, modern processing chips can generate a lot of heat. If the heat is not dissipated adequately, the temperature of the chip can increase significantly, possibly causing the chip to cease functioning properly or to become permanently damaged. The Cyrix 6x86-P166 chip generates heat at a rate of approximately 24 W, and Cyrix advises that the chip should not reach a temperature of over 70°C (see Cyrix (1998)).

One method of dissipating heat is to mount heat fins on the chip with a cooling fan to blow air between the fins. For the Cyrix chip, 12 heat fins are used with a fan mounted above them. Each heat fin has dimensions of width $w = 5.1$ cm, thickness $b = 1$ mm and length $\ell = 1.5$ cm. A schematic diagram of the heat fin assembly and fan for a Cyrix 6x86 CPU chip is shown in Figure 11.7.

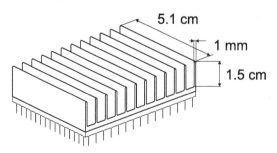

Figure 11.7: Heat dissipation assembly for a Cyrix CPU computer chip. It consists of a cooling fan and 12 heat fins attached to the chip.

The Cyrix 6x86 chip, for example, generates heat at a rate of $q = 24$ Watts, Cyrix (1998). If the only way the heat can dissipate is from the surface of the chip, then we can model the temperature of the chip using Newton's law of cooling.

Example 11.6: *Use Newton's law of cooling to calculate the equilibrium temperature of the chip if there are no cooling fins and the dimensions of the chip are 5.1 cm × 5.3 cm.*

Solution: *Newton's law of cooling states that the rate of heat loss Q is*

$$Q = hS\Delta U, \tag{11.30}$$

where ΔU is the temperature difference between the chip and the ambient temperature. For the Cyrix chip, with surface area $S = 5.1\,\text{cm} \times 5.3\,\text{cm} = 2.7 \times 10^{-3}\,\text{m}^2$ and using a value of $h = 5\,\text{W}\,\text{m}^{-2}\,{}^{\circ}\text{C}^{-1}$, we obtain

$$\Delta U \simeq \frac{24}{5 \times 2.7 \times 10^{-3}} \simeq 1{,}800^{\circ}\text{C} \tag{11.31}$$

for the temperature difference between the chip face and the surrounding air.

Of course, the chip would cease to function long before this temperature were reached. Clearly we need to dissipate the heat more efficiently.

We are interested in calculating the temperature of the chip with the heat fins attached. We are also interested in answering the question:

> *How does this temperature vary if we increase or decrease the length of the heat fins?*

Model assumptions and approach

We develop a model for a single heat fin where the geometry of the model is shown in Figure 11.8.

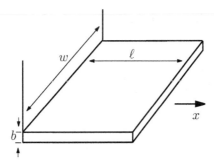

Figure 11.8: Diagram of model used for a single heat fin. Heat enters at $x = 0$ (left end) with flux q generated from the computer chip. We assume the heat flux from $x = \ell$ (right end) is zero.

We make some simplifying assumptions on which to build the mathematical model:

- We assume the heat loss is distributed evenly over the 12 heat fins.
- We assume the heat loss from the sides of the fin is negligible compared with the heat loss from the top and bottom.
- We also neglect any heat loss from the end of the heat fin.

The differential equation

We can calculate the heat flux entering a single heat fin. This heat flux, denoted by q, is obtained by taking the rate of heat loss for a single fin $24/12 = 2\,\text{W}$ and then dividing by the area of the cross-section, bw. This gives a heat flux of $q = 3.92 \times 10^4\,\text{W}\,\text{m}^{-2}2$.

In Chapter 9 (Section 9.7), we formulated an appropriate differential equation for the equilibrium temperature distribution, $U(x)$, along a single heat fin. This differential equation is

$$\frac{d^2U}{dx^2} = \beta(U - u_s), \qquad \text{where} \quad \beta = \frac{2h}{kb}. \tag{11.32}$$

Here $U(x)$ is the temperature, x the distance along the heat fin, k the conductivity, h the convective heat transfer coefficient and u_s the temperature of the surrounding air. This differential equation is an inhomogeneous, constant coefficient, second-order differential equation.

For simplicity we set the temperature of the heat fin relative to the temperature of the surrounding air. On this temperature scale, $u_s = 0$ and the differential equation (11.32) becomes

$$\frac{d^2U}{dx^2} = \beta U \qquad \text{where} \quad \beta = \frac{2h}{kb}, \tag{11.33}$$

which is now homogeneous. This is not a major restriction as it is simple, with a change of variable, to convert a problem with non-zero u_s to one with $u_s = 0$. Equation(11.33) represents a *heat balance*. The LHS term comes from heat conduction and the RHS comes from heat lost to the surrounding air.

Analytic solution

First we derive a general solution to the differential equation. Then we find the associated constants and hence the particular solution, using the boundary conditions.

Example 11.7: *Find the general solution to the differential equation (11.33).*

Solution: *The general solution can be found by substituting a trial solution of the form $U(x) = e^{\lambda x}$ into the differential equation (11.33). This gives rise to the characteristic equation $\lambda^2 = \beta$. Since β is a positive constant, the roots of the characteristic equation are $\pm\sqrt{\beta}$. A general solution of the differential equation (11.33) is then*

$$U(x) = C_1 e^{\sqrt{\beta}x} + C_2 e^{-\sqrt{\beta}x}, \tag{11.34}$$

where C_1 and C_2 are arbitrary constants.

It is also possible to express this solution in terms of the hyperbolic functions[1] $\cosh(\sqrt{\beta}x)$ and $\sinh(\sqrt{\beta}x)$. Often this form can lead to simpler algebra when applying certain types of boundary conditions. (Further details of hyperbolic functions can be found in Appendix B.4.)

---

[1]The hyperbolic functions cosh and sinh are defined by

$$\cosh(z) = \frac{1}{2}\left(e^z + e^{-z}\right), \quad \sinh(z) = \frac{1}{2}\left(e^z - e^{-z}\right).$$

The boundary conditions

We require two boundary conditions, one at the end $x = 0$, the origin, and one at the end $x = \ell$, to give us two simultaneous equations for the two arbitrary constants C_1 and C_2. At $x = 0$, we know the value of the heat flux to be q. We write this boundary condition as

$$J(0) = q. \tag{11.35}$$

At the end $x = \ell$, we have assumed we can neglect the heat loss. This is expressed mathematically as

$$J(\ell) = 0. \tag{11.36}$$

Example 11.8: *Apply the boundary conditions to obtain an expression for the temperature along the heat fin.*

Solution: *Both boundary conditions involve the heat flux J. From Fourier's law $(J(x) = -k\,dU/dx)$ and the general solution (11.34), we obtain*

$$J(x) = -k\sqrt{\beta}\left(C_1 e^{\sqrt{\beta}x} - C_2 e^{-\sqrt{\beta}x}\right). \tag{11.37}$$

Applying the first boundary condition $J(0) = q$ and evaluating this at $x = 0$ gives the equation

$$C_1 - C_2 = \frac{-q}{k\sqrt{\beta}}. \tag{11.38}$$

To apply the second boundary condition, $J(\ell) = 0$, we substitute $x = \ell$ into the expression for $J(x)$ to get

$$C_1 e^{\sqrt{\beta}\ell} - C_2 e^{-\sqrt{\beta}\ell} = 0. \tag{11.39}$$

We solve equations (11.38) and (11.39) simultaneously, giving an expression for C_2. Substituting this into (11.38) gives

$$\left(1 - e^{2\sqrt{\beta}\ell}\right) C_1 = \frac{-q}{k\sqrt{\beta}},$$

which allows us to solve for C_1 as

$$C_1 = \frac{q}{k\sqrt{\beta}}\left(\frac{1}{e^{2\sqrt{\beta}\ell} - 1}\right). \tag{11.40}$$

Substituting this back into our expression for C_2 we obtain

$$C_2 = \frac{q}{k\sqrt{\beta}}\left(\frac{e^{2\sqrt{\beta}\ell}}{e^{2\sqrt{\beta}\ell} - 1}\right). \tag{11.41}$$

With these expressions for C_1 and C_2 we have, from the general solution (11.34), an expression for the temperature along the heat fin, namely

$$U(x) = \frac{q}{k\sqrt{\beta}}\left(\frac{e^{\sqrt{\beta}x} + e^{2\sqrt{\beta}\ell}e^{-\sqrt{\beta}x}}{e^{2\sqrt{\beta}\ell} - 1}\right). \tag{11.42}$$

From this general solution for the temperature in equation (11.42), we can immediately write an expression for the temperature of the computer CPU chip, which is the quantity of interest. This is the temperature at $x = 0$ and is given by

$$U(0) = \frac{q}{k\sqrt{\beta}}\left(\frac{e^{2\sqrt{\beta}\ell} + 1}{e^{2\sqrt{\beta}\ell} - 1}\right). \tag{11.43}$$

Results and interpretation

Using some appropriate parameter values for copper heat fins we take $k = 386$ and $h = 32$ SI units. The value of the convective heat transfer coefficient h is higher than the one we used in the calculation in equation (11.31) to incorporate the effect of the fan.[2] We also have $b = 1\,\text{mm} = 10^{-3}\,\text{m}$, $w = 5.1 \times 10^{-2}\,\text{m}$ and $\ell = 1.5\,\text{cm} = 1.5 \times 10^{-2}\,\text{m}$. We then calculate $\beta = 174$ (in SI units). We have previously calculated $q = 3.92 \times 10^4\,\text{W}\,\text{m}^{-2}$ for each heat fin. Substituting these values into (11.43) we obtain

$$U(0) = 41.4°\text{C}.$$

This is the temperature rise above the ambient temperature, or the surrounding temperature, as our model assumed $u_s = 0$.

In Figure 11.9, we plot the temperature along the length of the heat fin. The temperature is highest at the chip surface and decreases along the length of the heat fin, as expected.

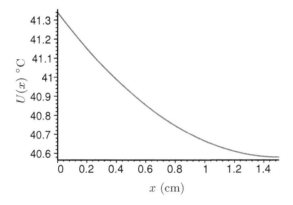

Figure 11.9: Graph of temperature in °C along the length of the heat fin.

It is more interesting to sketch a graph of the temperature of the computer chip for different lengths of the heat fin. This graph, illustrated in Figure 11.10, provides some insight into how long to make the heat fin for the best results. The temperature of the chip decreases as we increase the length of the heat fin. Physically this makes sense, as increasing the length of the heat fin increases the area through which heat is lost. However, $U(0)$ appears to approach a limiting value as ℓ (the length of the fin) increases.

From (11.43) we can calculate the observed limiting temperature as we let ℓ become very large. This will provide a lower bound for the coolest temperature that the chip can reach. To find this limit we let $\ell \to \infty$ in (11.43). Dividing both top and bottom by $e^{2\sqrt{\beta}\ell}$, to avoid infinite limits in the numerator and denominator, yields

$$U(0) = \frac{q}{k\sqrt{\beta}} \left(\frac{1 + e^{-2\sqrt{\beta}\ell}}{1 - e^{-2\sqrt{\beta}\ell}} \right). \qquad (11.44)$$

Hence, the limiting value as $\ell \to \infty$ is

$$U_{\min} = \frac{q}{k\sqrt{\beta}}. \qquad (11.45)$$

---

[2]To calculate h, a formula from Holman (1981) was used assuming an air speed of $1\,\text{m}\,\text{s}^{-1}$.

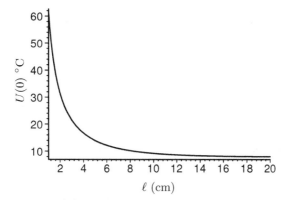

Figure 11.10: The temperature $U(0)$ in °C of the computer chip as a function of the length l of the heat fin.

For the same parameter values as above, this gives a minimum temperature for U of $U_{min} = 8.1$°C. Clearly, from Figure 11.10, the temperature is close to 8.1°C for a fin length of much less than $\ell = \infty$, and thus not much is gained by increasing the fin length beyond a certain point (say 10 cm). However, there is a substantial gain by increasing the length from 2 to 4 cm.

Summary of skills developed here:

- *Obtaining the general solution to the homogeneous heat fin equations.*
- *Applying boundary conditions to this general solution, including other combinations of boundary conditions, such as one or both boundaries at prescribed temperatures.*
- *Interpreting the solution for the temperature of a computer chip to find a theoretical minimum temperature.*

11.6 Case Study: Tumour growth

We have developed the theory of diffusive mass transport and now show how this mechanism can be used in medical research, with an application to the growth of tumours through oxygen availability. We do not include solution graphs, but focus instead on how diffusion theory can inform a model formulation. The case study is based on an early model of tumour growth by Greenspan (1972), the purpose of which was to explain why some tumours do not grow beyond a certain size. Although there are three distinct growth phases, our focus here is mainly on the first phase and the oxygen concentration inside the tumour. The reader is referred to Greenspan (1972) for further details and extensions.

Cancer is a distressing disease, but widespread with most people knowing someone who has been affected. Cancer cells proliferate (divide) at a much faster rate than normal cells and the resulting, often spherical, growths are called tumours, although not all tumours are cancerous. The earliest stages of tumour development are regulated by the availability of nutrients, in particular oxygen, from surrounding tissues.

It has been observed that some tumours grow to a fixed size of just a few millimetres across, while others grow very large. What causes some to be limited in their growth is a decline in available nutrients. In these tumours, nutrients such as oxygen diffuse into the tumour from the surface with declining oxygen availability towards the centre. Other tumours secrete chemicals that cause surrounding blood vessels to grow into them and supply oxygen equally to all parts of the tumour; these can grow much larger.

In this case study we focus on the former situation of tumours that only grow to a fixed size, and begin to investigate mechanisms that limit growth. This type of tumour is called nonvascular, or avascular. When the tumour is small oxygen can reach all parts easily. However, as it increases in size, live cells close to the surface absorb most of the oxygen with less oxygen available towards the centre. If the oxygen concentration falls below a critical level, then the cells stop dividing and a central quiescent core develops and the growth rate of the tumour declines markedly (Greenspan, 1972).

Greenspan (1972) described three phases of growth for avascular tumours: (1) where proliferating cells cause the tumour mass to grow quickly; (2) where a core of quiescent (not reproducing) cells form with a surrounding proliferating layer; and (3) where a necrotic core of dead and decomposing cells forms and is surrounded by an inner layer of quiescent cells and an outer layer of proliferating cells. These three phases are shown in Figure 11.11. In this case study we focus on calculating the oxygen concentration inside the tumour during the first phase of growth.

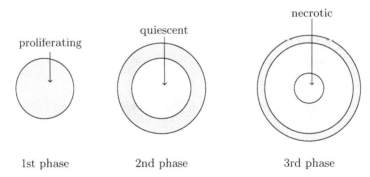

Figure 11.11: Diagram of the three phases of growth of an avascular tumour: (1) phase-1 has only proliferating cells (P); (2) phase-2, where a core of quiescent cells (Q) form with an outer proliferating layer (P); and (3) phase-3 where a core of necrotic cells (N) that are dead and decomposing.

First growth phase. *To model the first growth phase, we assume that a tumour is spherical and that oxygen diffuses into the tumour from the surface towards its centre. For any region inside the tumour, the equilibrium oxygen concentration $C(r)$ will be a function of the distance from the centre of the tumour, r. To simplify the problem we assume that the oxygen concentration quickly reaches equilibrium. This process will happen much more quickly than the tumour will grow in size so it is reasonable to think of the oxygen concentration $C(r)$ at equilibrium with respect to time, but the tumour radius $R(t)$ depending on time.*

Oxygen molecules diffuse inwards from high concentrations on the outside of the tumour towards lower concentrations at its centre. So the governing equation will be a diffusion equation, with an absorption term representing the consumption of oxygen by live cells as they diffuse inwards. From equation (9.39) in Section 9.8 the equilibrium oxygen concentration satisfies the diffusion equation

$$\frac{D}{r^2}\frac{d}{dr}\left(r^2\frac{dC}{dr}\right) + M(r) = 0, \tag{11.46}$$

but with mass production rate $M(r) < 0$ because the cells consume oxygen.

We can solve the differential equation for oxygen concentration ((11.46)) with constant oxygen consumption rate $M(r) = -A_0$ and boundary conditions for the region $0 < r < R(t)$. The solution is (see exercises, Exercise 11.17)

$$C(r) = c_1 - \frac{A_0}{6D}(R^2 - r^2), \tag{11.47}$$

where c_1 is the initial concentration on the surface $r = R(t)$.

This solution will be valid only while $C(r) > c_q$, a threshold concentration below which cells stop dividing and become quiescent. The quiescent core will start to form where the lowest concentration becomes c_q. This first occurs at the centre of the tumour. Setting $C(0) = c_q$ (at $r = 0$) and solving for R, we obtain the radius when cells first become quiescent, $R = r_q$, as

$$r_q = \sqrt{(c_1 - c_q)\frac{6D}{A_0}}, \qquad \text{where } c_1 > c_q. \tag{11.48}$$

To interpret this, note that r_q will become smaller when c_q is closer to c_1. Also, r_q gets larger as the absorption rate of oxygen, A_0, becomes smaller, which is what we would expect. To fully specify the solution we also need to find the radius $R(t)$ for the tumour as a function of time by modelling the growth dynamics. This is explained in Exercise 11.18.

Further growth phases. We mention briefly, without without providing a solution, how to model the next phase of growth. See exercises, Exercise 11.19 to solve the governing equations.

The quiescent core expands as more cells in the proliferating layer starve and become quiescent. Let $R_q(t)$ be the time-dependent radius of the quiescent core, with $R(t)$ the outer radius of the tumour. After the quiescent core forms, there are two regions (see Figure 11.11) with distinct dynamics and thus modelled with different diffusion equations.

Let $C_p(r)$ be the equilibrium oxygen concentration in the proliferating layer, and $C_q(r)$ the equilibrium oxygen concentration in the quiescent core. In the proliferating layer, cells still consume oxygen, but in the quiescent layer no consumption takes place even though oxygen can diffuse into the region with $C_q(r) < c_q$. So $M(r) = 0$ for $0 < r < R_q(t)$ and $M(r) = -A_0$ for $R_q(t) < r < R(t)$. Thus the two diffusion equations are: for the proliferating layer,

$$\frac{D}{r^2}\frac{d}{dr}\left(r^2\frac{dC_p}{dr}\right) - A_0 = 0, \qquad R_q(t) < r < R(t)$$

and for the quiescent core,

$$\frac{D}{r^2}\frac{d}{dr}\left(r^2\frac{dC_q}{dr}\right) = 0, \qquad 0 < r < R_q(t).$$

Two boundary conditions are the oxygen concentration at the outer surface

$$C_p(R) = c_1,$$

and zero oxygen flux at its centre,

$$J_q(0) = 0 \quad \Rightarrow \quad -D \left.\frac{dC_q}{dr}\right|_{r=0} = 0.$$

At the boundary between the two regions we assume the concentration is continuous and equal to c_q, the threshold concentration,

$$C_p(R_q) = C_q(R_q) = c_q,$$

and the mass fluxes must match, so $J_q(R_q) = J_p(R_q)$.

We conclude by noting that avascular tumours as discussed above are small, about 1 to 3 mm across, and cause few health complications. Usually they cease expansion altogether. However, when avascular tumours become vascular, that is, they modify blood vessels of the host forming new capillaries that extend into the tumour, then the availability of oxygen can lead to explosive growth and the spread of cells to other parts of the body through direct access to the blood supply. Thus understanding the process of vascularisation, the conditions under which it occurs, and its integration with avascular growth as examined here are important areas of current research.

There are many other factors that affect the growth of avascular tumours and the model presented here is only one of the early models. To read about some of the newer ideas and controversies, see Roose et al. (2007), and see Araujo and McElwain (2004) for a comprehensive historical perspective.

11.7   Exercises for Chapter 11

11.1. Solving a boundary value problem. *Consider the differential equation*

$$\frac{d^2U}{dx^2} = 1,$$

which satisfies the boundary conditions

$$U(0) = 1, \qquad U(2) = 0.$$

(a) *Find the general solution.*

(b) *Apply the boundary conditions to find the solution.*

(c) *Suppose the two boundary conditions are replaced with*

$$U(0) = 1, \qquad \frac{dU}{dx}(2) = 0.$$

   Apply these boundary conditions and find the solution.

(d) *Suppose the two boundary conditions are replaced with*

$$\frac{dU}{dx}(0) = 1, \qquad \frac{dU}{dx}(2) = U(2).$$

Apply these boundary conditions and find the solution.

11.2. Formulating a boundary value problem. Write, in mathematical form, boundary conditions for the following. Express them in terms of temperature U or heat flux J.

(a) A $1\,\text{m}^2$ section of a furnace wall gains heat from the end $x = 0$ at a fixed rate of $300\,\text{W}$, while the other end is maintained at temperature $30°\text{C}$.

(b) The outside of a wall of a house loses heat according to Newton's law of cooling, to the surrounding air at temperature $10°\text{C}$, while the inside gains heat according to Newton's law of cooling from the inside of the house, which is at a temperature of $25°\text{C}$.

(c) A slab of material has its right end held at temperature $80°\text{C}$ and the left end gaining heat according to Newton's law of cooling from the surroundings at temperature $100°\text{C}$.

11.3. Formulating boundary conditions. In the following, write boundary conditions involving the temperature U or the heat flux J. (It may be useful to draw a diagram.)

(a) In a spherical shell defined by $r = a$ to $r = b$, heat flows radially. The temperature on the outside is u_2 and the inside of the shell receives heat at a rate of $10\,\text{W}$.

(b) In a wall of length L the temperature at one end $x = L$ is u_1 and the temperature at the other end is u_2.

(c) In a cylindrical heat fin, of small radius a, heat is conducted along the length of the wire. The heat fin is attached to a surface whose temperature is maintained at $100°\text{C}$. The other end of the heat fin is assumed to lose no heat.

(d) A wall located between $x = 0$ and $x = \ell$ has temperature at $x = \ell$ fixed at u_1. The other side of the wall gains heat from the surroundings, at temperature u_2, according to Newton's law of heat exchange.

(e) The material inside a cylindrical insulating layer, of inner radius 1 and outer radius 2, gains heat from the surroundings at temperature $3u_1$, according to Newton's law of heat exchange. The temperature at the inner radius is maintained at u_1.

11.4. Heat flux through a wall. In an internal wall of thickness d, the temperature at $x = 0$ is maintained (by a thermostat) at temperature u_1 and the other side is maintained at temperature u_2.

(a) Find an expression for the temperature at any point inside the wall.

(b) Hence find an expression for the heat flux through the wall.

11.5. Heat flux through a window. Consider the flow of heat through a window pane of width ℓ. Assume that heat is gained from the inside and lost from the outside of the window according to Newton's law of cooling.

The temperature of the air on the inside is u_i and the temperature on the outside is u_o. The Newton cooling coefficients are h_i and h_o for the inside surface and outside surface of the window, respectively. You are given that the equilibrium temperature, inside the window, satisfies

$$\frac{d^2U}{dx^2} = 0.$$

(a) What is the general solution for the equilibrium temperature?

(b) What are the boundary conditions for this problem?

(c) Hence deduce the heat flux from the outer surface $x = \ell$ is given by the expression

$$\frac{k(u_i - u_o)}{\ell + \frac{k}{h_i} + \frac{k}{h_o}}.$$

(d) Interpret this formula in terms of heat resistances.

11.6. Heat flux and thermal resistance. *Suppose we have two different materials with widths d_1 and d_2, conductivities k_1 and k_2, and which are joined together at $x = 0$. The equilibrium temperatures $U_1(x)$ and $U_2(x)$ both satisfy the basic equilibrium heat equations*

$$\frac{d^2 U_1}{dx^2} = 0, \qquad \frac{d^2 U_2}{dx^2} = 0.$$

The temperature on the inside $(x = -d_1)$ is held (by a thermostat) at a temperature u_i and the temperature on the outside $(x = d_2)$ is held (by a thermostat) at a temperature u_o.

(a) *Assuming the temperature and heat flux are continuous where the materials join at $x = 0$, what are the boundary conditions?*

(b) *Show that the heat flux is given by*

$$J = \frac{u_i - u_0}{\frac{d_1}{k_1} + \frac{d_2}{k_2}}.$$

(c) *Interpret this in terms of thermal resistances.*

11.7. Internal floor heating. *Houses are often heated via a heating source located within a concrete slab floor (e.g., by attaching an electrical resistance wire to the reinforcing bars embedded within the concrete). Suppose such a heat source is located in a slab of thickness d and generates heat at a constant rate q (per unit volume). The equilibrium temperature $U(x)$ satisfies the differential equation*

$$k \frac{d^2 U}{dx^2} + q = 0,$$

where k is the thermal conductivity. If each side of the slab is maintained at the same temperature u_0, find an expression for $U(x)$, the temperature inside the slab.

11.8. Nuclear reactor. *In the shielding wall of a nuclear reactor the rate of heat generated by gamma rays interacting with the wall, the internal heat generation (per unit volume) $q(x)$, can be modelled by*

$$q(x) = q_0 e^{-ax},$$

where q_0 and a are positive constants. The equilibrium temperature inside the wall satisfies the differential equation

$$k \frac{d^2 U}{dx^2} + q(x) = 0.$$

(a) *Sketch $q(x)$ as a function of x.*

(b) *Determine the general solution for the equilibrium temperature.*

(c) *Given that the inside of the wall $x = 0$ is maintained at a constant temperature of u_1 and the outside of the wall is maintained at a constant temperature of u_2, find an expression for the equilibrium temperature.*

11.9. Combining heat resistances. *To prove that the individual heat resistances are additive, consider*

$$J = \frac{u_1 - u_a}{R_g} = \frac{u_a - u_b}{R_a} = \frac{u_b - u_2}{R_g}.$$

Find expressions for u_a and u_b and substitute them back into one of the equations for J. This will establish equation (11.15) of the double glazing case study (Section 11.3).

11.10. Double glazing. *Following the case study on double glazing (Section 11.3), assume that the inside room temperature is* $20°C$*, the outside environment temperature is* $0°C$*, the thickness of each of the two panes of glass is* $d = 4\,\mathrm{mm}$*, the conductivity of glass is* $k_g = 0.8\,\mathrm{J\,m^{-1}\,s^{-1}\,°C^{-1}}$ *and the conductivity of air is* $k_a = 0.05\,\mathrm{J\,m^{-1}\,s^{-1}\,°C^{-1}}$*. Take Newton cooling coefficient* $h = 10\,\mathrm{W\,m^{-2}\,°C^{-1}}$*.*

Calculate the heat flux for a gap width of twice the thickness of a single pane of glass, using each of the models corresponding to each of the Δ_0*,* Δ_1 *and* Δ_2 *models defined in the case study. Compare this to the heat flux through a single pane of glass with width* $2d$*.*

11.11. Comparing glazing models. *Using* `Maple` *or* `MATLAB`*, draw graphs of the three delta functions in the case study on double glazing (Section 11.3). Look at limits over time, and with the standard specifications of manufacturers how these three compare.*

11.12. Cooling lizards. *Lizards at rest are known to generate heat internally at a rate of* $q = 0.5\,\mathrm{W\,kg^{-1}}$*. We might model the lizard as a cylinder of radius* a *and length* ℓ*. The equilibrium temperature satisfies the differential equation*

$$\frac{k}{r}\frac{d}{dr}\left(r\frac{dU}{dr}\right) + \rho q = 0,$$

where ρ *is the (average) density of lizard tissue,* k *is the conductivity, and* r *measures radial distance from the centre of the lizard's body (a cylinder).*

(a) *Starting from a suitable word equation, outline a derivation of this differential equation.*

(b) *Find the general solution.*

(c) *For a solid cylinder (our lizard), we require the equilibrium temperature to be finite at* $r = 0$*. What does this tell you about the value of one of the arbitrary constants?*

(d) *Suppose heat is lost to the surroundings according to Newton's law of cooling, where the surroundings are at temperature* u_0*. Find an expression for the equilibrium temperature inside the lizard (cylinder). Where is the equilibrium temperature a maximum?*

11.13. Equilibrium temperature in a spherical shell. *The equilibrium temperature inside a spherical shell, of inner radius* a *and outer radius* b*, satisfies the differential equation*

$$\frac{d}{dr}\left(r^2\frac{dU}{dr}\right) = 0.$$

(a) *Find the general solution of the differential equation.*

(b) *Find the equilibrium temperature if the inner surface* $r = a$ *is maintained at temperature* u_1 *and the outer surface* $r = b$ *is maintained at temperature* u_2*.*

11.14. Hyperbolic functions. *Verify that*

$$U(x) = D_1 \cosh(\sqrt{\beta}x) + D_2 \sinh(\sqrt{\beta}$$

is a general solution of the differential equation

$$\frac{d^2U}{dx^2} = \beta U.$$

Now find the solution that satisfies the boundary conditions applied at $x = 0$*,*

$$U(0) = 1, \qquad \frac{dU(x)}{dx} = 0 \quad \text{at } x = 0.$$

For details on hyperbolic functions, see Appendix B.4.

11.15. Cooling an engine. *Heat fins are often attached to engines to cool them when in operation. Find an expression for the temperature inside a heat fin where one end of the heat fin is attached to a metal engine and is maintained at temperature u_1. You may neglect any heat loss from the other end of the heat fin.*

The equilibrium temperature satisfies the differential equation

$$\frac{d^2 U}{dx^2} = \beta U,$$

where β is a constant and the temperature of the surroundings is zero. Use the hyperbolic form of the general solution from Exercise 11.14.

11.16. Rectangular heat fin. *In Section 11.5 it was shown that the relative temperature in a rectangular heat fin satisfies the differential equation*

$$\frac{d^2 U}{dx^2} = a^2 U, \qquad a^2 = \frac{2h}{kb},$$

where b is the thickness of the heat fin, k is the conductivity, h the Newton cooling coefficient, and we have taken the temperature of the surrounding air to be zero (see Section 11.5, equation (11.32)).

(a) *Write the boundary conditions assuming that the temperature at the end attached to the heat source is u_1 and the heat lost from the other end is given by Newton's law of cooling.*

(b) *Hence obtain an expression for the equilibrium temperature. (Hint: You will find the algebra easier if you use the hyperbolic form of the general solution*

$$U(x) = C_1 \cosh(ax) + C_2 \sinh(ax),$$

where C_1 and C_2 are arbitrary constants.)

11.17. Proliferating tumour. *For a spherical tumour of radius R, in the first phase of growth, the equilibrium oxygen concentration $C_p(r)$ satisfies*

$$\frac{D}{r^2} \frac{d}{dr}\left(r^2 \frac{dC_p}{dr}\right) - A_0 = 0, \qquad 0 < r < R,$$

where A_0 is the oxygen absorption rate and r is radial distance from the centre of the tumour. Solve this equation assuming zero flux at the centre and fixed concentration c_1 on the tumour surface.

11.18. Radius of the proliferating tumour. *Consider a tumour in phase-1 of its growth where there is a sphere of proliferating cells.*

(a) *Given the cell population has a per-capita growth rate s, justify the equation*

$$3R^2 \frac{dR}{dt} = sR^3.$$

(b) *Solve the equation given $R(0) = r_0$.*

(c) *By solving this differential equation, find an expression for t_q, the time for the tumour to reach the end of the first phase of growth.*

11.19. Avascular tumour with phase-two growth. *Consider a tumour that has already reached the phase-two growth stage. Assume that the proliferating layer absorbs oxygen at a constant rate* $A = A_0$ *ml per second per unit volume of cells, and let* $c_p(r)$ *be the oxygen concentration here. The governing equation is*

$$\frac{D}{r^2}\frac{d}{dr}\left(r^2\frac{dC_p}{dr}\right) - A_0 = 0. \qquad R_q(t) < r < R(t)$$

In the proliferating core, with $C_q(r)$ *as the concentration, then*

$$\frac{D}{r^2}\frac{d}{dr}\left(r^2\frac{dC_q}{dr}\right) = 0, \qquad 0 < r < R_q(t).$$

The boundary conditions are

$$C_p(R_q) = C_q(R_q) = c_q, \qquad \frac{dC_q}{dr}(0) = 0. \qquad \frac{dC_q}{dr}(R_q) = \frac{dC_p}{dr}(R_q).$$

(a) *Give a brief explanation of each of the boundary conditions. Comment on the number of boundary conditions required for solution of the differential equations.*

(b) *Solve for the concentrations* $C_q(r)$ *and* $C_p(r)$.

   [Hint: It is suggested that you do not use the boundary condition $C_p(R_q) = c_q$ *in your solution. This can be set aside to be used after the solution is obtained to find an expression for* R_q *given* R].

11.20. Vascular tumour. *Consider a model for tumour cells that are growing around a blood vessel of radius* r_b, *as shown in Figure 11.12. State the governing differential equations for the proliferating layer and for the quiescent layer and give appropriate boundary conditions. Assume the oxygen concentration at the blood vessel surface is a fixed value* c_b.

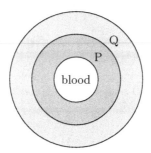

Figure 11.12: Diagram of model for growth of tumour cells around a blood vessel.

Hint: The geometry is not spherical for this problem. What is a suitable geometry to use?

11.21. Tumour growth in a test tube. *Scientific experiments grow tumour tissue in a test tube which is a small cylindrical glass tube where the growth can only proceed in the linear direction. Thus, when growing tumour cells in a test tube, the growth is in one direction instead of radially, and we assume that nutrients diffuse into the tumour from the outer end of the test tube. Consider an experiment where the growth has reached phase-2 with a quiescent layer of height* h_q *measured from the bottom of the test tube, and a proliferating layer of thickness* $h - h_q$.

(a) *Write suitable differential equations for the quiescent layer,* $0 < x < h_q$, *and the proliferating layer,* $h_q < x < h$, *and give suitable boundary conditions.*

(b) *Solve the equations to obtain the concentrations of each layer, $C_q(x)$ and $C_p(x)$.*

   Hint: Don't apply the boundary condition $C_p(R_q) = c_q$ — it would be needed later to solve for h_q in terms of h.

(c) *Verify that the surface concentration, c_1 is the maximum concentration inside the proliferating layer.*

Chapter 12

Introduction to partial differential equations

This chapter provides a brief introduction to the derivation of partial differential equations (PDEs). It is a natural extension of the previous models of heat and lake pollution, and indicates how they might be improved and generalised. A case study is included to illustrate how such equations are applied in practice.

12.1 The heat conduction equation

In previous chapters we have mentioned the dependence of certain functions on two independent variables: time and position. We now see how to derive a PDE for the temperature when the system is not in thermal equilibrium, that is, when there are changes in both of these independent variables.

Introduction

So far we have assumed thermal equilibrium so that, for all calculations, the heat flux is dependent on x (position) alone. Let us now consider what happens when the material is not in thermal equilibrium. We have to take account of the heat used in raising (or lowering) the temperature. The temperature will now depend on both position x and time t. Hence we let $U(x,t)$ be the temperature, and denote the heat flux as $J(x,t)$ at position x and time t.

Formulating a PDE

We return to the problem of heat flow through a wall, where the wall is now no longer in thermal equilibrium. From Section 9.5, the word equation describing the rate of change of heat content in a small section x to $x + \Delta x$ of the wall is

$$\left\{ \begin{array}{c} \text{rate of} \\ \text{change of} \\ \text{heat in section} \end{array} \right\} = \left\{ \begin{array}{c} \text{rate heat} \\ \text{conducted} \\ \text{in at } x \end{array} \right\} - \left\{ \begin{array}{c} \text{rate heat} \\ \text{conducted} \\ \text{out at } x + \Delta x \end{array} \right\}. \tag{12.1}$$

Since the temperature is no longer in thermal equilibrium, we must regard the temperature and heat flux as functions of two variables, $U(x,t)$ and $J(x,t)$, where x measures distance and t measures time.

The rate of change of heat is calculated from the rate at which heat is 'converted' into an increase in temperature. For the LHS of this equation we use equation (9.1), from Section 9.2, which relates heat to temperature as

$$\left\{ \begin{array}{c} \text{rate of} \\ \text{change of} \\ \text{heat in section} \end{array} \right\} = c \times \{\text{mass}\} \times \frac{\partial U}{\partial t}, \tag{12.2}$$

301

with c the specific heat. Note the partial derivative with respect to time, since U depends on both x and t. (For more details on partial derivatives, see Appendix B.2.) The mass of the section can be expressed in terms of the wall density ρ, as $\rho A \Delta x$, where A is the usual surface area.

The heat conduction terms can be expressed in terms of the heat flux

$$\left\{\begin{matrix} \text{rate heat} \\ \text{conducted} \\ \text{in at } x \end{matrix}\right\} = AJ(x,t), \qquad \left\{\begin{matrix} \text{rate heat} \\ \text{conducted} \\ \text{out at } x+\Delta x \end{matrix}\right\} = AJ(x+\Delta x,t). \tag{12.3}$$

Then, from equations (12.1) and (12.2),

$$c\rho \frac{\partial U}{\partial t} \Delta x = -J(x+\Delta x,t) + J(x,t). \tag{12.4}$$

Example 12.1:   For equation (12.4), take the limit as $\Delta x \to 0$ and then use Fourier's law to obtain a differential equation for the temperature.

Solution: We first divide through by Δx to obtain

$$c\rho \frac{\partial U}{\partial t} = -\left[\frac{J(x+\Delta x,t) - J(x,t)}{\Delta x} \right].$$

Recall the definition of a partial derivative,

$$\frac{\partial J}{\partial x} = \lim_{\Delta x \to 0} \frac{J(x+\Delta x,t) - J(x,t)}{\Delta x},$$

where we are looking at the change in J with respect to the x variable, while the t variable remains constant. If we now take the limit $\Delta x \to 0$, we obtain

$$c\rho \frac{\partial U}{\partial t} = -\frac{\partial J}{\partial x}. \tag{12.5}$$

Finally, we apply Fourier's law to express the equation in terms of temperature. For quantities that vary with both space x and time t, Fourier's law must be written as

$$J(x,t) = -k\frac{\partial U}{\partial x}. \tag{12.6}$$

Substituting Fourier's law (12.6) back into (12.5) gives

$$c\rho \frac{\partial U}{\partial t} = -\frac{\partial}{\partial x}\left(-k\frac{\partial U}{\partial x} \right).$$

For constant k this simplifies to

$$\frac{\partial U}{\partial t} = \alpha \frac{\partial^2 U}{\partial x^2}, \qquad \text{where} \quad \alpha = \frac{k}{c\rho}. \tag{12.7}$$

This second-order partial differential equation is called the *heat equation*, and the parameter α is known as the thermal diffusivity. Note that for equilibrium temperatures, that is, for $\partial U/\partial t = 0$, (12.7) reduces to

$$\frac{d^2 U}{dx^2} = 0,$$

which was the ordinary differential equation obtained in Section 9.5.

Other applications of the heat/diffusion equation

This heat equation (12.7) is widely known, arising in many other applications. In such contexts,where it models the changes in concentration due to the movement of mass by random motion, it is called the *diffusion equation*. One other area of application is finance, where another similar partial differential equation, called the Black–Scholes equation, is used to compute share option prices. The Black–Scholes equation can be transformed into the heat equation by a change of variables. See Wilmott (1998). Variants of this equation are also used in population dynamics with spatial dependence, as in Murray (1990), where the random movement of animals can be modelled by a diffusion term.

One application of mass diffusion is the diffusion of chemicals. Partial differential equations would provide a valuable extension to the model we developed in Chapter 2 on lake pollution. There we assumed that pollution concentration within the lake was uniformly distributed, while in reality it is a function of both time and distance from its source. We see how to do this later in this chapter.

Summary of skills developed here:
- *Formulate a PDE to describe the time-dependent conduction of heat.*
- *Distinguish cases where the use of an ODE is appropriate, and where a PDE is appropriate.*

12.2 Oscillating soil temperatures

We consider an interesting and fundamental property of heat processes, namely the oscillatory nature of temperatures of large masses, such as the ocean and the soil. It will provide a theoretical basis for the case study concerning the detection of land mines.

Problem description

It is well known that soil acts as an excellent thermal insulator for variations in temperature. Since wine demands an environment with as close to a constant temperature as possible, wine cellars are constructed underground, making use of this thermal property. The thermal mass of the ocean, which has a substantial impact on the milder climatic temperatures experienced in coastal regions, is another well-known case of this phenomenon.

The theory we develop concerns the soil mass and its temperatures over time. We intend to develop a model that reflects the oscillatory nature of this thermal mass.

Model assumptions and approach

Let us consider the problem of determining the temperature in the soil due to periodic heating by day and cooling by night. We consider an infinite soil depth, with the surface $x = 0$ corresponding to the soil surface and the x-axis pointing into the ground.

We let $U(x, t)$ denote the relative temperature variation in the ground, at some time t, and at a distance x into the ground. We measure the temperature relative to some mean temperature, in this case the mean surface temperature. From Section 12.1 the governing differential equation is the standard non-equilibrium heat conduction equation,

$$\frac{\partial U}{\partial t} = \alpha \frac{\partial^2 U}{\partial x^2}, \tag{12.8}$$

where α is the thermal diffusivity for soil. (Recall that $\alpha = k/c\rho$, with k the conductivity, ρ the density and c the specific heat.)

Boundary conditions for the PDE

On the surface of the ground, $x = 0$, and we have to specify the heat flux $Q(t)$ (for which we develop a model shortly). Due to heating by the sun during the day and then cooling at night, we require a suitable function that oscillates with time.

The change in temperature is driven by a diurnal cycle. Thus we model the surface heat flux $Q(t)$ with a simple trigonometric function, $\cos(\omega t)$, where ω is determined from the period being 1 day, that is,

$$Q(t) = q\cos(\omega t), \qquad \omega = \frac{2\pi}{60 \times 60 \times 24} \simeq 7.3 \times 10^{-5} \text{ seconds.} \tag{12.9}$$

Hence, using Fourier's law of heat conduction, we can write the boundary condition on the ground surface $(x = 0)$ as

$$-k\frac{\partial U}{\partial x} = q\cos(\omega t) \quad \text{at } x = 0. \tag{12.10}$$

(Alternatively, we could have considered a more complicated boundary condition by allowing Newton cooling/heating from the ground surface, but we do not do so here.)

The second boundary condition states that the temperature variation goes to zero at a long distance from the surface, that is,

$$-k\frac{\partial U}{\partial x} \to 0 \quad \text{as } x \to \infty. \tag{12.11}$$

Complex valued temperatures

To solve the PDE (12.8), with the two boundary conditions (12.10) and (12.11), we assume a suitable form for the solution. Bearing in mind PDE (12.8) has constant coefficients, we would assume exponential forms for both the x-dependence and the t-dependence. However, this would not match with the cosine term in the boundary condition (12.10). Similarly, assuming a form $U(x,t) = F(x)\cos(\omega t)$ and matching with the boundary condition (12.10), will not help to solve the PDE (12.8).

A useful trick is to use complex numbers. This technique allows us to simplify the problem we have to solve. We write the problem in terms of complex exponentials creating a problem which is easier to solve. Since our original problem is the real part of the complex problem, the solution to our problem is the real part of the complex problem solution. We represent the trigonometric functions as complex exponentials, using Euler's identity

$$e^{iz} = \cos(z) + i\sin(z), \qquad \text{where} \qquad i = \sqrt{-1}.$$

(See Appendix B.3 for an introduction to complex numbers.) We define a complex valued temperature $\hat{U}$, where U is the real part of $\hat{U}$,

$$U(x,t) = \text{Real}\left(\hat{U}(x,t)\right). \tag{12.12}$$

The problem we now solve is

$$\frac{\partial \hat{U}}{\partial t} = \alpha \frac{\partial^2 \hat{U}}{\partial x^2}, \quad -k\frac{\partial \hat{U}}{\partial x} = qe^{i\omega t} \text{ at } x = 0, \quad -k\frac{\partial \hat{U}}{\partial x} = 0 \text{ as } x \to \infty. \qquad (12.13)$$

We then take the real part of the solution, which corresponds to the solution of the original problem (12.8).

Analytic solution

Assume the solution for the complex valued temperature can be written in the form

$$\hat{U}(x,t) = F(x)e^{i\omega t}. \qquad (12.14)$$

Now substituting (12.14) into equation (12.13) we obtain

$$i\omega F(x)e^{i\omega t} = \alpha \frac{d^2 F(x)}{dx^2} e^{i\omega t}.$$

Note that the derivative is an ordinary derivative, which is the case because F is a function of x alone. Dividing each equation by $e^{i\omega t}$, we obtain an ODE for $F(x)$,

$$\alpha \frac{d^2 F}{dx^2} = i\omega F. \qquad (12.15)$$

We now find its general solution.

Example 12.2: *Find the general solution to the differential equation (12.15).*

Solution: *The equation is a constant coefficient differential equation. Hence, let $F(x) = e^{\lambda x}$. Substituting this solution form into equation (12.15),*

$$\lambda^2 = \frac{i\omega}{\alpha}, \qquad so \qquad \lambda = \pm\sqrt{\frac{\omega}{\alpha}} \times i^{1/2}.$$

From Appendix B.3 we see that

$$i^{1/2} = \pm\frac{(1+i)}{\sqrt{2}}.$$

So the solution is (substituting back into $F(x) = e^{\lambda x}$ and taking a linear combination of the solutions)

$$F(x) = C_1 e^{\sqrt{\frac{\omega}{2\alpha}}(1+i)x} + C_2 e^{-\sqrt{\frac{\omega}{2\alpha}}(1+i)x},$$

where C_1 and C_2 are arbitrary complex valued constants. Defining

$$b = \sqrt{\frac{\omega}{2\alpha}}(1+i), \qquad (12.16)$$

we can write

$$F(x) = C_1 e^{bx} + C_2 e^{-bx}. \qquad (12.17)$$

Using (12.14) and (12.17), the complex valued temperature is given by

$$\hat{U}(x,t) = \left(C_1 e^{bx} + C_2 e^{-bx}\right)e^{i\omega t}, \qquad \text{where} \quad b = \sqrt{\frac{\omega}{2\alpha}}(1+i). \qquad (12.18)$$

The arbitrary constants C_1 and C_2 are found by applying the boundary conditions in (12.13), as shown in the following example.

Example 12.3:   *Apply the boundary conditions to find values for C_1 and C_2 and hence obtain the complex valued temperature.*

Solution: *Applying the boundary condition at $x = 0$, $\hat{U} = qe^{i\omega t}$ gives*

$$-kb\left(C_1 - C_2\right)e^{i\omega t} = qe^{i\omega t}, \tag{12.19}$$

and hence

$$C_1 - C_2 = \frac{-q}{kb}.$$

Applying the other boundary condition, as $x \to \infty$, noting that $e^{-bx} \to 0$ and $e^{bx} \to \infty$ (see Exercise 12.4), the condition can only be satisfied if

$$C_1 = 0.$$

Hence we obtain $C_2 = q/kb$.

Substituting for C_1 and C_2 back into (12.18), the complex valued temperature $\hat{U}(x,t)$ is now given by

$$\hat{U}(x,t) = \frac{q}{kb}e^{-bx}e^{i\omega t}. \tag{12.20}$$

The actual temperature is obtained by taking the real part of this complex valued solution. Using the expression for b from (12.16), and Euler's identity $e^{i\theta} = \cos(\theta) + i\sin(\theta)$ (whence $(1+i)/\sqrt{2} = e^{i\pi/4}$), allows us to write (12.20) in the form

$$\hat{U}(x,t) = \left(\frac{q}{k}\sqrt{\frac{\alpha}{\omega}}\right)e^{-i\pi/4}e^{-\sqrt{\omega/2\alpha}(1+i)x}e^{i\omega t}$$

$$= \left(\frac{q}{k}\sqrt{\frac{\alpha}{\omega}}e^{-\sqrt{\omega/2\alpha}\,x}\right)e^{i(\omega t - \sqrt{\omega/2\alpha}\,x - \pi/4)}.$$

Taking the real part of $\hat{U}(x,t)$ (using Euler's identity) yields the expression for the real-valued temperature

$$U(x,t) = \left(\frac{q}{k}\sqrt{\frac{\alpha}{\omega}}e^{-\sqrt{\omega/\alpha}\,x}\right)\cos\left(\omega t - \sqrt{\frac{\omega}{\alpha}}x - \pi/4\right). \tag{12.21}$$

This can also be done using **Maple**, with the built-in functions **Re** and **evalc**. This is illustrated in the **Maple** code Listing 12.1.

Listing 12.1: Maple code: c_hp_soiltemp.mpl

```
restart:
interface(imaginaryunit=i);
assume(omega>0,alpha>0,k>0);
b:=(omega/alpha)^(1/2)(1+_i)/2^(1/2);
U:=-q/(k*b)*exp(-b*x)*exp(_i*omega*t);
evalc(Re(U));
```

We can summarise the solution technique described above for oscillatory problems involving linear PDEs:

- Introduce a complex valued function $\hat{U}(x,t) = U(x,t) + iV(t)$, where $U(x,t)$ is the solution we seek, and assume its form to be $\hat{U} = F(x)e^{i\omega t}$.

- Substitute this into the PDE, where all the $e^{i\omega t}$ terms cancel, leaving a constant coefficient ODE for $F(x)$.

- Substitute for $F(x) = e^{\lambda x}$ and hence obtain the general solution for $\hat{U}$.

- Apply the boundary conditions to find the arbitrary constants.

- Take the real part of $\hat{U}(x,t)$ to obtain $U(x,t)$ (using Euler's identity).

Interpreting the solution

A general oscillation can be expressed in the form

$$A\cos\left(\omega(t-\phi)\right),$$

where A is the *amplitude* of the oscillation and ϕ is the *phase-shift*. The amplitude measures the maximum extent of the oscillation and the phase-shift measures the time difference between the oscillatory behaviour of the above function and that of $A\cos(\omega t)$. The first term in brackets in (12.21), which is the coefficient of the cosine term, represents the amplitude of the oscillation. This term decreases exponentially with depth x. The phase-shift is proportional to the depth. This means the oscillation lags behind what is happening at the ground surface, by an amount which increases with depth. Furthermore, the temperature on the surface also lags behind the heat flux applied to the surface by an amount $\pi/(4\omega)$ (which is equivalent to $1/8$ of a day). Figure 12.1 provides a typical sketch of the temperature profiles at different times, which illustrates this decrease in amplitude, and an increase in phase-shift, with depth.

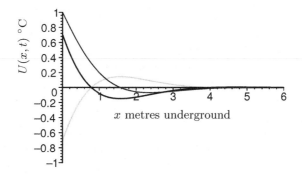

Figure 12.1: `Maple` graph of the ground temperature at different times resulting from an oscillating heat flux applied at the surface. (Solution to equation (12.21) with parameters $k = 1$, $\alpha = 1$ and $q = 1$.) The temperature is plotted at times $\omega t = 0$ (black), $\omega t = \pi/4$ (thin black) and $\omega t = \pi$ (grey) radians, with $\omega = 1$.

In practice, for soils, growth in the exponential term is sufficiently large for the amplitude to die off quickly. This implies that soil is a very good insulator from temperature variations caused by daily fluctuations. As was mentioned at the beginning of the section, wine requires a close to constant temperature for storage. Thus wine is stored in cellars, which do not have to be very deep to reduce the temperature variation to less than $1°C$.

Summary of skills developed here:

- *Write appropriate governing equations for a heat conduction problem with an oscillatory driving temperature.*
- *For a problem with oscillating temperatures, convert to a complex valued problem to simplify the analysis.*
- *Convert back from the complex valued temperatures to real valued temperatures.*

12.3   Case Study: Detecting land mines

In the following case study we use the above theory of oscillatory temperatures of a soil mass to detect non-soil objects buried beneath the earth's surface. The main idea is that such objects change the oscillatory pattern sufficiently to be detected and, furthermore, different buried objects such as land mines or rocks produce changes sufficiently different to allow us to distinguish between them. This case study is based on Miller (1995), Davies (1994) and discussions with Tony Richings.

Land mines are a serious problem. Decades after a war, land mines still maim and cripple civilians with an enormous impact on their communities. Large areas on the Falkland Islands are fenced off and unsafe for use as a result of a war fought more than 15 years ago. In November 1998, Hurricane Mitch together with floods and mud slides in Nicaragua and Honduras exposed long-buried land mines, carrying them to previously clear regions and creating further distress. Many countries around the world are affected by mines: El Salvador, Ethiopia, Peru, Argentina, Afghanistan, Angola, Mozambique, Iraq, Iran, Kuwait, Korea and Cambodia, to name some. Cambodia is probably one of the worst affected countries in that the mines are widespread and regularly moved by monsoon floods. Although the war there ended some six years ago, there were in 1999 an average of 350 accidents a month due to land mines, which represents a substantial proportion of the total Cambodian population of six million. The safe detection and removal of land mines thus assumes crucial importance.

A common method for spotting land mines is to use metal detectors; however, land mines are now often made of plastic to avoid detection. Another approach, being investigated by the Defence Science and Technology Organisation in Australia, is to use thermal infrared imaging. This works because the temperature of the ground just above a buried land mine is different from the temperature of the surrounding ground. The mine produces a 'hot spot' on the surface during the day and a 'cold spot' at night.

To gain a better understanding of this process, mathematical models were developed by CSIRO (Miller, 1995), which could be used to investigate such effects as the depth and size of the land mine, the thermal properties of land mine and soil, as well as the weather conditions. In fact, two models were produced: a full three-dimensional model used for quantitative predictions and a one-dimensional model for initial investigations. We consider a slightly simplified version of the one-dimensional model.

Consider a land mine buried at a depth d of $d = 5\,\text{mm} = 0.005\,\text{m}$. The ground is heated by the sun during the day and cools at night, and we make use of this oscillatory heat flux $Q(t)$ to model the heating and cooling processes.

We neglect any horizontal heat flow in the ground above the land mine and thus consider a one-dimensional model where the heat flow is only in the vertical direction. This is illustrated in Figure 12.2 below. We also let $U_1(x,t)$ denote the temperature of the soil and $U_2(x,t)$ denote the temperature of the land mine, where t is time and x denotes distance into the ground from the surface. The surface of the soil is taken at $x = -d$ and the interface between the soil and mine at $x = 0$.

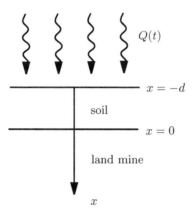

Figure 12.2: The 1-D model for thermal detection of a land mine. The soil surface is heated by day and cools at night; this is represented by an oscillatory heat flux $Q(t)$ applied to the soil surface.

The temperature at any point will be oscillating with time due to the diurnal cycle of heating and cooling of the sun; therefore the temperatures will not reach equilibrium but depend continuously on time and depth. The heat equation is used to model these temperatures of the soil above the land mine, as well as the land mine (with the configuration illustrated in Figure 12.2) resulting in the equation system

$$\frac{\partial U_1}{\partial t} = \alpha_1 \frac{\partial^2 U_1}{\partial x^2}, \qquad -d < x < 0, \tag{12.22}$$

$$\frac{\partial U_2}{\partial t} = \alpha_2 \frac{\partial^2 U_2}{\partial x^2}, \qquad 0 < x < \infty, \tag{12.23}$$

where

$$\alpha_1 = \frac{k_1}{\rho_1 c_1}, \qquad \alpha_2 = \frac{k_2}{\rho_2 c_2}. \tag{12.24}$$

The parameter values k_1 and k_2 are the conductivities of the soil and land mine, respectively, ρ_1 and ρ_2 the densities of the soil and land mine and c_1 and c_2 the specific heats of the soil and mine.

The associated boundary conditions are as follows. On the surface of the soil, $x = -d$ and the heat flux has a specified value $Q(t) = q\cos(\omega t)$, where q is the amplitude of the oscillatory heat flux. By Fourier's law this gives the boundary condition at $x = -d$,

$$-k_1 \frac{\partial U_1}{\partial x} = q\cos(\omega t) \quad \text{at } x = -d. \tag{12.25}$$

Where the soil and the top of the land mine join, $x = 0$ and we impose a condition for the continuity of temperature and heat flux,

$$U_1(0,t) = U_2(0,t), \qquad -k_1 \frac{\partial U_1}{\partial x} = -k_2 \frac{\partial U_2}{\partial x} \quad \text{at } x = 0. \tag{12.26}$$

A further boundary condition, specified for $x \to \infty$, states that the rate of flow of heat, or heat flux, is zero. Hence,

$$-k_2 \frac{\partial U_2}{\partial x} = 0, \quad \text{at } x = \infty. \tag{12.27}$$

The governing equations can be solved to obtain a solution for the soil surface temperature (at $x = -d$), given by the real part of the following complex function:

$$\hat{U}_1(-d, t) = \frac{q}{k_1 b_1} \left(\frac{(1+r) + (1-r)e^{-2b_1 d}}{(1+r) - (1-r)e^{-2b_1 d}} \right) e^{i\omega t}, \tag{12.28}$$

where

$$r = \frac{k_2 b_2}{k_1 b_1}, \qquad b_1 = \sqrt{\frac{\omega}{\alpha_1}} \frac{(1+i)}{\sqrt{2}}, \qquad b_2 = \sqrt{\frac{\omega}{\alpha_2}} \frac{(1+i)}{\sqrt{2}}.$$

Note that, in the limit as $d \to \infty$, this expression for the surface temperature reduces to the (complex valued) temperature for the surface temperature for a semi-infinite soil. (Semi-infinite soil here refers to deep soil with no buried land mine, as would be found in the vicinity of the land mine.) This is the same as setting $r = 1$, so that the thermal parameters of the land mine are the same as those of the surrounding soil.

One thing we notice about the solution (12.28) is that the temperature is proportional to q, the amplitude of the periodic heat flux term. In practice this quantity might be difficult to measure directly. However, it can be determined by measuring the temperature in soil where there are no buried objects.

We might then be interested in comparing the relative temperature of the soil surface above the land mine with the temperature in an infinite homogeneous soil. This is shown in Figure 12.3 for a MK5 land mine, buried under 5 mm of soil. We have taken

$$\alpha_1 = 0.6 \times 10^{-6}, \quad \alpha_2 = 0.1 \times 10^{-6}, \quad k_1 = 1.2, \quad k_2 = 0.26.$$

The temperature on the soil surface above the land mine (solid line) at approximately noon, $t = 0$ days, is greater than the surface temperature for soil without any buried mine (dotted line). Similarly, at night ($t = 0.5$ days) the temperature is lower than that for the soil without a land mine. (Actually, there is a small phase-lag in the soil temperatures without the land mine that can be observed by examining the graph carefully.) Thus, where the land mine is buried, we see a 'hot spot' on the soil surface during the day and a 'cold spot' at night.

A similar calculation for a buried rock ($\alpha_2 = 1.1 \times 10^{-6}$, $k_2 = 2.5$) shows the opposite, that is, a 'cold spot' during the day and a 'hot spot' at night. This makes thermal detection attractive from the point of view of discriminating land mines from other buried objects.

Our model predictions are not as accurate as we might have hoped: the predicted temperature variations are far too large. This is because the model is only one-dimensional and fails to account for heat conduction in the horizontal direction. When the model is extended to include this conduction, the temperature variations become far smaller.

Hopefully, such models will be relevant in the clearing of land mines in affected countries. There is a current political drive to ban land mines completely and most countries, including Australia, signed such an agreement at the Ottawa Convention in 1997. However, the USA, China and Vietnam refused. It is an interesting (and possibly unrelated) fact that while land mines are manufactured in several countries, the USA, China and Vietnam are the principal manufacturers of land mines today.

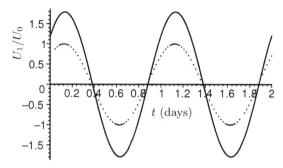

Figure 12.3: Normalised temperature of soil surface versus time in days t. The solid line is the normalised surface temperature for an MK5 land mine buried 5 mm below the surface. The dotted line is the normalised temperature of the soil surface when there is no buried object. Both temperatures have been expressed as multiples of the maximum surface temperature, $U_0 = q/(kb)$, for a soil with no buried land mine.

12.4 Lake pollution revisited

We return to the problem of modelling the levels of pollution in a lake from Section 2.5. There are a number of processes by which a mass of pollutant can travel through a lake. Previously we considered a lake in which the pollution was well mixed and was transported with the water flow. This means of transport is advection, which is equivalent to the convection of heat. Below, we relax the 'well mixed' assumption, allowing the pollution levels to be different at different locations within the lake, although again advection is responsible for the movement of pollution. Modelling this process leads to a PDE for the pollution concentration, which we derive and solve analytically using a technique known as the method of characteristics. We also indicate how `Maple` can be used to solve the equation, and generate solutions for given boundary conditions. In the final section, the much slower diffusive process of pollution transport is introduced and a PDE to model this is developed.

Model assumptions and approach

We assume pollution to travel solely by water flow through the lake, that is, by advection. We assume a unidirectional flow of water from its entrance into, to its exit from, the lake. Thus if x is in the direction of water flow, and y and z are the other axes in 3D space, then, for any fixed x, there is no change in pollution concentration in the y and z directions. Once again, we consider the volume V to be constant, as well as the flow of water into, and out of, the lake, and the pollution source will be from a feeder river at one end. However, unlike in our previous lake pollution model from Section 2.5, in this case pollution will not be assumed to be of uniform concentration throughout the lake. Alternatively, we model pollution levels at any location x within the lake and at any time t. To do this, we make an assumption about the shape of the lake: it is taken as rectangular! As justification for this simplification, in many lakes there is a main channel of flow from entrance to exit, where advection is the dominant means of pollution movement through the lake. This channel is often, roughly, rectangular in shape. As the formulation progresses below, you will see how the non-rectangular shape of real lakes might be dealt with, as well as changes in pollution concentration in the y and z directions.

The concentration C of pollutant depends on both position x (horizontal distance from its entry point) in the lake and time t. Hence, we let $C(x, t)$ be the concentration and examine how it varies throughout the lake with time. The pollution mass flux (across a unit of area per unit of time) will be denoted $J(x,t)$, at position x and time t. We consider the movement of pollution in one direction only: the direction of changing x.

Consider a lake, estimated by a rectangular shape of length $(b - a)$ and constant cross-sectional area A, as illustrated in Figure 12.4. Thus $a \le x \le b$. Here $F(x,t)$ is taken as the

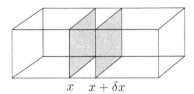

$x \quad x + \delta x$

Figure 12.4: Diagram of a 'rectangular' lake, from $x = 0$ to $x = b$.

horizontal water flow (across area A per unit of time), so that $F(a,t)$ is the horizontal water flow across area A at the entrance to the lake and $F(b,t)$ the horizontal water-pollution flow across area A leaving the lake.

Formulating the PDE

The input-output principle, or balance law, provides us with a basis on which to build the equations. Applied to the total mass of pollutant in a section of width δx,

$$\left\{ \begin{array}{c} \textit{rate of change} \\ \textit{of pollutant} \\ \textit{mass} \end{array} \right\} = \left\{ \begin{array}{c} \textit{rate pollutant} \\ \textit{mass enters} \\ \textit{at } x \end{array} \right\} - \left\{ \begin{array}{c} \textit{rate pollutant} \\ \textit{mass leaves} \\ \textit{at } x + \delta x \end{array} \right\}. \tag{12.29}$$

If we let $C(x,t)$ denote the pollutant concentration and let $J(x,t)$ denote the pollutant mass flux at position x and time t, and if A is the area of the cross-section (assumed constant), then equation (12.29) becomes

$$\frac{\partial}{\partial t}(CA\delta x) = J(x,t)A - J(x + \delta x, t)A. \tag{12.30}$$

Note that each of the three terms in (12.30) has units of mass divided by time. Furthermore, the rate of change with time is given by a partial derivative since the quantities involved depend on both time t and position x. We divide throughout by $A\delta x$ to obtain

$$\frac{\partial C}{\partial t} = -\left(\frac{J(x + \delta x, t) - J(x,t)}{\delta x} \right).$$

Letting $\delta x \to 0$ yields the equation

$$\frac{\partial C}{\partial t} = -\frac{\partial J}{\partial x}. \tag{12.31}$$

We have assumed that the pollutant is transported solely by the motion of the fluid, that is, via advection. We need an expression for the pollutant mass flux due to the motion of the water-pollutant mixture. F was defined as the flow rate (litres per second) of the mixture, which we assumed to be constant. So the rate of flow of pollutant mass (at the point x) is FC. The pollutant mass flux (rate of flow of mass per unit area per unit time) is thus given by

$$J(x,t) = \frac{FC(x,t)}{A}. \tag{12.32}$$

Substituting this into equation (12.31) yields a PDE for the pollutant concentration,

$$\frac{\partial C}{\partial t} + \frac{F}{A}\frac{\partial C}{\partial x} = 0. \tag{12.33}$$

On the boundaries of the lake we place conditions that allow the constants of integration to be established, just as initial conditions were required for the solution of differential equations in previous sections. (For further details on boundary conditions, see Section 11.1.) We assume that pollution is entering the lake at $x = a$, such that

$$C(a, t) = \frac{g(t)}{F}, \tag{12.34}$$

where $g(t)$ is a time-dependent function describing the total amount of pollution entering the lake at a. A second condition is placed on the pollution concentration at location x when $t = 0$, that is, the initial concentration at each location x within the lake:

$$C(x, 0) = P(x). \tag{12.35}$$

Analytical solution

We now need to solve the PDE (12.33), together with its boundary conditions, equations (12.34) and (12.35). In Exercise 12.6 we examine the simple case where the system has settled to some steady state, or equilibrium, and deal with the general case below.

We are interested in establishing the changing level of pollution over time, at any location in the lake, before any steady-state solution is reached. For this we introduce the *method of characteristics* in its simplest form. It is a method commonly used for the solution of PDEs, but the idea is not extended in this text. (The reader is referred to any text on PDEs for further details on this and other solution methods.)

Example 12.4: *Suppose pollution is entering a lake from a river at one end. Modelling the spread of pollution with (12.33), establish the level of pollution at any time and position (horizontal distance from the pollution source) in the lake.*

Solution: *We need to solve*

$$\frac{\partial C}{\partial t} + \frac{F}{A}\frac{\partial C}{\partial x} = 0$$

$$C(a, t) = \frac{g(t)}{F} \quad (\text{possibly}\ \ 0)$$

$$C(x, 0) = P(x). \tag{12.36}$$

Rearranging the PDE above,

$$\frac{\partial C}{\partial t} = -\frac{F}{A}\frac{\partial C}{\partial x}. \tag{12.37}$$

This implies that the rate of change of C with respect to t is a constant $(-F/A)$ times the rate of change of C with respect to x. Set constant $(F/A) = v$. Then,

$$\frac{\partial C/\partial x}{\partial C/\partial t} = -\frac{A}{F} = -\frac{1}{v} = -\frac{dt}{dx}.$$

So v is a velocity, and solving the simple differential equation on the RHS, $t = x/v - K_1$ (with K_1 an arbitrary constant), or $x = vt + K$ where $K = vK_1$. What we now show is that pollution concentration C is constant along these lines (curves) defined by x and t.

From the definition for the derivative of a function of two variables $C(x,t)$, we have that

$$\frac{dC}{dt} = \frac{\partial C}{\partial x}\frac{\partial x}{\partial t} + \frac{\partial C}{\partial t}\frac{\partial t}{\partial t}$$
$$= \frac{\partial C}{\partial x}\frac{\partial x}{\partial t} + \frac{\partial C}{\partial t}.$$

Let us return now to equation system (12.36) and apply this definition. With $x = vt + K$,

$$\frac{\partial C}{\partial t} + \frac{F}{A}\frac{\partial C}{\partial x} = 0 = \frac{dC}{dt}.$$

Thus along the lines $x = vt + K$, the pollutant concentration C is constant. This is a very powerful result (as will be seen below), and these lines are known as characteristics.

Next we establish the value of C on each characteristic (and thus at any specific time and location) using the boundary conditions. Consider the (x,t)-plane: The particular line passing through $(a,0)$ can be determined by setting $x = a$ and $t = 0$ in $x = vt + K$, and solving for K. The associated characteristic is

$$t = \frac{A}{F}(x - a).$$

Similarly a series of characteristics can be established. These are illustrated in Figure 12.5 in the positive (x,t)-plane.

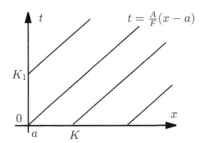

Figure 12.5: Diagram of characteristics.

Notice that when $t < A(x - a)/F$, the characteristics cut the x-axis at $(K,0)$, which implies that the value of C (which is constant along line $x = vt + K$) is $P(x) = P(K)$, from the second boundary condition in system (12.36). These values for $P(x)$ are the initial values of pollution at each x-location in the lake.

If $t > A(x - a)/F$, then $t = A(x - a)/F + K_1$ for some positive constant K_1. The point of intersection between this characteristic and the boundary $x = a$ is at $t = K_1$. On substitution into the first boundary condition of system (12.36),

$$C(a,t) = \frac{g(t - A(x - a)/F)}{F} = \frac{g(K_1)}{F}.$$

Thus we have as a solution for any location x, $a \le x \le b$, and with $v = F/A$,

$$C(x,t) = \begin{cases} g\left(t - \frac{A(x-a)}{F}\right)/F & \text{if } t > \frac{A(x-a)}{F} \\ P(x - \frac{Ft}{A}) & \text{if } t < \frac{A(x-a)}{F}. \end{cases} \tag{12.38}$$

We can now look at the cleaning time this model of a lake predicts, in the case where no further pollution enters the lake, and compare it with the cleaning time in our previous model from Section 2.5 where the automatic mixing of a pollutant was assumed. This is a particular case of the example above, when no pollution enters the lake, and thus the boundary condition at $x = a$ and $t > 0$ is $g(t) = 0$, so that $C(a, t) = 0$. The resulting solution is

$$C(x, t) = \begin{cases} 0 & \text{if } t > \frac{x-a}{v} \\ P(x - vt) & \text{if } t < \frac{x-a}{v} . \end{cases} \qquad (12.39)$$

Note that for $t > (b - a)/v$ (or $x > b$) there are no positive values for the pollution concentration in the lake, whence it is clean. The complete cleaning time is given by $(b - a)/v = A(b - a)/F$. Thus v is the velocity, in the x-direction, of the pollution through the lake.

Equation (12.39) describes a very important phenomenon: that of *wave motion*. Over time, the wave (of pollution) propagates in the x-direction, with velocity v. This notion can be understood by examining the equation solution as t increases. Notice that $P(a) = P(x - vt)$ for some x and t (from (12.39)), which implies that the pollution level at location $x - vt$ is the same as that at a. Alternatively stated, wherever $a = x - vt$ or $x = a + vt$. This latter equation implies that the pollution level at location a will have moved in the positive x-direction with velocity v in time t: the wave shifts to the right with velocity v. This moving wave phenomenon via advection is clarified in the sequence of diagrams of Figure 12.6.

The system is initialised with a 'wave' of pollution, $g(t)$, at the boundary $a = 0$, as in the first graph. For illustration, the lake is assumed pollution-free at this initial time. Each subsequent plot is the level of pollution in the lake for $a \le x \le b$, at a later time, with $t = t + nh$, where $n = 1, 2, \ldots$ and $h = \pi$ for the illustrated figure.

In order to calculate the time required for a polluted lake, with fresh water flowing into it, to have its pollution level reduced to 5% of its current level, we apply (12.39), with $g(t) = 0$ and $P(x, 0) \neq 0$ for some x, where $a \le x \le b$. To establish the time required for such a reduction in the amount of pollution in the entire lake, we convert concentration to mass $M(x, t)$, using $C = M/V(x)$, where $V = \Delta x \times \text{width} \times \text{depth}$ is the volume of water at location x. Since width and depth are assumed constant, and $C\Delta x$ is a measure of mass, the solution of t in

$$\int_a^b C(x, t)\, dx = 0.05 \int_a^b P(x)\, dx$$

is the answer sought. This can be simplified to

$$\int_a^b P(x - vt)\, dx = 0.05 \int_a^b P(x)\, dx,$$

$$\int_{a-vt}^{b-vt} P(s)\, ds = 0.05 \int_a^b P(x)\, dx.$$

Numerical solution

`Maple` can be used to solve PDE (12.33) and obtain a general solution. Then, with specified boundary conditions, we can examine how the system changes with time and space. The code for this is given in Listing 12.2. The inclusion of the **orientation** here allows you to plot t or x against pollution concentration. Alternatively, excluding **orientation** will result in a 3D plot with a time axis, a spatial axis, and a pollution concentration axis.

Listing 12.2: Maple code: c_hp_lakepde.mpl

```
restart:with(PDEtools):
pdelake := diff(C(x,t),t) + F/A*diff(C(x,t),x) = 0;
generalSolution:=pdsolve(pdelake);

#Plotting the solution for given conditions
PDEplot(pdelake,[y,5*(y-2),1+sin(5*(y-2))],
```

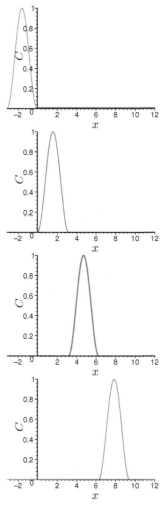

Figure 12.6: Illustration of solution (12.39), with a wave of pollution $P(x) = \sin^2(x - vt)$ arriving at an unpolluted lake. For simplicity, $v = 1$ and the 'rectangular' lake stretches from $a = 0$ to $b = 12$.

```
            y=0..50,orientation=[1,90]);
PDEplot(pdelake,[y,5*(y-2),1+sin(5*(y-2))],
            y=0..50,orientation=[-90,90]);
```

Extensions and diffusion

Although advection of pollution in the main central channel of a lake is the main contributor to the movement of pollution through it, there are many other processes at work; chemical processes, for example. Acting at a molecular level, certain nutrients react with chemicals to create further pollutants.

Another process operating at the microscopic level is diffusion. Diffusion may be the main process for reducing pollution in sheltered bays, away from the main advection channel, and as such is essential to understand in order to ascertain safety levels for food sources from, or recreational sports in, these regions. Diffusion operates in all directions. Thus the diffusion of chemicals is a slow process, compared with the advection modelled above, and occurring in all directions puts it beyond the scope of this book. However, we will develop the diffusion equation in a single direction to provide an understanding of how the process works.

For simplicity we will return to Figure 12.4, and imagine that diffusion may occur in the y direction only, at right angles to the direction of the advection in direction x. Using the same thin rectangular slice (in Figure 12.4) as before, let the face of the slice facing the y direction have area B. Thus $B = \Delta x \times$ depth and is comparable with A of the advection model. Let $J(y, t)$ be the pollution flux through a diffusive process across a unit of area per unit of time. Then, in the same way as PDE (12.33) was developed for $C(x, t)$, with pollution flux $J(x, t)$ via advection in the x-direction across area A, we can develop a PDE for diffusion in the y direction across area B, to give

$$\frac{\partial C}{\partial t} + \frac{\partial J}{\partial y} = 0. \tag{12.40}$$

Since pollution moves from regions of high concentration to regions of low concentration, the flux J is proportional to the negative gradient of pollution concentration C. Thus we write

$$J = -D\frac{\partial C}{\partial y},$$

where D is the constant of proportionality (called the diffusivity). Thus we have the diffusion equation in one direction,

$$\frac{\partial C}{\partial t} = D\frac{\partial^2 C}{\partial y^2}. \tag{12.41}$$

We will not solve this equation here, nor develop it to 3D space, but the interested reader will find details in any text on PDEs. However, the solution to the diffusion equation (12.41) will follow from the solution to the heat equation (see Section 12.2), which takes the same form, although with different boundary conditions.

Summary of skills developed here:

- *Formulate a PDE to describe the advection of a pollutant from some source in a lake, such as a river.*

- *Understand the differences between modelling pollution spread via advection and diffusion.*

- *Distinguish cases where the use of an ODE is appropriate, and where a partial differential equation is appropriate.*

12.5 Exercises for Chapter 12

12.1. Non-equilibrium temperature model. *For Exercise 9.2, from Chapter 9, show that the non-equilibrium temperature satisfies a PDE of the form*

$$\frac{\partial U}{\partial t} = \alpha\frac{\partial^2 U}{\partial x^2} + \beta,$$

where α and β are constants you must determine.

12.2. Non-equilibrium temperature for a cylinder. *Extend the formulation in Exercise 9.6 of Chapter 9 to obtain a PDE for the non-equilibrium temperature $U(r,t)$.*

12.3. Non-equilibrium temperature for a cylindrical heat fin. *In Exercise 9.8 of Chapter 9 you found an ODE for the equilibrium temperature. Modify the derivation to obtain the following PDE for time-dependent temperature,*

$$c\rho\frac{\partial U}{\partial t} = k\frac{1}{r}\frac{\partial}{\partial r}\left(r\frac{\partial U}{\partial r}\right) - 2h(U - u_s),$$

where c is the specific heat, ρ is the density, k is the conductivity and h is the convective heat transfer coefficient.

12.4. Using Euler's identity. *In Section 12.2 limits were required to establish values for the constants of integration, which led to equation (12.20). Given $b = a(1+i)$ where $a > 0$ is a positive constant, what are*

$$\lim_{x\to\infty} e^{bx} \qquad \text{and} \qquad \lim_{x\to\infty} e^{-bx} \text{?}$$

(Hint: Use Euler's identity.)

12.5. Oscillating surface temperature. *Consider the PDE*

$$\frac{\partial U}{\partial t} = \alpha\frac{\partial^2 U}{\partial x^2}$$

with the boundary conditions

$$U(0,t) = u_1\cos(\omega t), \qquad u(x,t) \to 0, \quad \text{as } x \to \infty.$$

These are the equations for the temperature in the ground due to a given oscillating temperature on the surface. Solve these equations, and express your answer as a real-valued function.

12.6. Lake pollution in steady state. *Suppose pollution has been entering a lake from a river at one end over a period of time, such that a stable steady-state situation has been reached. Modelling the spread of pollution with (12.33), establish the level of pollution at any time and position (horizontal distance from the pollution source) in the lake.*

12.7. Including a variable flow rate into the lake pollution model. *In the case of a non-constant F, with $F(x,t)/A = v(x,t)$, velocity, show that equation (12.33) becomes*

$$\frac{\partial C}{\partial t} + \frac{\partial}{\partial x}\left(C(x,t)\,v(x,t)\right) = 0\,.$$

Appendix A

Differential equations

A.1 Properties of differential equations

A *differential equation* is an equation that involves the derivative(s) of some unknown function. Such equations occur in many forms and thus specific terminology, outlined below, is used to classify them. Solving a differential equation involves finding an expression for the unknown function, for which there are a variety of analytical techniques available. Since not all equations can be solved analytically, numerical methods have been developed and are widely used. Some analytical techniques relevant to this text are introduced below, with a brief introduction to numerical methods covered in Chapter 4.

First some terminology relevant to the discussion of differential equations is introduced:

- The *order* of a differential equation is the highest derivative occurring in it. Note that the order of the differential equation is the same as the number of arbitrary constants occurring in the general solution. For example, the differential equation

$$\frac{dy}{dt} - y = 0$$

  is first-order and has a general solution $y = C_1 e^t$, which contains only one arbitrary constant C_1. The differential equation

$$\frac{d^2 y}{dt^2} - y = 0$$

  is second-order. Its general solution is $y = C_1 \cosh(t) + C_2 \sinh(t)$, which contains two arbitrary constants.

- A differential equation is *linear* if it can be expressed in the form

$$a_n(t)\frac{d^n y}{dt^n} + a_{n-1}(t)\frac{d^{n-1}y}{dt^{n-1}} + \ldots + a_1(t)\frac{dy}{dt} + a_0(t)y = F(t),$$

  where the a_i and F are functions of t alone. Otherwise it is *nonlinear*. For example,

$$\frac{dy}{dt} - y = 0 \quad \text{and} \quad t\frac{dy}{dt} - y = \sin t$$

  are linear, whereas

$$y^3 \frac{dy}{dt} - y = 0 \quad \text{and} \quad \frac{dy}{dt} - \sin y = 0$$

  are nonlinear.

- An equation is said to be *homogeneous* when it involves only terms that are functions of the dependent variable and/or its derivatives. That is, no term must be a constant term or a function of the independent variable alone. The equation is *inhomogeneous* if a term involving the independent variable is included. Thus, for example,

$$\frac{dy}{dt} - y = 0$$

  is homogeneous, while

$$\frac{dy}{dt} - y = 2t$$

  is inhomogeneous.

- A *system of differential equations* is a group of differential equations (more than one) that together describe some phenomenon. The equations are *coupled* if they depend on each other, such as in a predator-prey system,

$$\frac{dx}{dt} = \beta x - c_1 xy, \qquad \frac{dy}{dt} = c_2 yx - \alpha y.$$

They are *uncoupled* otherwise.

- The system is called a *dynamical system* when the independent variable is time and the solution used to predict future states. That is, the system describes how the modelled phenomenon changes with time. The coupled system above is thus a dynamical system.

- Differential equations are *ordinary differential equations* (ODEs) when there is only one independent variable, such as space or time. In the case of two independent variables, such as dependence on both space and time together, the equations are called *partial differential equations* (PDEs), in which case the derivatives are written with 'curly' d's, as for $y = y(x, t)$ an example PDE is

$$\frac{\partial y}{\partial t} = 2xyt + \sin t.$$

A.2   Solution by inspection

Some differential equations are sufficiently simple for us to spot the general solution by inspection. This requires a good knowledge of the derivatives of the elementary functions such as polynomials, exponential functions, hyperbolic functions and trigonometric functions.

Example A.1: *Find a solution to the differential equation*

$$\frac{d^2 y}{dt^2} = -b^2 y,$$

where b is a non-zero constant (hence $-b^2$ is negative).

Solution: *We look for functions y with the property that their second derivative is a constant squared (b^2), times the function itself (y), and is negative. Functions with this property are the trigonometric functions $\cos(bt)$ and $\sin(bt)$. Hence the general solution is*

$$y = C_1 \cos(bt) + C_2 \sin(bt),$$

where C_1 and C_2 are arbitrary constants.

A few other common examples of differential equations whose solutions may be obtained by inspection include:

- $\dfrac{dy}{dt} = ay$, with solution $y = Ce^{at}$

- $\dfrac{dy}{dt} = at$, with solution $y = \dfrac{a}{2}t^2 + C$

- $\dfrac{d^2 y}{dt^2} = b^2 y$, with solution $y = C_1 e^{bt} + C_2 e^{-bt}$ or, equivalently, with solution

$$y = c_1 \cosh(bt) + c_2 \sinh(bt).$$

A.3   First-order separable equations

These are first-order differential equations which can be put in the form

$$\frac{dy}{dt} = f(y)g(t),$$

where $f(y)$ is a function of y only and $g(t)$ is a function of t only. We can solve these equations by 'separating the variables', that is, by dividing the differential equation through by $f(y)$ and then integrating both sides with respect to the independent variable t.

Example A.2: *Solve the differential equation $\dfrac{dy}{dt} = 2ty + 2t$ using indefinite integrals.*

Solution: *We identify this differential equation as a separable differential equation by writing it in the form*

$$\frac{dy}{dt} = 2(y+1)t.$$

Thus, separating the variables, we obtain

$$\frac{1}{y+1}\frac{dy}{dt} = 2t.$$

Integrating both sides with respect to t, we obtain

$$\int \frac{1}{y+1}\frac{dy}{dt}\,dt = \int 2t\,dt.$$

On the LHS the change of variables rule for integration allows us to change the integration variable from t to y, giving

$$\int \frac{1}{y+1}\,dy = \int 2t\,dt.$$

Carrying out the integrations, we obtain

$$\ell n\,|y+1| = t^2 + C,$$

where C is an arbitrary constant. Hence, solving for y,

$$|y+1| = e^{t^2 + C},$$

or

$$y(t) = Ae^{t^2} - 1,$$

where $A = \pm e^C$ is a convenient relabelling of the arbitrary constant.

Alternatively, the above example could be solved on some fixed interval using definite integrals, and thus not requiring the arbitrary constants, as follows.

Example A.3: *Solve the differential equation $\dfrac{dy}{dt} = 2ty + 2t$ on the interval $[t_0, t]$.*

Solution: *As above, this differential equation is a separable differential equation and can be written in the form*

$$\frac{dy}{dt} = 2(y+1)t.$$

Separating the variables, we obtain

$$\frac{1}{y+1}\frac{dy}{dt} = 2t.$$

Integrating both sides with respect to t

$$\int_{t_0}^{t} \frac{1}{y+1} \frac{dy}{dt} \, dt = \int_{t_0}^{t} 2t \, dt.$$

On the LHS the change of variables rule for integration allows us to change the integration variable from t to y, giving

$$\int_{y_0}^{y} \frac{1}{y+1} \, dy = \int_{t_0}^{t} 2t \, dt,$$

where $y(t_0) = y_0$. *Carrying out the integrations we obtain*

$$\ell n \, |y+1| - \ell n \, |y_0+1| = t^2 - t_0^2,$$

and then

$$\ell n \left| \frac{y+1}{y_0+1} \right| = t^2 - t_0^2.$$

From here y can be solved for as before

$$\left(\frac{y+1}{y_0+1} \right) = \pm e^{t^2 - t_0^2}.$$

This can be written as

$$y(t) = Ae^{t^2} - 1,$$

where $A = \pm(y_0+1)e^{-t_0^2}$ *is a convenient relabelling of the constant.*

A.4   First-order linear equations and integrating factors

These are differential equations that can be put into the form

$$\frac{dy}{dt} + p(t)y = q(t),$$

where functions p and q are functions of t only. To solve these, we find an *integrating factor* that helps us integrate the equation directly. The integrating factor $R(t)$ is given by the formula

$$R(t) = e^{\int_a^t p(t) \, dt}.$$

If we then multiply throughout by this integrating factor, the differential equation can be converted into the form

$$\frac{d}{dt} \left(R(t)y(t) \right) = R(t)q(t),$$

and then the integral (with respect to t) of the LHS is simple to find, with the function on the right a function of t alone.

Example A.4: Solve the differential equation $\dfrac{dy}{dt} = 2t(y + 1)$.

Solution: First we see that the differential equation can be written in the appropriate form:

$$\frac{dy}{dt} - 2ty = 2t.$$

Thus $p(t) = -2t$, $q(t) = 2t$ and the integrating factor $R(t)$ is given by

$$R(t) = e^{\int_a^t p(t)\,dt} = e^{\int_a^t -2t\,dt} = e^{-t^2 + a^2} = A e^{-t^2},$$

where $A = e^{a^2}$ is constant. Hence, multiplying throughout by this factor, the differential equation becomes

$$\frac{d}{dt}\left(e^{-t^2} y\right) = 2t e^{-t^2}.$$

Integrating both sides with respect to t, we obtain

$$e^{-t^2} y = -e^{-t^2} + C,$$

(where C is the arbitrary constant of integration) and so

$$y = -1 + C e^{t^2}.$$

A.5 Homogeneous equations

We consider here *constant coefficient equations*. These are linear differential equations where the coefficients of each derivative, including that of zero order, are constants and the coefficient of y^0 is 0. This last condition defines the equation as homogeneous. Thus they are a linear combination of derivatives of the dependent variable and include no functions of the independent variable alone.

For first-order equations, they are of the general form

$$\frac{dy}{dt} + a_0 y = 0.$$

For second-order equations, they are of the general form

$$\frac{d^2 y}{dt^2} + a_1 \frac{dy}{dt} + a_0 y = 0,$$

where a_1 and a_0 are constants.

The characteristic equation

To solve such equations we use prior knowledge about their solutions; we look for a solution of the form $y(t) = e^{\lambda t}$, where the constant λ is to be determined. This always leads to an auxiliary equation, or characteristic equation, for λ. The general solution is then a linear combination of the solutions obtained.

Example A.5: *Solve* $\dfrac{d^2y}{dt^2} - \dfrac{dy}{dt} - 2y = 0.$

Solution: *Substitute* $y = e^{\lambda t}$ *into the differential equation to obtain*

$$\lambda^2 e^{\lambda t} - \lambda e^{\lambda t} - 2e^{\lambda t} = 0 \qquad so \qquad (\lambda^2 - \lambda - 2)e^{\lambda t} = 0.$$

We obtain the characteristic (auxiliary) equation

$$\lambda^2 - \lambda - 2 = (\lambda + 1)(\lambda - 2) = 0.$$

This quadratic equation has roots $\lambda = -1$ *and* $\lambda = 2$, *and thus the solutions are* e^{-t} *and* e^{2t}. *The general solution is a linear combination of* e^{-t} *and* e^{2t},

$$y(t) = C_1 e^{-t} + C_2 e^{2t}.$$

Classification of roots

For second-order equations the characteristic equation is a quadratic equation and thus we do not always obtain distinct real roots. The form of the solution depends on the nature of the roots, for which there are three possible cases:

- Real roots: $\lambda = m_1$ and $\lambda = m_2$. Then,

$$y = C_1 e^{m_1 t} + C_2 e^{m_2 t}.$$

- Complex roots $\lambda = \alpha \pm \beta i$. Then,

$$y = e^{\alpha t} \left(C_1 \cos(\beta t) + C_2 \sin(\beta t) \right).$$

- Equal roots: $\lambda = m_1$. A second independent solution can be obtained by multiplying by t. Then,

$$y = e^{m_1 t} \left(C_1 + C_2 t \right).$$

Two special cases of the above are worth noting:

- Real perfect squares: $\lambda = \pm m_1$. Then,

$$y = C_1 e^{m_1 t} + C_2 e^{-m_1 t},$$

  or an equivalent form is

$$y = c_1 \cosh(m_1 t) + c_2 \sinh(m_1 t),$$

  where the arbitrary constants c_1 and c_2 are linearly related to the arbitrary constants C_1 and C_2.

- Purely imaginary roots: $\lambda = \pm m_1 i$. Then,

$$y = C_1 \cos(m_1 t) + C_2 \sin(m_1 t).$$

A.6   Inhomogeneous equations

We consider again *constant coefficient equations*. This time we include a forcing term, that is, include a function dependent on t alone, or independent of the dependent variable. The inclusion of forcing defines the equation as inhomogeneous. For first-order equations, the general form is

$$\frac{dy}{dt} + a_0 y = F(t),$$

and for second-order equations,

$$\frac{d^2y}{dt^2} + a_1\frac{dy}{dt} + a_0y = F(t),$$

where a_1 and a_0 are constants and F is a non-zero function called the forcing term. (When $F(t) = 0$ the equation is homogeneous. Alternatively, when $F(t) \neq 0$, the equation is inhomogeneous.)

Splitting the solution

We solve this equation by splitting the solution into a homogeneous solution $y_h(t)$ and a *particular solution* $y_p(t)$, hence

$$y(t) = y_h(t) + y_p(t).$$

Here y_h is the *general solution* of the homogeneous equation (with $F(t) = 0$), and y_p is any particular solution satisfying the original equation.

Finding particular solutions

A particular solution of a differential equation is any function that satisfies the differential equation. If the RHS is a simple elementary function, then it is usually possible to find a particular solution by making an educated guess based on the form of the RHS and its derivatives.

Example A.6: Find a particular solution of $\dfrac{d^2y}{dt^2} - \dfrac{dy}{dt} - 2y = 3$.

Solution: Since the RHS is constant, we seek a solution of the form $y_p = A$ where A is a constant. Substituting into the differential equation, we obtain $0 + 0 - 2A = 3$. Hence $A = 2/3$ and $y_p = 2/3$.

Example A.7: Find a particular solution of $\dfrac{d^2y}{dt^2} - \dfrac{dy}{dt} - 2y = t^2$.

Solution: We look for a solution of the form

$$y_p = A + Bt + Ct^2,$$

where A, B and C are constants. Substituting this into the differential equation, we get $2C - (2Ct + B) - 2(Ct^2 + Bt + A) = t^2$. Equating coefficients of the powers of t (noting that the RHS can be expressed as $t^2 + 0t + 0$), we obtain the set of simultaneous equations

$$-2C = 1,$$
$$-2C - 2B = 0,$$
$$2C - B - 2A = 0.$$

Solving these, $C = -1/2$, $B = 1/2$ and $A = -3/4$. Hence,

$$y_p = -\frac{1}{2}t^2 + \frac{1}{2}t - \frac{3}{4}.$$

Example A.8: Find a particular solution of $\dfrac{d^2y}{dt^2} - \dfrac{dy}{dt} - 2y = 2e^{5t}$.

Solution: We look for a solution of the form

$$y_p = Ae^{5t}.$$

Substituting into the differential equation we obtain

$$25Ae^{5t} - 5Ae^{5t} - 2Ae^{5t} = 2e^{5t}.$$

Hence

$$(25 - 5 - 2)A = 2$$

and $A = 1/9$. Thus

$$y_p = \frac{1}{9}e^{5t}.$$

Example A.9: Find a particular solution of $\dfrac{d^2y}{dt^2} - \dfrac{dy}{dt} - 2y = 2\sin(2t)$.

Solution: We try a solution in the form

$$y_p = A\cos(2t) + B\sin(2t).$$

Substituting this form into the differential equation, we obtain

$$-4A\cos(2t) - 4B\sin(2t) - [-2A\sin(2t) + 2B\cos(2t)] - 2[A\cos(2t) + B\sin(2t)]$$
$$= 2\sin(2t) + 0\cos(2t).$$

Equating the coefficients of $\cos(2t)$ and $\sin(2t)$, we obtain

$$-6A - 2B = 0,$$
$$-6B + 2A = 2.$$

Solving for A and B gives $A = 1/10$ and $B = -3/10$ and hence,

$$y_p = -\frac{1}{10}\cos(2t) + \frac{3}{10}\sin(2t).$$

Thus we have some general rules we can apply:

- For constant RHS, try $y_p = A$, where A is a constant.
- For polynomial RHS, try a general polynomial up to the order of the RHS polynomial.
- For exponential RHS, try a general exponential with the same argument.
- For trigonometric RHS, try a linear combination of sin and cos.

There are some exceptions that should be noted. When the particular solution is the same as the homogeneous solution, the usual approach for finding the general solution will fail. In the context of oscillations, this is called the resonance condition. In this case, try the appropriate choice for a particular solution multiplied by t, as illustrated in the following example.

Example A.10: *Find a particular solution of $\dfrac{d^2y}{dt^2} - \dfrac{dy}{dt} - 2y = 3e^{2t}$.*

Solution: *Note that the homogeneous solution is $C_1 e^{2t} + C_2 e^{-t}$. For the particular solution, rather than try $y_p = Ae^{2t}$, try $y_p = Ate^{2t}$.*

Substituting this into the differential equation yields

$$(4Ate^{2t} + 4Ae^{2t}) - (2Ate^{2t} + Ae^{2t}) - 2(Ate^{2t}) = 3e^{2t}.$$

Equating coefficients of te^{2t}, we gain no information ($0 = 0$). However, from the coefficients of e^{2t}, we obtain $2A = 3$. Hence a particular solution is

$$y_p(t) = \frac{3}{2} te^{2t}$$

and the general solution is

$$y(t) = C_1 e2t + C_2 e^{-t} + \frac{3}{2} te^{2t}.$$

Appendix B

Further mathematics

B.1 Linear algebra

This appendix on linear algebra presents only the theory and concepts required to support the text. A general text on linear algebra, such as Lay (1994), is recommended for further background and a complete presentation of the theory.

Matrix notation and algebra

Square matrices, that is, matrices with the same number of rows and columns (in our case, 2×2), can be added (subtracted), multiplied, or multiplied by a constant as follows:

(a) $\begin{bmatrix} a_1 & b_1 \\ a_2 & b_2 \end{bmatrix} + \begin{bmatrix} c_1 & d_1 \\ c_2 & d_2 \end{bmatrix} = \begin{bmatrix} a_1 + c_1 & b_1 + d_1 \\ a_2 + c_2 & b_2 + d_2 \end{bmatrix}$

(b) $\begin{bmatrix} a_1 & b_1 \\ a_2 & b_2 \end{bmatrix} \times \begin{bmatrix} c_1 & d_1 \\ c_2 & d_2 \end{bmatrix} = \begin{bmatrix} (a_1 c_1 + b_1 c_2) & (a_1 d_1 + b_1 d_2) \\ (a_2 c_1 + b_2 c_2) & (a_2 d_1 + b_2 d_2) \end{bmatrix}$

(c) $k \times \begin{bmatrix} a_1 & b_1 \\ a_2 & b_2 \end{bmatrix} = \begin{bmatrix} ka_1 & kb_1 \\ ka_2 & kb_2 \end{bmatrix}.$

They can also be post-multiplied by a vector with an equal number of rows (in our case, 2):

(d) $\begin{bmatrix} a_1 & b_1 \\ a_2 & b_2 \end{bmatrix} \begin{bmatrix} x \\ y \end{bmatrix} = \begin{bmatrix} a_1 x + b_1 y \\ a_2 x + b_2 y \end{bmatrix}.$

For this last case we use the form

$$\mathbf{A}x = \begin{bmatrix} a_1 & b_1 \\ a_2 & b_2 \end{bmatrix} \begin{bmatrix} x \\ y \end{bmatrix}.$$

Matrices and vectors provide a very useful way of expressing systems of equations. With this above notation, the system

$$a_1 x + b_1 y = 0,$$
$$a_2 x + b_2 y = 0,$$

can be given as

$$\mathbf{A}x = \mathbf{0},$$

where

$$\mathbf{A} = \begin{bmatrix} a_1 & b_1 \\ a_2 & b_2 \end{bmatrix}, \qquad x = \begin{bmatrix} x \\ y \end{bmatrix}.$$

It should be noted that while the above notation is given for 2×2 matrices, the concepts can be extended to $n \times n$ matrices. There is a wealth of linear algebra theory that can then be applied to these systems, some of which is given below. (For a full understanding of the concepts, a text on linear algebra, such as Lay (1994), is recommended.)

Determinants

Suppose $\mathbf{A}$ is the 2×2 matrix $\begin{bmatrix} a_1 & b_1 \\ a_2 & b_2 \end{bmatrix}$ and $\boldsymbol{x} = \begin{bmatrix} x \\ y \end{bmatrix}$ is a vector.

Then $\mathbf{A}\boldsymbol{x} = \boldsymbol{0}$ is

$$\begin{bmatrix} a_1 & b_1 \\ a_2 & b_2 \end{bmatrix} \begin{bmatrix} x \\ y \end{bmatrix} = \begin{bmatrix} 0 \\ 0 \end{bmatrix} \tag{B.1}$$

or

$$a_1 x + b_1 y = 0,$$
$$a_2 x + b_2 y = 0.$$

The *trivial solution* to this is $x = y = 0$. The *nontrivial solution* excludes this above possibility, and thus

$$x = -\frac{b_1}{a_1} y \qquad \text{if} \quad a_1 \neq 0.$$

Then,

$$-\frac{a_2 b_1}{a_1} y + b_2 y = 0$$

so that

$$y \left(b_2 - \frac{a_2 b_1}{a_1} \right) = 0$$

and then, since $y \neq 0$, for a nontrivial solution to the vector equation (B.1), we must have

$$a_1 b_2 - a_2 b_1 = 0.$$

This value $(a_1 b_2 - a_2 b_1)$ is called the *determinant* of A and is denoted $\det A$ or $|A|$. It is the difference between the diagonal products of A in the case of a 2×2 matrix. The important conclusion is that for a nontrivial solution to a vector equation $A\boldsymbol{x} = \boldsymbol{0}$, we require $|A| = 0$. (Such notions can be extended to $n \times n$ matrices.)

Eigenvectors and eigenvalues

An *eigenvector* of a 2×2 matrix $\mathbf{A}$ is a non-zero vector $\boldsymbol{x}$ such that $\mathbf{A}\boldsymbol{x} = \lambda \boldsymbol{x}$ for some scalar λ. A scalar λ is an *eigenvalue* of A if there is a nontrivial solution $\boldsymbol{x}$ to $\mathbf{A}\boldsymbol{x} = \lambda \boldsymbol{x}$ such that $\boldsymbol{x}$ is the eigenvector corresponding to the eigenvalue λ.

Note that, using the results of the previous section, the existence of a nontrivial (i.e., non-zero) solution $\boldsymbol{x}$ to

$$\mathbf{A}\boldsymbol{x} = \lambda \boldsymbol{x} \qquad \text{or} \quad \mathbf{A}\boldsymbol{x} - \lambda \boldsymbol{x} = (\mathbf{A} - \lambda \mathbf{I})\boldsymbol{x} = \boldsymbol{0}$$

means that the determinant of the matrix $(\mathbf{A} - \lambda \mathbf{I})$ is zero. That is, $|\mathbf{A} - \lambda \mathbf{I}| = 0$. (Here I is the identity matrix, and

$$I = \begin{bmatrix} 1 & 0 \\ 0 & 1 \end{bmatrix}$$

for 2×2 matrices.) Recall,

$$\mathbf{A} - \lambda \mathbf{I} = \begin{bmatrix} a_1 & b_1 \\ a_2 & b_2 \end{bmatrix} - \begin{bmatrix} \lambda & 0 \\ 0 & \lambda \end{bmatrix} = \begin{bmatrix} a_1 - \lambda & b_1 \\ a_2 & b_2 - \lambda \end{bmatrix}$$

and then

$$\begin{aligned} |\mathbf{A} - \lambda \mathbf{I}| &= (a_1 - \lambda)(b_2 - \lambda) - a_2 b_1 \\ &= a_1 b_2 - \lambda b_2 - \lambda a_1 + \lambda^2 - a_2 b_1 \\ &= \lambda^2 - \lambda(a_1 + b_2) + (a_1 b_2 - a_2 b_1). \end{aligned}$$

Notice that this is $\lambda^2 - \lambda(\text{trace}\mathbf{A}) + (\det \mathbf{A})$, where trace $\mathbf{A}$ (or tr $\mathbf{A}$) is the sum of the diagonal terms of A, and $\det \mathbf{A} = |\mathbf{A}| = a_1 b_2 - a_2 b_1$.

We call

$$|\mathbf{A} - \lambda \mathbf{I}| = \lambda^2 - \lambda(a_1 + b_2) + (a_1 b_2 - a_2 b_1)$$

the *characteristic polynomial* of $\mathbf{A}$, and $|\mathbf{A} - \lambda \mathbf{I}| = 0$ is called the characteristic equation of $\mathbf{A}$. This characteristic equation can be solved to find the eigenvalues. Clearly, in this case, there can be at most two distinct eigenvalues, with the possibility of complex valued eigenvalues.

Linear dependence and independence

Any point in the (x, y)-plane can be expressed as a linear combination of x and y, which have direction vectors $[1, 0]^T$ and $[0, 1]^T$ respectively.[1] We say that $[1, 0]^T$ and $[0, 1]^T$ *span* $\mathbb{R}^2$, or that they form a *basis* for $\mathbb{R}^2$: that is, every point in the plane can be expressed in terms of them. Further,

$$\begin{bmatrix} 1 \\ 0 \end{bmatrix} \neq c \begin{bmatrix} 0 \\ 1 \end{bmatrix}$$

and

$$\begin{bmatrix} 0 \\ 1 \end{bmatrix} \neq c \begin{bmatrix} 1 \\ 0 \end{bmatrix}$$

for any constant c. This property is called *linear independence* of the vectors. Thus we say that two vectors are linearly independent if neither can be expressed as a linear combination of the other. Another way to say this (the typical mathematical definition) is

a set of vectors is linearly independent if the vector equation

$$c_1 v_1 + c_2 v_2 + \ldots + c_k v_k = 0$$

has only the trivial solution $c_1 = c_2 = \ldots = c_k = 0$.

Theorem 4 (Linear independence of eigenvectors.) *Suppose $v_1, \ldots, v_k$ are eigenvectors that correspond to k distinct eigenvalues $\lambda_1, \ldots, \lambda_k$ of some matrix A. Then the set of eigenvectors $\{v_1, \ldots, v_k\}$ is linearly independent.*

    Proof *Suppose $\{v_1, \ldots, v_k\}$ is not linearly independent (i.e., linearly dependent), then there exists some p such that v_p is a linear combination of the eigenvectors $\{v_1, \ldots, v_{p-1}\}$, with constants c_i,*

$$v_p = c_1 v_1 + \ldots + c_{p-1} v_{p-1}. \tag{B.2}$$

Now, multiplying by A,

$$A v_p = c_1 A v_1 + \ldots + c_{p-1} A v_{p-1}$$

and then

$$\lambda_p v_p = c_1 \lambda_1 v_1 + \ldots + c_{p-1} \lambda_{p-1} v_{p-1}, \tag{B.3}$$

since $A v_i = \lambda_i v_i$ by the definition of eigenvalues and eigenvectors. Also, multiplying (B.2) by λ_p gives

$$\lambda_p v_p = c_1 \lambda_p v_1 + \ldots + c_{p-1} \lambda_p v_{p-1} \tag{B.4}$$

and then subtracting (B.4) from (B.3)

$$0 = c_1 (\lambda_1 - \lambda_p) v_1 + \ldots + c_{p-1}(\lambda_{p-1} - \lambda_p) v_{p-1}.$$

Since $\{v_1, \ldots, v_{p-1}\}$ are linearly independent and non-zero, the expressions $(\lambda_i - \lambda_p)$ for $i = 1, \ldots, p-1$ are all zero. However, this is a contradiction because the eigenvalues are all distinct. This means that $\{v_1, \ldots, v_k\}$ are linearly independent.

    What we have shown is that, in the case of A, a 2×2 matrix with two distinct eigenvalues and two associated eigenvectors (u and v), the eigenvectors are linearly independent. And if we have two linearly independent vectors in $\mathbb{R}^2$, they must necessarily span $\mathbb{R}^2$, that is, form a basis for $\mathbb{R}^2$. Thus x and y can be expressed as a linear combination of the eigenvectors

$$x = z_1 u + z_2 v.$$

---

[1]The superscript T denotes the transpose. Thus,

$$[1, 0]^T = \begin{bmatrix} 1 \\ 0 \end{bmatrix}.$$

Matrix invertibility

An $n \times n$ matrix $\mathbf{A}$ is *invertible* if there exists a matrix $\mathbf{A}^{-1}$ such that $\mathbf{A}^{-1}\mathbf{A} = \mathbf{I} = \mathbf{A}\mathbf{A}^{-1}$, where $\mathbf{I}$ is the identity matrix.

Theorem 5 (Matrix invertibility.) *Let $\mathbf{A}$ be an $n \times n$ matrix (a square matrix). Then the following are equivalent; that is, if any one of them is true, then so are the rest.*

(a) *$\mathbf{A}$ is invertible*

(b) *The columns of $\mathbf{A}$ are linearly independent*

(c) *The columns form a basis for $\mathbb{R}^n$*

(d) *$|\mathbf{A}| \neq 0$ (i.e. $\det \mathbf{A} \neq 0$)*

(e) *$\mathbf{A}\mathbf{x} = \mathbf{0}$ has no solutions other than the trivial one,*

(f) *The rows of $\mathbf{A}$ are linearly independent*

(g) *The rows of $\mathbf{A}$ form a basis for $\mathbb{R}^n$*

Proof. *The details are omitted, but for those who are interested, see any text on linear algebra, for example Lay (1994).*

B.2   Partial derivatives and Taylor expansions

Partial derivatives

Suppose z is dependent on two variables x and y. We say $z = f(x,y)$. If we keep y constant ($y = y_1$ say), then z can be considered a function of x alone and if the derivative of $f(x,y_1)$ exists; then it is called the partial derivative of f with respect to x. Mathematically, it is described with 'curly' d's as

$$\frac{\partial f}{\partial x} = f_x \qquad - \qquad \text{a first-order partial derivative.}$$

Similarly, with x fixed ($x = x_1$ say), we have (if the derivative exists)

$$\frac{\partial f}{\partial y} = f_y \qquad - \qquad \text{the other first-order partial derivative.}$$

Example B.1: $z = f(x,y) = x^2 y + x \sin y$

$$\frac{\partial f}{\partial x} = 2xy + \sin y, \qquad \frac{\partial f}{\partial y} = x^2 + x \cos y,$$

and

$$\frac{\partial^2 f}{\partial x^2} = \frac{\partial}{\partial x}\left(\frac{\partial f}{\partial x}\right) = 2y, \qquad\qquad \frac{\partial^2 f}{\partial y^2} = \frac{\partial}{\partial y}\left(\frac{\partial f}{\partial y}\right) = -x \sin y,$$

$$\frac{\partial^2 f}{\partial y \partial x} = \frac{\partial}{\partial y}\left(\frac{\partial f}{\partial x}\right) = 2x + \cos y, \qquad\qquad \frac{\partial^2 f}{\partial x \partial y} = \frac{\partial}{\partial x}\left(\frac{\partial f}{\partial y}\right) = 2x + \cos y.$$

These are the four second-order partial derivatives.

The partial derivatives represent slopes in the x and y-directions, respectively (see Figure B.1).

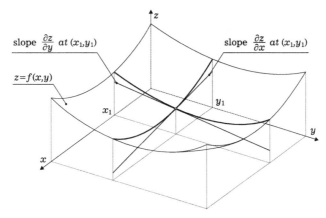

Figure B.1: Diagram showing the interpretation of the partial derivatives as the slopes of the graph of a function of two variables in the directions of the x and y axes.

The notation for these partial derivatives is given below, where it is worth noting the order of x and y in the subscripts.

$$\frac{\partial}{\partial y}\left(\frac{\partial f}{\partial x}\right) = \frac{\partial^2 f}{\partial y \partial x} = f_{xy},$$

$$\frac{\partial}{\partial x}\left(\frac{\partial f}{\partial y}\right) = \frac{\partial^2 f}{\partial x \partial y} = f_{yx},$$

$$\frac{\partial}{\partial y}\left(\frac{\partial f}{\partial y}\right) = \frac{\partial^2 f}{\partial y \partial y} = f_{yy},$$

$$\frac{\partial}{\partial x}\left(\frac{\partial f}{\partial x}\right) = \frac{\partial^2 f}{\partial x \partial x} = f_{xx}.$$

It is not always true that $f_{xy} = f_{yx}$. However, if all the derivatives concerned are continuous, we do have $f_{xy} = f_{yx}$ and the order of differentiation does not matter. This was the case in Example B.1 above.

For the partial derivatives then, y fixed at $y = y_1$ represents a vertical plane that intersects the surface $z = f(x, y)$ in a curve, and $\partial z/\partial x$ at the point (x_1, y_1) is the slope of the tangent to this curve at (x_1, y_1). (Similarly for x fixed at $x = x_1$.) This is illustrated in Figure B.1.

Taylor series expansions

We recall Taylor's Theorem for a single variable, and then extend the definition to functions of two variables.

Taylor's Theorem for a single variable. If F and its first n derivatives F', F'', ..., $F^{(n)}$ are continuous on some interval $[a, b]$ and $F^{(n)}$ is differentiable on (a, b), then there exists a number c_{n+1}, between a and b, such that

$$F(b) = F(a) + F'(a)(b - a) + \frac{F''(a)}{2!}(b - a)^2 + \ldots$$
$$+ \frac{F^{(n)}(a)}{n!}(b - a)^n + \frac{F^{(n+1)}}{(n+1)!}(c_{n+1})(b - a)^{n+1}.$$

This equation, in the single-variable case, provides an extremely accurate polynomial approximation for a large class of functions that has derivatives of all orders. With reference to Figure B.2, close to $x = \hat{x}$ we can approximate the curve $f(x)$ with the line tangent to the curve at $x = \hat{x}$. For a linear approximation, $F(b) \approx F(a) + F'(a)(b - a)$ with an error $e(b)$ having $a < c < b$,

$$|e(b)| \leq \frac{\max |F''(c)|}{2}(b - a)^2.$$

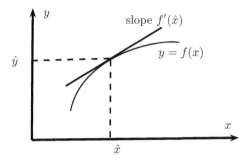

Figure B.2: Diagram showing the first-order Taylor expansion in the vicinity of the point $\hat{x}$.

For a quadratic approximation, $F(b) \approx F(a) + F'(a)(b-a) + \frac{F''(a)}{2}(b-a)^2$ with (again $a < c < b$)

$$|e(b)| \leq \frac{\max|F''(c)|}{6}|b-a|^3.$$

Taylor's Theorem can be extended to the two-variable case. As illustrated in Figure B.1, close to $(x, y) = (x_1, y_1)$ we can approximate the surface $z = f(x, y)$ with the tangent plane to the surface at $(x, y) = (x_1, y_1)$.

We extend the definition of Taylor's Theorem to functions of two variables in the following way. Let R be some region containing $P(a, b)$. Let $S(a + h, b + k)$ be in R such that the line PS is in R.

Let f have continuous first- and second-order partial derivatives in R. Describe PS parametrically as $x = a + th$, $y = b + tk$, with $0 \leq t \leq 1$. Now we study the values of $f(x, y)$ on PS by considering the function

$$F(t) = f(a + ht, b + kt).$$

We have

$$F'(t) = \frac{\partial f}{\partial x}\frac{dx}{dt} + \frac{\partial f}{\partial y}\frac{dy}{dt} = h\frac{\partial f}{\partial x} + k\frac{\partial f}{\partial y}$$

and F' is continuous and differentiable on $[0, 1]$ because f_x and f_y are continuous and differentiable (by definition above). Also,

$$F''(t) = h\frac{\partial F'}{\partial x} + k\frac{\partial F'}{\partial y}$$

$$= h\left(h\frac{\partial^2 f}{\partial x^2} + k\frac{\partial^2 f}{\partial y \partial x}\right) + k\left(h\frac{\partial^2 f}{\partial x \partial y} + k\frac{\partial^2 f}{\partial y^2}\right)$$

$$= h^2\frac{\partial^2 f}{\partial x^2} + 2hk\frac{\partial^2 f}{\partial x \partial y} + k^2\frac{\partial^2 f}{\partial y^2}.$$

Since F has its first two derivatives continuous on the interval $[0, 1]$, it satisfies the hypothesis for the Taylor series expression in a single variable with $n = 2$, so

$$F(1) = F(0) + F'(0)(1-0) + F''(c)\frac{(1-0)^2}{2!} \qquad (\text{for} \quad c \in [0, 1])$$

$$= F(0) + F'(0) + \frac{1}{2}F''(c). \tag{B.5}$$

But $F(1) = f(a + h, b + k)$ and $F(0) = f(a, b)$ by definition. Also $F'(0) = hf_x(a, b) + kf_y(a, b)$ and $F''(c) = (h^2 f_{xx} + 2hk f_{xy} + k^2 f_{yy})$ evaluated at $(a + ch, b + ck)$.

Now substituting these expressions into (B.5) above,

$$f(a + h, b + k) = f(a, b) + hf_x(a, b) + kf_y(a, b) + (\text{terms of higher order}),$$

and the *tangent-plane approximation* is

$$f(a + h, b + k) \approx f(a, b) + hf_x(a, b) + kf_y(a, b).$$

B.3   Review of complex numbers

Complex numbers have the form $z = a + ib$, where $i = \sqrt{-1}$. The number a is called the real part of z, $R(z)$, and the number b is called the imaginary part of z, $I(z)$.

Complex numbers can be added, subtracted, multiplied and divided, as shown by the following examples:

- $(7 + 3i) + (-3 + 2i) = (7 - 3) + (3 + 2)i = 4 + 5i$
- $(2 + 3i)(3 - i) = 2 \times 3 - 2i + 3 \times 3i - 3i^2 = 6 + 7i - 3 \times (-1) = 9 + 7i$
- $\dfrac{2 + i}{1 + i} = \dfrac{2 + i}{1 + i} \times \dfrac{1 - i}{1 - i} = \dfrac{2 - 2i + i - i^2}{1 - i^2} = \dfrac{2 - i + 1}{1 + 1} = \dfrac{3 - i}{2}$

Also useful when dealing with complex numbers is *Euler's identity*,

$$e^{i\theta} = \cos\theta + i\sin\theta, \tag{B.6}$$

which implies that

$$e^{-i\theta} = \cos\theta - i\sin\theta. \tag{B.7}$$

When solving ODEs, the function $e^{-i\theta}$ is often more convenient to deal with than $\cos\theta$ and $\sin\theta$. Furthermore, the identity is useful for obtaining roots of complex numbers. For example, for $\theta = \pi/2$,

$$e^{i\pi/2} = \cos\left(\frac{\pi}{2}\right) + i\sin\left(\frac{\pi}{2}\right) = i.$$

So

$$i^{1/2} = (e^{i\pi/2})^{1/2} = e^{i\pi/4}$$

and from (B.6),

$$i^{1/2} = e^{i\pi/4} = \cos\frac{\pi}{4} + i\sin\frac{\pi}{4} = \frac{1}{\sqrt{2}}(1 + i).$$

Note that $i = \cos\left(\frac{3\pi}{2}\right) + i\sin\left(\frac{3\pi}{2}\right)$, since $\cos\left(\frac{3\pi}{2}\right) = 0$ and $\sin\left(\frac{3\pi}{2}\right) = 1$, and thus we also have

$$i^{1/2} = \cos\left(\frac{3\pi}{4}\right) + i\sin\left(\frac{3\pi}{4}\right) = -\frac{1}{\sqrt{2}}(1 + i).$$

B.4   Hyperbolic functions

From Euler's identity (equation (B.6), in Appendix B.3) we can define $\sin x$ and $\cos x$ as

$$\sin x = \frac{e^{ix} - e^{-ix}}{2i}, \qquad \cos x = \frac{e^{ix} + e^{-ix}}{2}. \tag{B.8}$$

And recall that $\sin^2 x + \cos^2 x = 1$ indicates that in the (x, y)-plane, the point $(\sin x, \cos x)$ lies on the unit circle and the functions $\sin x$ and $\cos x$ are called circular functions. In the same way we can define the *hyperbolic functions*, where $(\sinh x, \cosh x)$ lies on the rectangular hyperbola with equation $\sinh^2 x - \cosh^2 x = 1$.

Formally they are defined in terms of e^x as

$$\sinh x = \frac{e^x - e^{-x}}{2}, \qquad \cosh x = \frac{e^x + e^{-x}}{2}, \tag{B.9}$$

hence

$$\tanh x = \frac{e^x - e^{-x}}{e^x + e^{-x}}.$$

These functions are commonly used to simplify expressions of exponentials and thus often appear in the integral solutions of the text.

To get an idea of how they behave, they are graphed below using `Maple` in Figures B.3 to B.5.

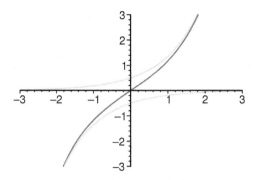

Figure B.3: Graph of sinh x (black) with $e^{-x}/2$ (grey) and $-e^{-x}/2$ (grey).

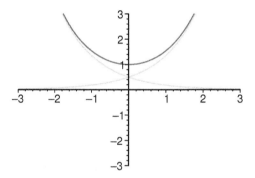

Figure B.4: Graph of cosh x (black) with $e^{-x}/2$ (grey) and $e^{x}/2$ (grey).

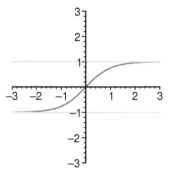

Figure B.5: Graph of tanh x (black) with upper and lower bounds at ± 1 (grey).

B.5   Integration using partial fractions

Integrating a complicated fraction can be simplified by writing it as the sum of simpler fractions. This process is useful in many applications, but in particular for integration where the integral of an original function becomes the sum of integrals, which are much easier to find.

Example B.2: Write

$$\frac{x+4}{x^2 - 5x + 6}$$

as a sum of simpler fractions and hence find its integral.

Solution: *The denominator of the function can be factorised and written as two partial fractions,*

$$\frac{x+4}{x^2 - 5x + 6} = \frac{x+4}{(x-2)(x-3)} = \frac{A}{x-2} + \frac{B}{x-3}.$$

Now

$$\frac{A}{x-2} + \frac{B}{x-3} = \frac{A(x-3) + B(x-2)}{(x-3)(x-2)}$$

so that equating the coefficients in the numerator with those of the original function gives two simultaneous equations, $A + B = 1$ and $-3A - 2B = 4$. Solving these, we have $A = -6$ and $B = 7$. Thus the original fraction can be written as the sum of partial fractions

$$\frac{x+4}{x^2 - 5x + 6} = \frac{-6}{x-2} + \frac{7}{x-3}.$$

While integration of the original function is clearly not simple, the latter expression is easy to integrate:

$$\int \frac{-6}{x-2} dx = -6 \, \ell n \, |x-2| + C_1 \quad \text{and} \quad \int \frac{7}{x-3} dx = 7 \, \ell n \, |x-3| + C_2.$$

Thus,

$$\int \frac{x+4}{x^2 - 5x + 6} dx = -6 \, \ell n \, |x-2| + 7 \, \ell n \, |x-3| + C_3,$$

where $C_3 = C_1 + C_2$.

Example B.3: *Write the fraction*

$$\frac{1}{x^2 - a^2}$$

as a sum of simpler fractions.

Solution: *First, the denominator of the original fraction can be factorised*

$$\frac{1}{x^2 - a^2} = \frac{1}{(x-a)(x+a)}$$

and we would like to write this as

$$\frac{A}{x-a} + \frac{B}{x+a}.$$

But

$$\frac{A}{x-a} + \frac{B}{x+a} = \frac{A(x+a) + B(x-a)}{(x-a)(x+a)},$$

from which, by equating the numerators,

$$1 = Ax + Aa + Bx - Ba.$$

Thus $A + B = 0$ (*equating the coefficients of* x) *and* $a(A - B) = 1$ (*equating the constants*), *which implies that* $A = 1/2a$ *and* $B = -1/2a$. *So the original fraction can be written as*

$$\frac{1}{x^2 - a^2} = \frac{1}{2a(x - a)} - \frac{1}{2a(x + a)}.$$

In general, fractions can be separated into partial fractions as follows:

$$\frac{f(x)}{x(x + a)(x - a)} = \frac{A}{x} + \frac{B}{x + a} + \frac{C}{x - a},$$

$$\frac{f(x)}{x^2 + a} = \frac{A}{x} + \frac{Bx + C}{x^2 + a},$$

$$\frac{f(x)}{x(x - a)^2} = \frac{A}{x} + \frac{B}{x - a} + \frac{C}{(x - a)^2}.$$

Note that when splitting a fraction into partial fractions, the numerator of each partial fraction is a polynomial with degree one less than the denominator. (For further details see, for example, Adams (1995).)

Appendix C

Notes on Maple and MATLAB

C.1 Brief introduction to Maple

`Maple` is an extensive symbolic software package, able to carry out many sophisticated calculations in many areas of mathematics. As an extension to the calculator, it is able to cope with extremely long numbers, up to about 500,000 digits and can solve differential equations symbolically and numerically, graph numerical solutions, do complex algebra, and much more. Since this book is concerned with differential equations, and their solution, not all of which are solvable analytically, `Maple` or `MATLAB` (see Section C.3), or some other equivalent software package, is essential for a complete understanding of the models developed and discussed in this book.

This brief introduction is far from adequate for those who are new to such packages, but it serves to illustrate the `Maple` environment, the manner in which statements and functions operate, and how to get help on any topic.

Basics

Below (Figure C.1) is a `Maple` screen dump of some basics in value assignments and algebra. Note the use of semicolons at the end of each statement; a semicolon allows the results to be displayed on the worksheet, while a colon suppresses information to the screen, although it exists and can be accessed at any time by typing the variable name. Note too that **restart** resets all variables to blanks, and it is good practice to start each independent code segment with this command.

Getting help

At any time, a question mark in front of a command will provide full information on that topic, such as, `?plot` for information, options and examples on plotting functions. Information on the commands required for a variety of functions pertaining to equations, plotting procedures and solution finding, typically used in this text, can also be found under 'Maple code', in the text index.

'Further information' (below) suggests a website for help on particular topics, as well as suggesting a basic text (of which there are many) for those requiring a detailed introduction to `Maple`.

C.2 Solving differential equations with Maple

This book is concerned with both the numerical and symbolic solution of differential equations, and the plotting of these results. Figure C.2 illustrates how to find the analytical and numerical solutions to a differential equation, as well as how to use the plotting and display functions to graph the results. These routines are typical of those that are required repeatedly throughout the book. Note the use of `with(plots)` in the opening line to access routines in the plotting library of `Maple`. (This code segment is taken from Section 2.5.)

Further information

There are any number of excellent books on introducing `Maple`, such as Holmes et al. (1993). More advanced texts are also available, such as Heck (1996).

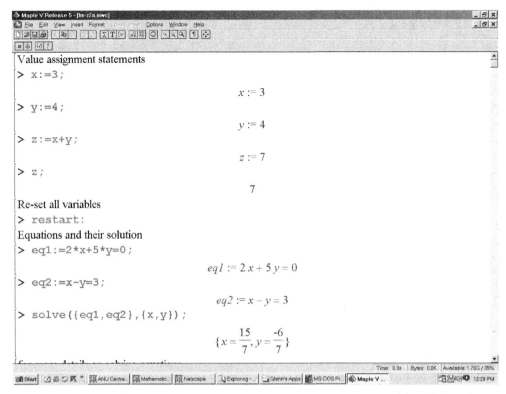

Figure C.1: **Maple** screen dump illustrating some basic assignment statements, and introducing the **Maple** environment.

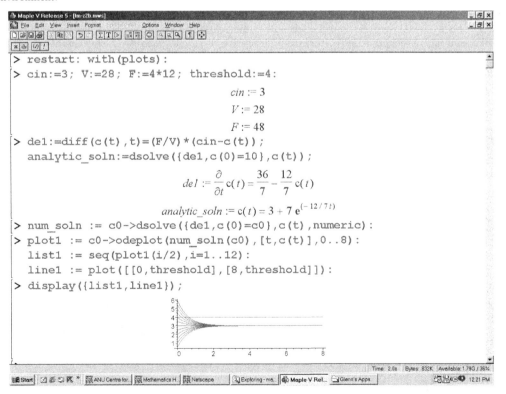

Figure C.2: **Maple** screen dump illustrating the solution to a differential equation, and a plotting routine.

C.3 Brief introduction to MATLAB

MATLAB is a high-level programming language for mathematics and engineering, in particular. The name MATLAB is short for 'MATrix LABoratory'. The characterising feature of MATLAB is that it makes dealing with and manipulating matrices both easy and fast. Originally Maple did not do symbolic algebra, but now there is a toolbox available (the symbolic toolbox, included with student versions of MATLAB) that provides the ability to do symbolic algebra.

Basic commands for vectors and matrices

The basic element (data-type) in MATLAB is the array (or matrix). Square brackets are used to denote arrays, with commas (or blanks) to separate elements and semicolons to separate rows. Thus the following commands are used to store in memory the mathematical constructs

```
a = [1 2 3];
b = [4; 5; 6];
A = [7 -8 9; -4 5 -6];
```

These produce the following mathematical vectors and matrices,

$$a = \begin{bmatrix} 1 & 2 & 3 \end{bmatrix}, \qquad b = \begin{bmatrix} 4 \\ 5 \\ 6 \end{bmatrix} \qquad \mathbf{A} = \begin{bmatrix} 7 & -8 & 9 \\ -4 & 5 & -6 \end{bmatrix}.$$

Note that a = [1, 2, 3] is equivalent to a = [1 2 3]. Also note that the semicolon at the end of each statement is used to end the statement and suppress the output (as opposed to when a semicolon occurs within the square brackets where it indicates a new line of the array).

Scalar variables are 1×1 matrices but do not require the square brackets; for example, c = 1 and d = 2.34 are used to store the scalars $c = 1$ and $d = 2.34$ in memory. Note that MATLAB normally regards all variables as real numbers, so c=1 is the same as c=1.0. However, integer-valued variables and complex-valued variables can also be defined if needed.

The usual addition, subtraction, multiplication and division operations apply to scalar variables. For array variables, addition and subtraction work for arrays of the same size and also it is possible to add or subtract a scaler to an array. For multiplication, * also corresponds to the usual matrix multiplication.

Additionally, the combinations .*, ./, .^ correspond to *element-by-element* operations; for example,

[1, 2, 3] .* [4, 5, 7]

results in the array $[1 \times 4, 2 \times 5, 3 \times 7] = [4, 10, 21]$. Similarly, [1, 2, 3] .^ 2 produces $[1^2, 2^2, 3^2] = [1, 4, 9]$.

Also very useful is colon notation, which is used to extract sub-arrays from arrays, and particularly to extract individual columns or rows from a matrix. For example, A(:,2) gives the second column of the matrix $\mathbf{A}$, A(1,:) gives the first row, and A(1:2,2:3) gives the sub-matrix corresponding to rows 1 to 2 and columns 2 to 3.

Elements of arrays can be accessed by round brackets. For the above definitions, b(2) has the value 5 and A(2,3) has the value -6. MATLAB also provides the usual constructs for-loops (for sequential operations). In particular, the *for-loop* is used when you know in advance how many sequential operations are needed. The syntax is for k=1:5 which means repeat for a variable k taking values 1 through to 5. Note the use of colon notation here is similar to that used above for the range of values for the rows or columns in a matrix. As an example of a for-loop, the following code constructs a matrix $\mathbf{A}$, where $A_{ij} = i + 2j$, by sequentially setting each element of the matrix, working along each row (in the inner loop) and then each column in the outer loop. The following is an example of two nested for-loops.

```
N = 10;
for i = 1:N
    for j = 1:N
```

```
        A(i,j) = i+2*j;
    end
end
```

MATLAB also provides other constructs, such as a while-loop (for when the number of times through a loop is not known in advance, but is determined by some condition) and the usual if-elseif and logical statements for decisions and branching in programs. The reader is referred to the MATLAB documentation, or some textbooks on MATLAB programming, such as Higham and Higham (2007); Hanselman and Littlefield (2005), for more information on how to use these. They are not required for the examples in this book.

The final important programming construct in MATLAB that we mention here is the MATLAB *function*. Functions are used to help structure complicated code, with each function performing a certain task and with the variables in one function protected from the other functions. A function takes a number of input arguments and returns a number of output arguments (often one output value, which can also be an array, but can be more than one). Some functions can even have a variable number of input and output arguments, for example, the built-in MATLAB function max(a, b) and \max(a, b, c, d) that returns the maximum value of the variables in the brackets.

MATLAB provides a substantial number of built-in functions, the simplest being functions to calculate standard mathematical operations, such as sin(x) to calculate $\sin(x)$, where x can be a scalar variable or an array. It also has functions for certain properties of arrays, such as length(a) for returning the length of an array; size(A) which returns the size of an array as a 1×2 array; and eig(A) for returning the eigenvalues of a square matrix. It is also possible to define one's own functions. The syntax for this is shown in the following code for a function that returns the sum of the cubes of two scalar numbers x and y.

```
function zout = myfcn(x, y)
zout = x^3+y^3;
```

In this the keyword *function* denotes that the following code is a separate function, and the variable zout is the output variable, which must be assigned a value inside the function. Here, myfcn is the name of the function. The number xin and yin here are the inputs for the function, and generally, can be scalars or arrays. To call the function, that is, to use the function inside some other function or code, a line of code, for example, is myfcn(3,2) which here returns the value $3^3 + 2^3 = 34$.

The function can be placed into a separate file, in which case it should have the same name as what you have called the function with the extension .m, that is, the name of the file here would be myfcn.m. However, it is also sometimes convenient to package up all the functions into one file, provided everything in that file is itself a function. A small modification of the above code for the function myfcn, changing the second line to zout = x.^3 + y.^3, would allow the function to take arrays as arguments provided the arrays were of the same size.

C.4   Solving differential equations with MATLAB

MATLAB provides functions for solving initial value problems. The standard ODE solver is the ode45 function. This function needs to be provided with an input argument corresponding to another function for the right-hand side of the differential equation. It also requires an argument for the range of values of the independent variable on which to solve the differential equation and an argument for the initial condition.

Let us consider the initial value problem

$$\frac{dx}{dt} = \frac{x^2}{10^3} - 3t, \qquad x(0) = 2.5.$$

The entire MATLAB code to solve this problem numerically and plot the results is given in Listing C.1.

Listing C.1: MATLAB code: c_app_solveode.m

```
function c_mainprogram; %define main program as function
trange = [0 2]; % set range of values of t to solve for
x0 = 2.5; % set initial value ofDE
[tsol, xsol] = ode45(@rhs, trange, x0); %solve the DE
plot(tsol, xsol); %plot the solution

function dxdt = rhs(t, x) % function definition
dxdt = x^2/10^3-3*t; % this defines RHS of dx/dt
```

Let us now examine the various parts of this code. The first line defines the main program as a function. This should have the same name as the filename, that is, **c_app_solveode.m**. The second line **trange=[0 2]** defines the range of t values to be input as an input argument of **ode45**. The line **x0=2.5** defines a variable to store the initial value $x(0)$. The next line is the one that does the work of solving the differential equation. Here the **ode45** function needs to input another function, called here **rhs**, which defines the RHS of the differential equation. Here **@rhs** is a *function handle*, that is, a "pointer" or "handle" to the actual function, which is what is needed to make a function an argument of another function. Note that the **ode45** function returns two equal sized arrays, called here **tsol** and **xsol**, which contain the solution $x(t)$ evaluated at a series of time steps contained in **tsol** with the values of the solution contained in the array **xsol**. Finally, the line **plot(tsol, xsol)** creates a figure and plots the array of values **xsol**, corresponding to $x(t)$ on the vertical axis, against the array of values **tsol**, corresponding to t, on the horizontal axis.

Let us consider the second block of code in Listing C.1. This defines the differential equation. This is a separate function that can be placed in the same file as the main program, as in Listing C.1, or placed in a separate file that has the same name as the function, called here **rhs.m**. Note that the function must always have both the arguments (t, x), in that order, even if the differential equation does not depend explicitly on t. The second line of this block then defines the RHS of the differential equation.

Solving a system of differential equations

A similar approach is taken for solving systems of differential equations. In this case we need to think of the system as a vector-differential equation. Consider

$$\frac{dx}{dt} = 3x - by, \qquad x(0) = 1,$$

$$\frac{dy}{dt} = 3x + ay^2 + 1, \qquad y(0) = 2,$$

where a and b are parameters in the model, that is,

$$\frac{d\boldsymbol{u}}{dt} = \boldsymbol{F}(t, \boldsymbol{u}), \qquad \boldsymbol{u} = \begin{bmatrix} x \\ y \end{bmatrix} = \begin{bmatrix} 3x - by \\ 3x + ay^2 + 1 \end{bmatrix}, \qquad \boldsymbol{u}(0) = \begin{bmatrix} x(0) \\ y(0) \end{bmatrix}.$$

The entire code to solve this is given in Listing C.2.

Listing C.2: MATLAB code: c_app_systemsolve.m

```
function c_app_systemsolve
global a b;     % define a,b,
trange = [0 1]; % range of values of t
a = 1; b = 3;    % parameter values
u0 = [1; 2];      % define initial values
[tsol, usol] = ode45(@rhs, [0 1], u0); %solve system

xsol = usol(:,1); ysol=usol(:,2); % extract x(t) and y(t)l
plot(tsol, xsol);      % plot x(t)
hold on;            % cause the plot to not be overwritten
plot(tsol, ysol, 'r:'); % plot y(t)
hold off;              % new plots replace the previous plot

function udot = rhs(t, u)
```

```
global a b;              % allows access to variables in main function
x = u(1); y = u(2);      % define x,y as components of vector u
xdot = 3*x - b*y;        % define the RHS of DE for x(t)
ydot = 3*x + a*y + 1;    % define the RHS of DE for x(t)
udot = [xdot; ydot];     % assemble the DEs into a column vector
```

The first line, of the first block of code, defines the name of the main program. The second line defines the variables **a** and **b** as *global variables*. The **global** command allows us to define the variables corresponding to the parameter values in the main program, and a similar statement in the function defining the RHS of the differential equation then allows that function access to the values stored in those variables. The third line defines the values of the parameters corresponding to a and b. The fourth line defines the range of t values on which to solve the differential equation. The fifth line of the 1st block defines the initial values. Note the initial values are specified as a column vector.

The first line of the second block of code solves the differential equation that is specified in the third block of code. Finally, we need to plot the solution, but to do this we need to extract the separate solution vectors from the system solution array **usol**, which is done on the second line. The solution array **usol** is an $N \times M$ array, where N is the number of time steps the solution was calculated at, and N is the number of differential equations in the system. To extract the separate arrays for plotting, a command **usol(:,1)** takes the first column of the array **usol**, with the colon(:) denoting all values of the column. The remaining lines plot the two solutions $x(t)$, $y(t)$. In this code the **hold on;** command keeps the plot open so that the second plot command can be included in the same plot without erasing the first command. The third argument of the plot command causes the second plot to be in the colour red with a dotted line.

The third block of code defines the system of differential equations. The first line defines the function **rhs** that is referenced by the handle **@rhs** in the **ode45** command in the second block of code. The second line is the global statement that makes the memory containing the parameter values a and b available to the function **rhs**. The variable **u**, in the function definition in the first line, is an array containing both the variables x and y, as a 2×2 column vector, in this case. So the third line assigns each component of **u** to the variables **x** and **y**. Then in the fourth and fifth lines we define the RHS for the differential equations for $x(t)$ and $y(t)$. Finally, in the sixth line we assign a 2×1 column vector, **udot**, to the two differential equations.

We could have written the differential equations in terms of $u(1)$ and $u(2)$ instead of defining the variables **x** and **y**; however, the use of **x** and **y** makes it easier to read and to check that the code for the RHS of the differential equations is correct.

In terms of programming style, it is often not a good idea to use global statements as we have done here. This is because for large, complicated programs there is a risk of forgetting which variables are global and accidentally putting the wrong values into them. It is then better to make these variables additional arguments of the **rhs** function; see the **MATLAB** documentation for examples of this. For simple programs, as are used in this book, however, it is convenient and simple to use the global statement.

Code for direction fields
In Chapter 6 plots are produced of the phase-plane using **MATLAB**. The following **MATLAB** function can be used to plot direction fields, similar to that produced by **Maple**. The code is given in Listing C.3.

Listing C.3: MATLAB code: c_dirplot.m

```
function c_dirplot(rhsfcn, xmin, xmax, ymin, ymax, Ngrid)
% function to plot direction fields given the RHS of a system of 2 DEs
% xmin, xmax, ymin, ymax scalar values defining plot region
% Ngrid= number of points on grid for arrows

lflag = 1; %a flag for whether arrows of same length (flag=1) or not

%make a grid
disp(xmin)
x = linspace(xmin, xmax, Ngrid);
```

```
y = linspace(ymin, ymax, Ngrid);
[Xm, Ym] = meshgrid(x,y);

%insert the function values at each point of the grid
%into the arrays xd and yd
for i = 1:Ngrid
    for j=1:Ngrid
        uvec = rhsfcn(0, [x(i), y(j)]);
            xd(j,i) = uvec(1)/norm(uvec);
            yd(j,i) = uvec(2)/norm(uvec);
            sfactor=0.6; %length of arrow
    end
end
%use Matlab quiver function to do the dirfield plot
quiver(Xm, Ym, xd, yd, sfactor,'r');
```

Appendix D

Units and scaling

D.1 Scaling differential equations

When dealing with models involving differential equations with several parameters, it can be useful to reduce the number of independent parameters in the system by making all the variables *dimensionless*. This process is called *scaling* the equation. The basic idea is to define new variables that are the old variables divided by some constant in the problem that has the same units as the variables.

Example D.1: *For the population model*

$$\frac{dN}{dt} = rN\left(1 - \frac{N}{K}\right), \quad N(0) = N_0,$$

where $N(t)$ is the population at time t, and K is the carrying capacity of the population (measured in number of people), n_0 the initial population (measured in number of people) and r is the intrinsic growth rate measured in years^{-1} (persons/person per year). Determine new variables n and τ that are dimensionless, to replace N and t.

Solution: *The variable N measures population size. There are two possible scales with the same dimensions: the carrying capacity K and the initial population N_0. We will choose the constant K. The variable t measures time (in hours). The constant r is measured in hours^{-1}, therefore r^{-1} has the same dimensions as t. We can thus define new dimensionless variables*

$$n = \frac{N}{K}, \quad \tau = \frac{t}{r^{-1}} = rt.$$

We can use the chain rule to express any derivatives in a differential equation in terms of scaled variables. Suppose X and T are dimensioned variables, and x and τ are scaled variables, where x_s and y_s are the constants used to scale the variables, so that

$$x = \frac{X}{x_s}, \quad y = \frac{Y}{y_s}, \quad \text{or} \quad X = x_s x, \quad Y = y_s y.$$

Then we have the following scaling law:

$$\frac{dX}{dY} = \left(\frac{x_0}{y_0}\right)\frac{dx}{dy}. \tag{D.1}$$

Equation D.1 follows directly from the chain rule,

$$\frac{dx}{dy} = \frac{d}{dy}(x_0 X) = x_0\frac{dX}{dy} = \frac{dX}{dY}\frac{dY}{dy} = x_0\frac{dX}{dY}\frac{1}{y_0}.$$

For second-order, and more generally, for nth-order derivatives, the scaling laws are

$$\frac{d^2 X}{dY^2} = \left(\frac{x_0}{y_0^2}\right)\frac{d^2 x}{dy^2}, \quad \frac{d^n X}{dY^n} = \left(\frac{x_0}{y_0^n}\right)\frac{d^n x}{dy^n}. \tag{D.2}$$

Example D.2: *Using the above scaling laws we now scale the model in Example D.1.*

Solution: *Write the model*

$$\frac{dN}{dt} = rN\left(1 - \frac{N}{K}\right), \quad N(0) = N_0,$$

in terms of the scaled variables $n = N/K$, $\tau = rt$. The scales are K for the variable N and r^{-1} for the variable t. Using equation (D.1) we can write the differential equation as

$$\frac{K}{r^{-1}}\frac{dn}{dt} = rn\left(1 - \frac{Kn}{K}\right)$$

which simplifies to

$$\frac{dn}{d\tau} = n(1 - n).$$

To scale the initial condition $N(0) = N_0$ we write this as $N = n_0$ when $t = 0$. In dimensionless variables, the initial condition becomes $Kn = N_0$ when $r^{-1}\tau = 0$, which simplifies to $n = N_0/K$ when $\tau = 0$. Hence we can write the dimensionless differential equation and initial condition as

$$\frac{dn}{dt} = n(1 - n), \quad n(0) = n_0, \quad \text{where} \quad n_0 = \frac{N_0}{K}.$$

This is a simpler form. Where we previously had three parameters, K, r and N_0, now we have just the one independent parameter $n_0 = N_0/K$ (which is a dimensionless combination of N_0 and K).

Example D.3: *Scale the SIR epidemic model*

$$\frac{dS}{dt} = -\beta SI,$$

$$\frac{dI}{dt} = \beta SI - \gamma I.$$

Solution: *We scale the variables using*

$$s = \frac{S}{s_0}, \quad i = \frac{I}{s_0}, \quad \tau = \frac{t}{\gamma^{-1}},$$

and so

$$S = s_0 s, \quad I = s_0 i, \quad t = \gamma^{-1}\tau.$$

Using the scaling law (D.1), the system becomes

$$\frac{s_0}{\gamma^{-1}}\frac{ds}{d\tau} = -\beta s_0^2 si,$$

$$\frac{s_0}{\gamma^{-1}}\frac{di}{d\tau} = \beta s_0^2 si - \gamma s_0 i,$$

which can be simplified to

$$\frac{ds}{d\tau} = -R_0 si,$$

$$\frac{di}{d\tau} = R_0 si - i,$$

where $R_0 = \beta s_0/\gamma$ is a single dimensionless parameter.

Advantages and disadvantages of scaling

There are some obvious advantages in scaling the equations of a model. First, there is the advantage of having fewer parameters to deal with. On the other hand, scaling a set of equations abstracts the model to some extent. It can put an additional obstacle (even a minor one) in interpreting the results or graph of a model when it is desired to obtain directly the original quantities.

For example, if the important question in a population model is the number of years until the population reaches a certain value, then it is more appropriate to produce a graph of dimensioned quantities. Furthermore, it can sometimes be useful to retain the physical interpretation of the coefficients of various terms, which can be lost when using a scaled model.

Generally speaking, if the parameter values for a model are well known, then a dimensional model may be more appropriate; but if several parameters are not known, and we need to explore the model, then there can be a big advantage to scaling the equations.

D.2   SI Units

Table D.1: Fundamental Units of Primary Quantities

Primary Quantity	SI Unit	cgs Unit
mass	kilogram, kg	gram, g
length	metre, m	centimetre, cm
time	second, s	second, s
temperature	Kelvin, K	Kelvin, K

Table D.2: Secondary Units Comprised of Fundamental Units

Quantity	SI Units
density ρ	$\mathrm{kg\,m^{-3}}$
velocity v	$\mathrm{m\,s^{-1}}$
acceleration a	$\mathrm{m\,s^{-2}}$
force F	Newtons $\mathrm{N = kg\,m\,s^{-2}}$
pressure p	Pascal $\mathrm{Pa = N\,m^{-2} = kg\,m^{-1}\,s^{-2}}$
energy E	Joule $\mathrm{J = kg\,m^2\,s^{-2}}$
power $\dot{E}$	Watt $\mathrm{W = J\,s^{-1}}$
heat flux J	$\mathrm{W\,m^{-2}}$
thermal conductivity k	$\mathrm{W\,m^{-1}\,K^{-1}}$
specific heat c	$\mathrm{J\,kg^{-1}\,K^{-1}}$
thermal diffusivity α	$\mathrm{m^2\,s^{-1}}$
Newton cooling coefficient h	$\mathrm{W\,m^{-2}\,K^{-1}}$

Appendix E

Parameters

E.1 Parameters

Practically all of the models in this book depend on parameters. In this appendix we provide a brief discussion of model parameters. First we look at approaches for estimating the values of the parameters. Then we look at approaches for assessing how sensitive the model outputs are to changes in parameter values as well as uncertainty analysis, which tells us about the variation in the model outputs given the variation in the parameter values.

We use an example problem to illustrate the main ideas, the SIR model for an infectious disease, as discussed in Section 5.2. The equations for this model were

$$\frac{dS}{dt} = -\beta_f S \frac{I}{N}, \qquad \frac{dI}{dt} = \beta_f S \frac{I}{N} - \gamma I. \tag{E.1}$$

This model depends on two parameters: β_f the transmission coefficient, and γ, the recovery rate. It also depends on initial conditions, $S(0) = s_0$ and $I(0) = i_0$, which can be also be regarded as parameters for the problem. We have chosen the frequency-dependent transmission form of the equations (5.5) because the parameter β_f is more convenient.

Estimating parameters

Estimating the values of parameters is often a difficult task. Without accurate parameter estimates a model is less useful as a predictive tool, although it may still be possible to infer the general behaviour and therefore add insight about the possible processes underlying the model.

One approach for estimating parameters is to directly measure the value of the parameter. For example, for the SIR model the parameter γ has an interpretation that γ^{-1} is the time an average infected individual is infectious. Based on knowledge of influenza we know that it has an infectious period of 1–3 days, so we can take $\gamma = 1/2$ as our estimate. It is feasible to measure the mean infectious period directly if sufficiently detailed records on the beginning time of infection and the times of recovery were kept for all individual patients; however, it is rare to have complete records of this type, and incomplete records can produce a biased estimate.

Estimating the parameter β_f is more difficult. We can use the definition of the basic reproduction number R_0, where $R_0 = \beta N/\gamma = \beta_f/\gamma$ and hence $\beta_f = R_0 = \gamma R_0$. There were three new infections in the first two days so this would give $\beta_f \simeq 3 \times 1/2 = 1.5$.

A different approach is to use data on some other aspect of the disease, for example the number of infections at given times or the number of recoveries. The idea is to then 'fit' suitable parameter values to the available data. This process of estimating the parameters is sometimes called *calibrating* the model. A number of approaches are available; perhaps the most common being the least squares method.

The least squares method (also known as regression) involves forming the sum of squares of the differences between the data and the model predictions for a given set of parameters. For example, suppose data were available for the number of infections, for N data points denoted by $\hat{I}_i$, $i = 1 \ldots N$. Some data for influenza infections in an English boarding school in 1978 were obtained by digitising the epidemic curve given in Anonymous (1978). The data are tabulated in Table E.1.

Table E.1: Number of infected students in an English boarding school collected in 1978 from 22 January to 4 February. These numbers have been digitised from the epidemic curve given in Anonymous (1978).

day	1	2	3	4	5	6	7	8	9	10	11	12	13	14
infected	3	7	22	78	233	300	256	233	189	128	72	33	11	6

Let $s(\beta, \gamma)$ be the sum of square of the differences between the data and the model at each time measurement,

$$s(\boldsymbol{p}) = \sum_{i=1}^{N}(I(t_i) - \hat{I}_i)^2.$$

This defines a function of the parameters $\boldsymbol{p} = (\beta_f, \gamma)$. Here $I(t_i)$ is the number of infectives predicted by the model at the times t_i, $i = 1 \dots N$, by solving the differential equations. The method of least squares minimises this function. Given that we need to solve the differential equations numerically, it also follows that we must use some numerical minimisation method. These are built-in to many software packages. For example, in Listing E.1, code for forming and minimising the least squares function is given at the end of this section. In Figure E.1 the fitted model is shown along with the model for our initial guess at the parameter values $\beta_f = 1.5$ and $\gamma = 1/2$. The least squares procedure has calibrated the model with the estimated parameters $\beta_f = 1.67$ and $\gamma = 0.44$.

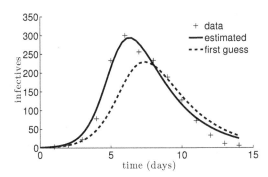

Figure E.1: Model fit using least squares minimisation procedure to estimate parameters β_f and γ with starting guess of $\beta_f = 1\,\text{day}$ and $\gamma_0 = 1/2\,\text{days}$. The estimated values from the least-squares procedure are $\beta_f = 1.67$ and $\gamma = 0.44$.

We had assumed the value of the initial numbers of infectives to be one, however we could also treat this as a parameter. The least squares procedure would then most likely estimate I_0 as non-integer. Of course it might not make sense for this parameter to have a non-integer value (but possible if I represents population density rather than population numbers). However, using more sophisticated optimisation techniques, it would be possible to constrain the value of I_0 to be an integer. Similarly it would be possible to constrain the parameters to not exceed a certain value or to not be negative, but this requires knowledge of integer and nonlinear programming techniques that are beyond the scope of this book.

While the technique appears like it works well, here it is prudent to point out that difficulties can occur. The technique depends on having a good starting guess for the parameters, which is sometimes hard to achieve. For more complicated problems it is possible to have local minima and the minimisation algorithm can get stuck at one of these. One way around this is to try a range of initial guesses. Another problem is that the search for a minimum over the parameter space may result in unphysical parameter values, such as negative values where only positive values have meaning and for negative values the model outputs might tend towards infinite values, causing overflow errors. A further problem can be that the estimation gives a good fit to the data but the parameter values are unrealistic. This however suggests that the original model is unrealistic and needs modifying.

Based on ideas from statistics, it is also possible to take a more sophisticated approach to estimating the parameters by defining a likelihood function. The likelihood is related to the probability of obtaining the data given the parameters. In this case we need to maximise the likelihood function. Maximum likelihood is equivalent to least squares estimation when the difference between the data and the model is assumed to be normally distributed, so maximum likelihood potentially gives more flexibility by specifying that this distribution is not normal (log-normal, for example, to ensure the difference is never negative) or binomial for small counts.

Furthermore, the Bayesian approach to parameter estimation specifies prior distributions for the parameters and then uses the data and the likelihood function to produce posterior distributions for the parameters. Specifying prior distributions can reflect both the uncertainty and our beliefs about the parameter values. Obtaining a posterior distribution for the parameters has the advantage that it can tell us about the uncertainty in the parameter values that is modified by the data.

Both maximum likelihood and Bayesian estimation are beyond the scope of the book.

Parameter sensitivity analysis

An important question to ask about a model is how sensitive the model is to changes in the parameter values. This tells us how important a parameter is and how accurately it needs to be measured.

We need a good output measure to compare results of different parameters. A bad choice in the SIR example would be the number of infections at 10 days. This is bad because as the parameters change, the dynamics change the epidemic curve so the number of infections at 10 days is probably not meaningful. Better measures would be the maximum number of infections over the time period, or the cumulative number of infections at the end of the time period, for example.

One measure of how sensitive the model output is to a given parameter is the derivative of the output with respect to that parameter. We can numerically approximate this by changing the parameter by a small amount, calculate the change in the output and then compute the ratio of the output change to the parameter change.

A better measure of sensitivity is the elasticity, which is defined for the relative changes in the model output with a relative change in the parameters. This allows a better comparison between sensitivities due to different parameters when the parameters are of different magnitude. The elasticity of a parameter p_i for a model output M is defined by

$$e(p_i) = \frac{\Delta M(p_i)/M(p_i)}{\Delta p_i/p_i}, \qquad \Delta M(p_i) = M(p_i + \Delta p_i) - M(p_i).$$

For the SIR example here, the elasticities of the parameters β_f and γ_f, for an output measure of the maximum number of infectives at any time, are $e(\beta_f) = 0.87$ and $e(\gamma) = -0.90$. Both are of similar magnitude, indicating they are equally important in the model. The negative sign for $e(\gamma)$ indicates that as γ is increased, the maximum number of infectives decreases.

Parameter uncertainty analysis

A different approach to studying the effect of changes of parameters on model output is *uncertainty analysis*. Here the parameters are assumed to follow a probability distribution where the distribution reflects our prior knowledge or beliefs about the range of values of the parameters.

Many simulations of the model are run with random selections of parameters from their distributions. The output will also be a probability distribution showing the possible range output that may occur and we can calculate probabilities of certain outcomes.

For the SIR example, using uniform distributions of the parameters, for β_f a range of $[1.5, 1.7]$ and for γ, use knowledge that the infection period γ^{-1} has a range of $[1, 3]$. Then the output distribution for the output measure of the maximum number of infectives is shown in Figure E.2, using 5,000 random draws of the β_f and γ. From this we can calculate that there is about a 48% chance of having more than 100 infectives at any one time but only a 3% change of having more than 140 infectives at any one time. This type of information could be useful for healthcare planning.

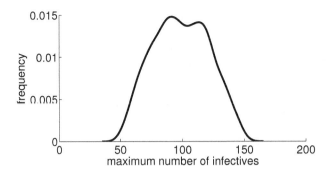

Figure E.2: Probability distribution of maximum number of infectives over the time interval 0 to 25 days with parameters β and α having uniform distributions with β having a range $[1, 5, 1.6]$ and γ^{-1} having a range $[1, 3]$.

Listing E.1: MATLAB code: c_z_parest.m

```matlab
function c_z_parest
% read the data points from an excel file
data = xlsread('school_flu_data.xlsx');
yd = data(:,2);
tv = data(:,1);

%set initial conditions
N = 763;
i0 = 1;
s0 = N-i0;
y0 = [s0; i0];

beta0 = 1.0;
gamma0 = 1/2;
pGuess = [beta0; gamma0];

myf = @(p) (ssq(tv, yd, p, y0));
pmin = fminsearch(myf, pGuess)
disp(pmin)


%% subfunctions %%%%%%%%%%%%%%%%%%%%%%%%%%%%%%%%%%%%%%%%%%%%

function ss = ssq(tdata, ydata, p, y0)
% function to compute sum of squares
% of differences between function and data
% Input:
%  beta,gamma = param values
%  tdata = vector of t values
%  ydata = vector of data values

% solve the ode
tvalues = [0; tdata];
[ts, y1] = ode45(@derhs, tvalues, y0, [], p);
ymodel = y1(2:end,2);
ss = sum( (ymodel-ydata).^2 );

function ydot = derhs(t, y, p)
% RHS of the ODE
N = 763;
beta = p(1);
gamma = p(2);
ydot = zeros(2,1);
S = y(1);
```

```
I = y(2);
ydot(1) = -beta*S*I/N;
ydot(2) = beta*S*I/N - gamma*I;
```

Appendix F

Answers and hints

Chapter 2

2.1 (a) No. Is a solution for all c.

(b) $y(t) = 5e^{2t}$.

2.3 (a) $p(h) = 1\,013\,e^{-0.1504h}$.

(b) Approximately 0.55 millibars.

(c) Approximately 0.79 kilometres.

2.5 4,103 years.

2.7 Approximately 284 days.

2.9 $\dfrac{dC}{dt} = \dfrac{F}{V}c_i - \dfrac{F}{V}C - rC$.

2.11 (b) Approximately 10 hours.

2.13 (b) Dividing by zero if $k_1 = k_2$.
$y(t) = k_1 x_0 t e^{-k_1 t}$.

(c) Hint: DE for the GI tract doesn't contain k_2.

2.15 (d) $x(t) = I - k_1 x$, $\quad x(0) = 0$,
$y(t) = k_1 x - k_2 y$, $\quad y(0) = 0$,
$z(t) = k_2 y$, $\quad z(0) = 0$.

2.17 (a) $y(t) = -k_3 t + y_0$, $\quad y(0) = y_0$.

(b) $y(t) = y_0 e^{-(k_3/M)t}$, $\quad y(0) = y_0$.

2.19 $\dfrac{d}{dt}(V_b C_1) = I - F_1 \times \dfrac{C_1 V_b}{V_g}$,

$\dfrac{d}{dt}(V_b C_2) = \alpha F_1 \times \dfrac{C_1 V_b}{V_g} - F_2 \times C_2$.

2.23 (a) $y_e = 1$ which is unstable.

(b) $C_e = c_i$. $F'(C) = -(F/V) < 0$.
$C_e = c_i$ is **stable** for all values of parameters.

Chapter 3

3.1 (a) $\dfrac{dX}{dt} = 0.7X - 0.2X - 300$, $\quad X(0) = X_0$.

(b) 1,176,000   (c) $X_e = 600$

3.3 (a) $N_e = 0$ and $N_e = a/b$.
$a > 0$ and $b > 0$ for $N > 0$.

(b) $N_e = 0$ is unstable for all positive a and b.
$N_e = a/b$ is stable for all positive a and b.

(c) $a = r$ and $b = r/K$. Claim is true.

3.5 $X' = (\beta - \alpha)X(1 - X/K)$

3.7 $X_e = \dfrac{K}{2}\left(1 + \sqrt{1 - \dfrac{4h}{rK}}\right)$, and

$X_e = \dfrac{K}{2}\left(1 - \sqrt{1 - \dfrac{4h}{rK}}\right)$

3.9 $X(t) = \dfrac{200}{1 + e^{-t/5}}$, so approx. 2 months.

3.11 (a) $\dfrac{dx}{dt} = (\beta - \alpha)x + i - e$.

(b) $\dfrac{dx}{dt} = \beta_x x - \alpha_x x + \gamma(x - y)$,

$\dfrac{dy}{dt} = \beta_y y - \alpha_y y - \gamma(x - y)$.

3.13 (a) Hint: See Appendix D.1.

(b) $\dfrac{dn}{dt} = rn(1 - n) - \epsilon n$, with $n = N(t)/K$.

3.15 The solutions are almost indistinguishable for both $r = 1.9$ and $r = 2.5$, and the populations move together rapidly.

With $r = 2.8$, the solutions move apart as time progresses.

3.17 (a) $N_e = 0$ is unstable.

(b) $N_e = 0$ is stable.

3.19 For $r \leq 1$, extinction.

For $r > 1$ a stable non-zero equilibrium.

3.21 $N_e = 0$ is unstable for $r > 0$.

$N_e = K$ is stable, for $r > 0$ and $r < 2$.

3.23 $X(t) = e^{mt}$, $m \approx 0.5671$.

Chapter 4

4.1 (a) 0.011   (b) 0.

4.4 When Digits is set to 20, the error decreases as the stepsize increases.

With Digits set to 10, however, it appears that the error is larger when a stepsize of 0.005 is used than when as stepsize of 0.01 is used.

Chapter 5

5.1 (a) Approximately 55 never infected. Maximum number of infectives was 120 on day 10.

(b) Approximately 7 never infected. Maximum number of infectives was 478 on day 5. Total infected was 993.

(c) Maximum infected approximately doubles (but not exactly).

5.3 $S' = -\beta SI$, $\quad E' = \beta SI - \sigma E$, $I' = \sigma E - \gamma I$, $\quad R' = \gamma I$.

5.5 (a) $S' = -\beta SI - vS$, $\quad I' = \beta SI - \gamma I$, and $V' = vS$.

(b) $S' = -\beta S - \nu S + \mu V$, $\quad I' = \beta SI - \gamma I$, and $V' = \nu S - \mu V$.

5.7 (a) $S'(t) = -c\dfrac{BS}{k_{50} + B} + bN - aS$,

$I'(t) = c\dfrac{BS}{k_{50} + B} - \gamma I - aI$,

$I'(t) = \gamma I - aR$,

$B'(t) = eI + (n_b - m_b)B$.
with $N = S + I + R$.

(b) $b = a$ for constant N.
DF equilibrium: $(S, I, R) = (N, 0, 0)$.

(c) $R_0 = \dfrac{ces_0}{(\gamma + a)k_{50}(m_b - n_b)}$.

5.9 (a) mean densities of predators and prey are higher.

(d) DDT increases the equilibrium prey population from 100 to 120 and decreases the equilibrium predator population from 100 to 90.

5.11 $X' = c_1 XY - \alpha_1 X$, $\quad Y' = c_2 XY - \alpha_2 Y$,

5.13 (a) $K_1 = 1.5$, $\quad K_2 = 1.2$.

(b) $X \to 0.5$, $\quad Y \to 1$.

5.15 $L'(t) = b_1 A - d_1 L - \sigma_1 L$,
$P'(t) = \sigma_1 L - d_2 P - \sigma_2 P$,
$A'(t) = \sigma_2 P - d_2 A$.

5.17 For $\epsilon = 0.5$, 8 days.
For $\epsilon = 0$, after 9 days.

5.19 $L' = -\beta LC + f\mu P$,
$C' = \beta LC - \sigma C + (1 - f)\mu P$,
$P' = \gamma C - \mu P$.

5.21 (b) $R' = -a_1 B$, $\qquad B' = -a_2 RB$.

(c) $R' = -a_1 B + r_1$, $\qquad B' = -a_2 RB + r_2$.

5.23 $S_h' = -\beta_1 S_h I_m$, $\quad S_m' = -\beta_2 S_m I_h$,
$I_h' = \beta_1 S_h I_m$, $\quad I'm = \beta_2 S_m I_h$

5.25 $U' = -bMU$, $\quad C' = bMU - aC$, $\quad M' = aC$

5.27 $R_0 = \beta/(\gamma + a)$.

5.29 (a) For frequency-dependent transmission: equations unchanged.

(b) For density-dependent transmission: require $\beta \to A\beta$.

Chapter 6

6.1 $(x, y) = (0, 0)$.

6.2 (a) $(X, Y) = (0, 0)$, $(1, 3/2)$
(b) $(X, Y) = (0, 0)$, $(1, 2)$
(c) $(X, Y) = (1/2, 1/2)$, $(X, 0)$, with X any real number.

6.4 (a) From chain rule $I = -S + K$, where K is a constant.

(b) The only nullclines in this system are $S = 0$ and $I = 0$.

(c) Yes. Following the phase-plane trajectory, as $S \to 0$, then I must tend to 0.

6.5 (a) $I = K - S$.

(b) $I = 0$ and $S = \gamma/\beta$.
With $S < \gamma/b$, $S' > 0$ and $I' < 0$.
With $S > \gamma/b$, $S' < 0$ and $I' > 0$.

6.6 (a) $(X, Y) = (0, 0)$, $(K, 0)$, $\left(\dfrac{\alpha_2}{c_2}, \dfrac{\beta_1}{c_1}\left(1 - \dfrac{\alpha_2}{c_2 K}\right)\right)$.

(b) Solutions appear to be spirals that tend to the coexisting steady state.

6.7 (a) $(0, 0, 0)$, $\quad \left(\dfrac{b_3}{a_3}, 0, \dfrac{a_1}{c_1}\right)$, $\quad \left(\dfrac{b_2}{a_2}, \dfrac{a_1}{b_1}, 0\right)$.

Coexistance of all three species is not possible in equilibrium here.

(b) Introducing an extra predator causes either this new predator or the existing predator to become extinct.

6.9 $(0, 0)$, $\quad (\beta_2/c_2, \beta_1/c_1)$.

6.11 $(X, Y) = (0, 0)$, $\quad \left(K(1 - \dfrac{p_1}{\beta_1}), 0\right)$, and $\left(\dfrac{\alpha_2 + p_2}{c_2}, \dfrac{\beta_1}{c_1}\left(1 - \dfrac{\alpha_2 + p_2}{Kc_2}\right) - \dfrac{p_1}{c_1}\right)$

6.12 (a) Substitute $Q = S/(\nu N)$.

(b) $N = K(r - a)/r$ and $S = \nu K(r - a)/r$.

6.15 $R(t) = r_0 \cosh(\alpha t) - \dfrac{b_0}{\nu}\sinh(\alpha t)$,
$B(t) = b_0 \cosh(\alpha t) - r_0\nu\sinh(\alpha t)$,
where $\alpha = \sqrt{a_1 a_2}$ and $\nu = \sqrt{a_2/a_1}$.

6.17 (a) $R = \dfrac{c_1}{2a_2}B^2 + K, \quad K = r_0 - \dfrac{c_1}{2a_2}b_0^2.$

(b) 500 red army soldiers.

(c) Red army hidden.

6.21 (a) $(N, 0, 0)$,
$\left(\dfrac{\gamma + a}{\beta}, \dfrac{a}{\beta}\left(\dfrac{\beta N}{\gamma + a} - 1\right), \dfrac{\gamma}{\beta}\left(\dfrac{\beta N}{\gamma + a} - 1\right)\right).$

(b) $\left(\dfrac{1}{R_0}, \dfrac{a}{\beta}(R_0 - 1), \dfrac{\gamma}{\beta}(R_0 - 1)\right).$

Chapter 7

7.1 (a) $(x, y) = (0, 0)$ is a centre.

(b) $x' = x - 5y, \ y' = x - y.$

7.4 (a) $\begin{bmatrix} x' \\ y' \end{bmatrix} = \begin{bmatrix} 1 & -1 \\ -y_e & -x_e \end{bmatrix} \begin{bmatrix} x - x_e \\ y - y_e \end{bmatrix}.$

(b) $(x, y) = (-4, -1)$ is an unstable node, $(x, y) = (1, 4)$ is a saddle point.

7.5 (a) $\begin{bmatrix} X' \\ Y' \end{bmatrix} = \mathbf{J} \begin{bmatrix} X - x_e \\ Y - y_e \end{bmatrix}$ where

$$\mathbf{J} = \begin{bmatrix} \beta_1 - 2d_1 x_e - c_1 y_e & -c_1 x_e \\ -c_2 y_e & \beta_2 - 2d_2 y_e - c_2 x_e \end{bmatrix}$$

(b) $(0, 0)$ is an unstable node, $(0, 1.25)$ and $(1.071, 0)$ are saddle points, $(1.111, 0.278)$ is a stable node.

The species will coexist.

7.7 (b) If $\gamma^2 < \beta\mu/4$ the equilibrium point is a stable node.
If $\gamma^2 > \beta\mu/4$ it is a stable spiral.

(c) For $\gamma > \sqrt{\beta\mu/4}$ solution has damped oscillations.
For $\gamma > \sqrt{\beta\mu/4}$ solution tends to equilibrium, (either monotonic increasing or decreasing depending on initial condition.)

7.9 The only equilibrium point is at $(R, B) = (0, b)$.
The eigenvalues of the system are $\lambda_1 = 0$ and $\lambda_2 = -a_1 b$.

No single equilibrium 'point' but a line of equilibrium points along $R = 0$.

The trajectories are parallel in the vicinity of this line.

Chapter 8

8.1 (b) $(0, 0)$, $\left(\dfrac{\epsilon_1}{\alpha_{11}}, 0\right)$, $\left(0, \dfrac{\epsilon_2}{\alpha_{22}}\right)$, and

$\left(\dfrac{\alpha_{22}\epsilon_1 - \alpha_{12}\epsilon_2}{\alpha_{11}\alpha_{22} - \alpha_{12}\alpha_{21}}, \dfrac{\alpha_{11}\epsilon_2 - \alpha_{21}\epsilon_1}{\alpha_{11}\alpha_{22} - \alpha_{12}\alpha_{21}}\right).$

If $\alpha_{11}\alpha_{22} - \alpha_{12}\alpha_{21} < 0$, either one or both species must die out.

8.3 Hint: Substitute

$$\dfrac{dX}{dt} = r_2 K \dfrac{dx}{d\tau} \quad \text{and} \quad \dfrac{dY}{dt} = r_2 c_3 K \dfrac{dy}{d\tau}.$$

8.5 (a) $F(V) = c_1 \left(\dfrac{V}{V + D}\right)$, and

$N(V, H) = r_2 \left(1 - \dfrac{JH}{V}\right).$

8.8 $J = \begin{bmatrix} b - a - \beta I & b - \beta S \\ \beta I & \beta S - a - \alpha_d \end{bmatrix},$

$T = \dfrac{-b(b - a)}{\alpha_d + a - \alpha}.$

Chapter 9

9.1 (a) 1.08×10^6 J.

(b) $Q = \rho c \Delta U, \ \rho = 8,390 \text{kg m}^{-3}$ and so $Q = 269$ J.

(c) $U = 14°C$

9.3 (a) Yes.

(b) No. The temperature U depends on x and so the conductivity also depends on x.

9.5 $\dfrac{d^2 U}{dx^2} + \dfrac{q(x)}{k} = 0$

9.7 $\dfrac{d}{dr}\left(r^2 \dfrac{dU}{dr}\right) = 0.$

9.9 $\dfrac{D}{r^2}\dfrac{d}{dr}\left(r^2 \dfrac{dC}{dr}\right) + M(r) = 0.$

9.10 $\dfrac{D}{r}\dfrac{d}{dr}\left(r \dfrac{dC}{dr}\right) + M(r) = 0.$

Chapter 10

10.1 (b) $\lambda \approx 0.011515 \text{ min}^{-1}, \ t = 26$ minutes.

10.3 $U = \dfrac{\beta}{\alpha} - \dfrac{\beta}{\alpha}e^{-\alpha t} + u_0 e^{-\alpha t} = u_0 + \dfrac{\beta}{\alpha}(1 - e^{-\alpha t}).$

10.5 $U(t) = \dfrac{q}{cm}t + u_0.$ So $t \simeq 3.6$ hours.

10.7 $M_1 \dfrac{dU_1}{dt} = q_0 + \dfrac{S_{10}}{R_{10}}(u_s - U_1) + \dfrac{S_{12}}{R_{12}}(U_2 - U_1)$

$M_2 \dfrac{dU_2}{dt} = \dfrac{S_{20}}{R_{20}}(u_s - U_2) + \dfrac{S_{12}}{R_{12}}(U_1 - U_2)$

10.8 (a) $\theta^{(1)}$ is stable, $\theta^{(2)}$ is unstable and $\theta^{(3)}$ is stable.

(b) Suppose θ_0 is small, then $\theta \to \theta^{(1)}$ and if θ_0 is sufficiently large then $\theta(t) \to \theta^{(3}$.

10.10 (b) $\theta \approx 0.38, \quad 0.67.$ The critical ambient temperature is $\theta_a \approx 0.22$

Chapter 11

11.1 (a) $U(x) = \frac{1}{2}x^2 + Ax + B$.

(b) $U(x) = \frac{1}{2}x^2 - \frac{3}{2}x + 1$.

(c) $U(x) = \frac{1}{2}x^2 - 2x + 1$.

(d) $U(x) = \frac{1}{2}x^2 + x - 1$.

11.3 (a) $J(a) = 10$ and $U(b) = u_2$.

(c) $U(0) = 100$ and $J(\ell) = 0$.

11.5 (a) $U(x) = Ax + B$, where A and B are arbitrary constants.

(c) $J = \dfrac{k(u_i - u_o)}{\frac{k}{h_i} + \frac{k}{h_0} + \ell}$.

(d) $J = (u_i - u_o)/R$ with $R = 1/h_i + 1/h_0 + \ell/k$.

11.7 $U(x) = \dfrac{q}{2k}x(l - x) + u_0$.

11.9 Hint: Solve

$$\frac{u_1 - u_a}{R_g} = \frac{u_a - u_b}{R_a}$$

for u_a and substitute into

$$\frac{u_a - u_b}{R_a} = \frac{u_b - u_2}{R_g}$$

to obtain u_b.

11.11 Δ_2 is the least efficient.

11.13 (a) $U(r) = -C_1 r^{-1} + C_2$.

(b) $U(r) = \dfrac{ab(u_2 - u_1)}{a - b}r^{-1} + \dfrac{u_1 a - u_2 b}{a - b}$.

11.15 $U(x) = u_1(\cosh(\sqrt{\beta}x) - \tanh(\sqrt{\beta}\ell)\sinh(\sqrt{\beta}x))$.

11.17 $C(r) = c_1 - \dfrac{A_0}{6D}(R^2 - r^2)$.

11.19 $C_q(r) = c_q$,

$C_p(r) = c_1 - \dfrac{A_0}{6D}(R^2 - r^2) + \dfrac{A_0 R_q^3}{3D}\left(\dfrac{1}{r} - \dfrac{1}{R}\right)$.

11.21 (a) $D\dfrac{d^2 C_q}{dx^2} = 0$, $\quad D\dfrac{d^2 C_p}{dx^2} - A_0 = 0$.

$C_q(0) = 0$, $\quad C_q(h_q) = c_q$,
$C_p(R) = c_1$, $\quad C_p(R_q) = c_q$, and
$C_p'(R_q) = C_q'(R_q)$.

(b) $C_q(x) = c_q$,

$C_p(x) = c_1 - \dfrac{A_0}{2D}(h^2 - x^2) + \dfrac{A_0}{D}h_q(h - x)$.

(c) Rearrange $C_p(x)$ so that

$C_p(x) = c_1 - \dfrac{A_0}{2D}(h^2 - x^2) + \dfrac{A_0 h_q}{D}(h - x)$.

Chapter 12

12.1 $\alpha = k/(c\rho)$ and $\beta = Q_0/(c\rho)$.

12.2 $\dfrac{\partial U}{\partial t} = \dfrac{k}{c\rho r}\dfrac{\partial}{\partial r}\left(r\dfrac{\partial U}{\partial r}\right) + \dfrac{Q_0}{c\rho}$.

12.5 $U(x,t) = u_1 e^{-\beta x}\cos(\omega t - \beta x)$, $\qquad \beta = \sqrt{\dfrac{\omega}{2\alpha}}$.

12.6 $C(x) = K$, where K is an arbitrary constant.

References

Adams, R. A. (1995). *Calculus: A Complete Course*. (3rd ed.). Don Mills, Ontario, Canada: Addison-Wesley.

Allen, L. (2003). *An Introduction to Stochasitic Processes with Applications to Biology*. Englewood Cliffs, NJ: Prentice-Hall.

Amaku, M., R. A. Dias and F. Ferreira (2010). Dynamics and control of stray dogs populaitons. *Mathematical Population Studies 17*, 69–78.

Anderson, A. (1996). Was *rattus exulans* in New Zealand 2000 years ago? AMS radiocarbon ages from the Shag River Mouth. *Archaeology in Oceania. 31*, 178–184.

Anderson, R. M. and R. M. May (1991). *Infectious Diseases of Humans*. Oxford, UK: Oxford University Press.

Anonymous (1978). Influenza in a boarding school. *British Medical Journal 1*, 587.

Araujo, R. P. and D. L. S. McElwain (2004). A history of the study of solid tumor growth: The contribution of mathematical modelling. *Bulletin of Mathematical Biology 66*, 1039–1091.

Banks, R. (1994). *Growth and Diffusion Phenomena: Mathematical Frameworks and Applications*. Berlin: Springer-Verlag.

Barlow, N. D. (2000). Non-linear transmission and simple models for bovine tuberculosis. *Journal of Animal Ecology 60*, 703–713.

Barnes, B., H. Sidhu and D. Gordon (2007). Host gastro-intestinal dynamics and the frequency of antimicrobial production by *Escherichia coli*. *Microbiology 153*, 2823–2827.

Barro, R. J. (1997). *Determinants of Economic Growth: A Cross-Country Empirical Study*. Cambridge, MA: MIT Press.

Beck, A. (2002). *The Ecology of Stray Dogs: A Study of Free-Ranging Urban Animals*. West Layfayette, IN: Purdue University Press.

Beck, M., A. Jakeman and M. McAleer (1993). *Construction and Evaluation of Models of Environmental Systems*, Chapter 1, pp. 3–35. New York: Wiley.

Beddington, J., C. Free and J. Lawton (1978). Characteristics of successful natural enemies in models of biological control of insect pests. *Nature 273*, 513–519.

Begon, M., M. Bennett, R. G. Bowers, N. P. French, S. M. Hazel and J. Turner (2002, Aug). A clarification of transmission terms in host-microparasite models: Numbers, densities and areas. *Epidemiol. Infect. 129*(1), 147–153.

Begon, M., J. Harper and C. Townsend (1990). *Ecology, Individuals, Populations and Communities* (2nd ed.). Oxford, UK: Blackwell.

Borelli, R. and C. Coleman (1996). *Differential Equations; A Modelling Perspective*. New York: Wiley.

Brauer, F. and C. Castillo-Chàvez (2001). *Mathematical Models in Population Biology and Epidemiology*. New York: Springer-Verlag.

Braun, M. (1979). *Differential Equations and Their Applications* (2nd ed.). Berlin: Springer-Verlag.

Burgess, J. and L. Olive (1975). Bacterial pollution in Lake Burley Griffin. *Water (Australian Water and Wastewater Association) 2*, 17–19.

Caulkins, J., R. Marrett and A. Yates (1985). Peruvian anchovy population dynamics. *U.M.A.P. Journal 6*(3), 1–23.

Codeço, C. (2001). Endemic and epidemic dynamics of cholera: The role of the aquatic reservoir. *BMC Infectious Diseases 1*, 1–15.

Costantino, R., J. Cushing, B. Dennis and R. Desharnis (1995). Experimentally induced transitions in the dynamic behaviour of insect populations. *Nature 375*, 227–375.

Costantino, R., R. Desharnais, J. Cushing and B. Dennis (1997). Chaotic dynamics in an insect population. *Science 275*, 389–391.

Cramer, N. F. and R. May (1972). Interspecific competition predation and species diversity: A comment. *Journal of Theoretical Biology 34*, 289–293.

Cyrix (1998). Technical documentation for Cyrix 6x86-P-166 processor.

Daley, D. and J. Gani (1999). *Epidemic Modelling. An Introduction.* Cambridge: Cambridge University Press.

Davies, P. (1994). *War of the Mines: Cambodia, Landmines and the Impoverishment of a Nation.* London: Pluto Press.

Dekker, H. (1975). A simple mathematical model of rodent population cycles. *Journal of Mathematical Biology 2*, 57–67.

Diekmann, O., J. A. P. Heesterbeek and T. Britton (2012). *Mathematical Tools for Understanding Infectious Disease Dynamics.* Princeton, NJ: Princeton University Press.

Dodd, A. (1940). The biological campaign against prickly-pear. Technical report, Brisbane Government Printer, Australia.

Edelstein–Keshet, L. (1988). *Mathematical Models in Biology.* New York: Random House.

Feichtinger, G., C. V. Forstand and C. Piccardi (1996). A nonlinear dynamical model for the dynastic cycle. *Chaos, Solitons and Fractals 7*, 257–271.

Frank, S. A. (1994). Spatial polymorphism of bacteriocins and other allelopathic traits. *Evolution Ecology 8*, 369–386.

Fulford, G. R. and P. Broadbridge (2000). *Industrial Mathematics: Case Studies in Heat and Mass Transport.* Cambridge: Cambridge University Press.

Fulford, G. R., P. Forrester and A. Jones (1997). *Modelling with Differential and Difference Equations.* Cambridge: Cambridge University Press.

Fulford, G. R., M. G. Roberts and J. A. P. Heesterbeek (2002). The metapopulation dynamics of an infectious disease: Tuberculosis in possums. *Theoretical Population Biology 61*, 15–29.

Fung, I. C.-H. (2014). Cholera transmission dynamic models for public health practitioners. *Emerg. Themes Epidemiol. 11*(1), 1.

Gani, J. (1972). Model-building in probability and statistics. In O. Sheynin (Ed.), *The Rules of the Game*, pp. 72–84. London: Tavistock Publications.

Gani, J. (1980). The role of mathematics in society. *Mathematical Scientist 5*, 67–77.

Gordon, D. M. and M. A. Riley (1999). A theoretical and empirical investigation of the invasion dynamics of colicinogeny. *Microbiology 145*, 655–661.

Gotelli, N. J. (1995). *A Primer of Ecology.* Sunderland, MA: Sinauer Associates.

Grad, Y. H., J. C. Miller and M. Lipsitch (2012). Cholera modeling: Challenges to quantitative analysis and predicting the impact of interventions. *Epidemiology 23*(4), 523–530.

Greenspan, H. (1972). Models for the growth of a solid tumor by diffusion. *Studies in Applied Mathematics 52*, 317–340.

Grenfell, B. and A. Dobson (1995). *Ecology of Infectious Diseases in Natural Populations.* Cambridge: Cambridge University Press.

Hanselman, D. and B. Littlefield (2005). *Mastering MATLAB 7.* Upper Saddle River, NJ: Pearson.

Hassell, M. (1978). *The Dynamics of Arthropod Predator-Prey Systems.* Princeton, NJ: Princeton University Press.

Hassell, M. P. (1976). *The Dynamics of Competition and Predation.* London: Edward Arnold.

Hearn, C. (1998). Private communication.

Heck, A. (1996). *An Introduction to Maple* (2nd ed.). New York: Springer-Verlag.

Higham, D. J. and N. J. Higham (2007). *MATLAB Guide* (2nd ed.). Philadelphia: SIAM.

Holman, J. (1981). *Heat Transfer.* New York: McGraw-Hill.

Holmes, M., J. Ecker, W. Boyce and W. Siegmann (1993). *Exploring Calculus with Maple.* New York: Addison-Wesley.

Hurewicz, W. (1990). *Lectures on Ordinary Differential Equations.* New York: Dover.

Jones, J. C. (1993). *Combustion Science. Principals and Practice.* Brisbane: Milennium Books.

Keeling, M. J. and P. Rohani (2008). *Modelling Infectious Diseases in Humans and Animals.* Princeton, NJ: Princeton University Press.

Keeton, W. T. (1972). *Biological Science* (2nd ed.). New York: W W Norton.

Kermack, W. and A. McKendrick (1927). A contribution to the mathematical theory of epidemics. *Proceedings of the Royal Society. A 115*, 700–721.

Kerr, B., M. A. Riley, M. W. Feldman and B. J. Bohannan (2002). Local dispersal promotes biodiversity in a real-life game of rock-paper-scissors. *Nature 418*, 171–174.

Kincaid, D. and W. Cheney (1991). *Numerical Analysis.* Belmont, CA: Brooks Cole.

Kirkup, B. C. and M. A. Riley (2004). Antibiotic-medicated antagonism leads to a bacterial game of rock-paper-scissors in vivo. *Nature 428*, 412–414.

Kormondy, E. K. (1976). *Concepts of Ecology* (2nd ed.). Englewood Cliffs, NJ: Prentice-Hall.

Krebs, C., M. Gaines, B. Keller, J. Myers and R. Tamarin (1973). Population cycles in small rodents. *Science 179*, 35–41.

Lay, D. C. (1994). *Linear Algebra and its Applications.* Reading, MA: Addison-Wesley.

Levin, B. R. (1988). Frequency dependent selection in bacterial populations. *Philosophical Transactions of the Royal Society of London. Series B, Biological Sciences 319*, 2823–2827.

Lewin, R. (1983a). Predators and hurricanes change ecology. *Science 221*, 737–740.

Lewin, R. (1983b). Santa Rosalia was a goat. *Science 221*, 636–639.

Lignola, P. and F. D. Maio (1990). Some remarks on modelling CSTR combustion processes. *Combustion and Flame 80*, 256–263.

Loyn, R. H., R. G. Runnalls, G. Y. Forward and J. Tyers (1983, Sep). Territorial bell miners and other birds affecting populations of insect prey. *Science 4618*, 1411–1413.

Lucas, R. (1988). On the mechanics of economic development. *Journal of Monetary Economics 22*, 3–42.

MacArthur, R. and J. Connell (1966). *The Biology of Populations.* New York: Wiley.

May, R. (1981). *Theoretical Ecology: Principles and Applications* (2nd ed.). Oxford: Blackwell Scientific Publications.

Mesterton-Gibbons, M. (1989). *A Concrete Approach to Mathematical Modelling.* Reading, MA: Addison-Wesley.

Michaelis, L. and M. Menten (1913). Die kinetik der invertinwirkung. *Biochemische Zeitschrift 49*, 333–369.

Microsoft (1995). Microsoft Encarta.

Miller, T. (1995). Mathematical model of heat conduction in soil for landmine detection. Technical Report DMS-C95/95, CSIRO Division of Mathematics and Statistics.

Monro, J. (1967). The exploitation and conservation of resources by populations of insects. *Journal of Animal Ecology 36*, 531–547.

Murray, J. (1990). *Mathematical Biology.* New York: Springer-Verlag.

Myers, J. and C. Krebs (1974). Population cycles in rodents. *Scientific American 230*, 38–46.

Oakley, R. and C. Ksir (1993). *Drugs, Society and Human Behaviour* (6th ed.). New York: McGraw-Hill.

Parrish, J. and S. Saila (1970). Interspecific competition, predation and species diversity. *Journal of Theoretical Biology 27*, 207–220.

Przemieniecki, J. (1994). *Mathematical Methods in Defence Analysis.* Washington, DC: American Institute of Aeronautics and Astronautics.

Quammen, D. (1997). *The Song of the Dodo: Island Biogeography in an Age of Extinctions.* London: Pimlico.

Renshaw, E. (1991). *Modelling Biological Populations in Space and Time.* Cambridge: Cambridge University Press.

Rivers, C., G. Wake and X. Chen (1996). The role of milk powder in the spontaneous ignition of moist milk powder. *Mathematics and Engineering in Industry 6*, 1–14.

Roberts, M. G. (1992). The dynamics and control of bovine tuberculosis in possums. *IMA Mathematics Applied to Medicine and Biology 9*, 19–28.

Roberts, M. G. (1996). The dynamics of bovine tuberculosis in possum populations, and its eradication by culling or vaccination. *Journal of Animal Ecology 65*, 451–464.

Roberts, M. G. and J. A. P. Heesterbeek (1993). Bluff your way in epidemic models. *Trends in Microbiology 1*, 343–348.

Roberts, M. G. and M. I. Tobias (2000). Predicting and preventing measles epidemics in New Zealand. *Epidemiology and Infection 124*, 279–287.

Rolls, E. (1969). *They All Ran Wild: A Story of Pests on the Land in Australia*. Melbourne: Angus and Robertson.

Romer, P. (1986). Increasing returns and long run growth. *Journal of Political Economy 94*, 1002–1037.

Roose, T., J. S. Chapman and P. K. Maini (2007). Mathematical models of avascular tumor growth. *SIAM Review 49*(2), 179–208.

Sansgiry, P. and C. Edwards (1996). A home heating model for calculus students. *College Mathematics Journal 27*(5), 394–397.

SBS (1998). *SBS World Guide*. Sydney: Australian Broadcasting Commission.

Smedley, S. I. and G. C. Wake (1987). Spontaneous ignition: Assessment of cause. In *Annual Meeting of the Institute of Loss Adjustersof NZ*, pp. 1–14.

Solow, R. M. (1956). A contribution to the theory of economic growth. *Quarterly Journal of Economic Growth 70*, 65–94.

Stewart, I. (1989). Portraits of chaos. *New Scientist 1689*(4 Nov.), 22–27.

Swan, T. W. (1956). Economic growth and capital accumulation. *Economic Record 32*, 334–361.

Taylor, J. G. (1980). *Force-on-Force Attrition Modelling*. Alexandria, VA: Operations Research Society of America.

Tung, K. K. (2007). *Topics in Mathematical Modeling*. Princeton, NJ: Princeton University Press.

van de Koppel, J. Huisman, J. van der Wal and H. Olff (1996). Patterns of herbivory along a productivity gradient: An empirical and theoretical investigation. *Ecology 77*(3), 736–745.

Vivaldi, F. (1989). An experiment with mathematics. *New Scientist 1689*(28 Oct.), 30–33.

Wake, G. (1995). Percy possum plunders. *CODEE Winter*, 13–14.

Wilmott, P. (1998). *Derivatives. The Theory and Practice of Financial Engineering*. Chichester, UK: Wiley.

Index

Printed in the United States
by Baker & Taylor Publisher Services